Topics in Applied Physics   Volume 15

W0106714

Topics in Applied Physics   Founded by Helmut K.V. Lotsch

Radiationless Processes

in Molecules and Condensed Phases

Edited by F. K. Fong

With Contributions by
D.J. Diestler   F.K. Fong   K.F. Freed
R. Kopelman   J.C. Wright

With 67 Figures

Springer-Verlag Berlin Heidelberg GmbH 1976

Professor FRANCIS K. FONG

Department of Chemistry, Purdue University
West Lafayette, IN 47907, USA

ISBN 978-3-662-30902-5   ISBN 978-3-540-38209-6 (eBook)
DOI 10.1007/978-3-540-38209-6

Library of Congress Cataloging in Publication Data. Main entry under title: Radiationless processes in molecules and condensed phases. (Topics in applied physics; v. 15). Includes bibliographical references. 1. Molecular theory. 2. Crystal lattices. 3. Relaxation (Nuclear physics). I. Fong, Francis K., 1938–. II. Diestler, D. J., 1941–. QD461.R2. 541'.22. 76-22773.

© by Springer-Verlag Berlin Heidelberg 1976
Originally published by Springer-Verlag Berlin Heidelberg New York in 1976
Softcover reprint of the hardcover 1st edition 1976

Preface

The present volume deals with a systematic treatment of radiationless processes in molecules and crystals. The subject matter broadly embraces molecular relaxation in the gaseous state and relaxation of active molecules or molecular aggregates in host lattices. Among topics discussed are energy transfer between rare-earth ions in crystals, energy migration in mixed crystals, molecular photodissociation, relaxation of electronically and vibrationally excited molecules, activated rate processes, multi-phonon processes in solids, and the primary processes in photosynthesis.

The book has been designed for researchers and graduate students in physical chemistry, solid-state physics and materials science. A conscious effort has been made at maintaining a balance between theory and experiment so that the reader will find a ready source of references to current experimental work and, if he so wishes, also obtain an overview of the latest developments in relaxation theory.

One central theme is the treatment of the experimental behavior in terms of the adiabatic approximation in which the physical mechanisms responsible for the relaxation process can be unambiguously identified. The primary goal is to integrate the diversified topics and to relate solid-state and molecular phenomena within a single theoretical and experimental framework.

It is hoped that the present effort does justice to the fundamental importance of the problems treated. In view of the fact that practically all dynamical processes in physics and chemistry are in some way describable as or related to some radiationless phenomena, this work merely touches upon certain aspects of the enormous subject. As such, it is intended that the book will serve as a starting point for the adventurous reader to leapfrog into areas as yet inadequately understood and largely unexplored.

I am grateful to Dr. HELMUT LOTSCH for his helpful editorial comments, and to Miss NANCY KOPANSKI for her dedicated assistance.

Lafayette, Indiana, March 1976   FRANCIS K. FONG

Contents

3. Vibrational Relaxation of Molecules in Condensed Media.
By D.J. DIESTLER (With 9 Figures)

5. Exciton Percolation in Molecular Alloys and Aggregates.
By R. Kopelman (With 18 Figures)

Contributors

DIESTLER, DENNIS J.
  Department of Chemistry, Purdue University
  West Lafayette, IN 47907, USA

FONG, FRANCIS K.
  Department of Chemistry, Purdue University
  West Lafayette, IN 47907, USA

FREED, KARL F.
  The Department of Chemistry. James Franck Institute
  University of Chicago, Chicago, IL 60637, USA

KOPELMAN, RAOUL
  Department of Chemistry, University of Michigan
  Ann Arbor, MI 48109, USA

WRIGHT, JOHN C.
  Department of Chemistry, University of Wisconsin
  Madison, WI 53706, USA

1. Introduction

F. K. FONG

In the past decade, there has been a rapid growth in the number of basic studies on radiationless transitions. This growth may be attributed, in part at least, to the expanding technology of lasers and related optical devices whose efficiencies are dependent on the extent to which non-radiative processes can compete with radiative decay. A perhaps more important reason for the widespread interest is the diversity of areas pervaded by the radiationless relaxation effects. The present volume deals with several topics of current interest, and attempts to outline a systematized approach to the study of a wide variety of radiationless phenomena.

1.1 A Systematic Approach to Radiationless Relaxation

In radiative relaxation, the decay of an excited state is accompanied by the emission of photons of energy commensurate with the difference between the initial and the final states of the transition. The system undergoing radiative relaxation is said to be coupled to the electromagnetic field. In radiationless transitions, no photons are emitted. A successful description of these transitions must therefore provide an alternative account for the mechanisms that allow for the coupling of the adiabatic states and for the energy conservation requirement in the relaxation process.

A convenient starting point in the treatment of radiationless relaxation is the adiabatic approximation [1.1]. In this approximation, the various molecular degrees of freedom are considered in terms of fast and slow subsystems whose variables are assumed to be separable. The total stationary wavefunction of the molecular system is written as a product of the wavefunctions of the fast and the slow subsystems. Relaxation properties are then calculated in terms of the non-adiabatic perturbation arising from interactions that are not diagonal in the adiabatic basis.

One important non-adiabatic perturbation is the nuclear kinetic energy of the slow subsystem [1.2–7] whose off-diagonal elements have not been included in the adiabatic basis set. Depending on the choice of the adiabatic Hamiltonian operator, different non-adiabatic operators may be responsible for radiationless transitions. Examples include the spin-orbit coupling operator in the case of adiabatic bases in which spin is a good quantum number [1.8] and the Coulomb interaction operator in the case of energy transfer processes [1.9–10].

Different models of adiabatic separation lead to different physical situations. In electronic relaxation, the electrons are taken to be fast compared to the nuclei. Likewise, in vibrational relaxation of guest molecules in a host lattice [1.11–14] the localized nuclear motions of the guest molecule might be considered fast with respect to the phonon modes of the lattice. In either case, it is the kinetic motions of the slow nuclei that induce transitions between the stationary states. In describing the re-orientation of a dipolar molecule in a rigid lattice, on the other hand, one might regard the librational motion of the dipole to be slow and the lattice phonon modes fast, so that it is the libron kinetic energy operator that makes possible transitions between different orientation states [1.5].

The adiabatic states are parametrically dependent on the nuclear coordinates. The corresponding eigenenergies are accordingly given by potential surfaces prescribed by the various normal coordinates. Couplings between adiabatic states of the system are therefore functionally dependent on the vibrational modes. Moreover, the nuclear re-organization in the final state accompanies the energy re-distribution in the transition. In the weak coupling limit the nuclear re-organization results in only a minor displacement of the adiabatic surface. In this limit, the rate constant will be sharply dependent on the gap between the initial and final states as well as on the frequencies of the phonon modes that accept the initial excitation energy [1.2, 6, 8]. In the strong-coupling limit, the relaxation process induces a large displacement in the adiabatic surface. The corresponding experimental behavior yields the Arrhenius-type rate expressions that display the usual exponential dependence on temperature and on the activation barrier to the transitions [1.5, 15–17].

On the following pages, we outline the theory of adiabatic separation and illustrate the variety of ways in which this theory may be applied. This outline serves as an introduction to a number of discussions detailed in later chapters of this book.

1.2 The Electronic Adiabatic Approximation

In quantum mechanics, the adiabatic approximation usually means the separation of the fast electronic subsystem from the slow nuclear vibrational subsystem. In anticipation of later generalizations, however, we specify this type of separation as *electronic* adiabatic approximation. The total molecular Hamiltonian is given in the form

$$\mathscr{H} = -(\hbar^2/2m) \sum_i \partial^2/\partial r_i^2 + V(r, R) - (\hbar^2/2) \sum_j \partial^2/M_j \partial R_j^2 \tag{1.1}$$

where the first term arises from the kinetic energy of the electrons, m and r_i being, respectively, the electronic mass and coordinates. The r, R are collective coordinates for the electrons and the nuclei, respectively. The $V(r, R)$ includes all the electron-electron, the nucleus-nucleus, and the electron-nucleus interactions. The last term in (1.1) is the nuclear kinetic energy operator. The sum over j includes all the normal modes, with the effective mass of the j-th mode given by M_j.

In the adiabatic approximation, we seek the solution to the Schrödinger equation

$$\left[-(\hbar^2/2m) \sum_i \partial^2/\partial r_i^2 + V(r, R) \right] |\alpha(r, R)\rangle = \varepsilon_\alpha(R) |\alpha(r, R)\rangle \qquad (1.2)$$

for fixed values of the nuclear coordinates R. The Born-Oppenheimer eigenstates $|\alpha(r, R)\rangle$ and eigenvalues $\varepsilon_\alpha(R)$ depend parametrically on R. We now use $\varepsilon_\alpha(R)$ as the effective adiabatic potential for the nuclear motion. Replacing the first two terms in (1.1) by $\varepsilon_\alpha(R)$, the Schrödinger equation for the slow nuclear subsystem is given by

$$\left[\varepsilon_\alpha(R) - (\hbar^2/2) \sum_j \partial^2/M_j \partial R_j^2 \right] |\chi_{\alpha, n}(R)\rangle = E_{\alpha, n} |\chi_{\alpha, n}(R)\rangle \qquad (1.3)$$

where $E_{\alpha, n}$ is the eigenvalue for the n-th vibrational level in the α-th electronic state, and $|\chi_{\alpha, n}(R)\rangle$ is the corresponding eigenstate. A total wavefunction $\psi_{\alpha, n}$ in the adiabatic approximation can be written as the product

$$|\psi_{\alpha, n}\rangle = |\alpha(r, R)\rangle |\chi_{\alpha, n}(R)\rangle . \qquad (1.4)$$

Applying the completeness relation, we may express the Hamiltonian operator in the adiabatic representation

$$H = \sum_{\alpha, \alpha', n, n'} |\psi_{\alpha, n}\rangle \langle \psi_{\alpha, n} | \mathcal{H} | \psi_{\alpha', n'} \rangle \langle \psi_{\alpha', n'} |. \qquad (1.5)$$

Substituting (1.1) and (1.4) into (1.5), we write

$$H = \sum_{\alpha, \alpha', n, n'} \langle \alpha | \langle \chi_{\alpha, n} | \left[-(\hbar^2/2m) \sum_i \partial^2/\partial r_i^2 \right.$$
$$\left. + V(r, R) - (\hbar^2/2) \sum_j \partial^2/M_j \partial R_j^2 \right] |\chi_{\alpha', n'}\rangle |\alpha'\rangle |\alpha\rangle |\chi_{\alpha, n}\rangle \langle \chi_{\alpha', n'} | \langle \alpha' | \qquad (1.6)$$

where the parametric dependences of $|\alpha\rangle$ and $|\chi_{\alpha, n}\rangle$ on r and R have been suppressed for the sake of simplicity. Now both $|\psi_{\alpha', n'}\rangle$ and $|\alpha'\rangle$ depend on R, so that the term corresponding to the nuclear kinetic energy operator in (1.6) is a sum of three terms according to

$$-(\hbar^2/2)\sum_j \frac{\partial^2}{M_j \partial R_j^2}|\chi_{\alpha',n'}\rangle|\alpha'\rangle = -|\alpha'\rangle\sum_j (\hbar^2/2M_j)\frac{\partial^2}{\partial R_j^2}|\chi_{\alpha',n'}\rangle$$

$$-\sum_j (\hbar^2/M_j)\frac{\partial}{\partial R_j}|\alpha'\rangle\frac{\partial}{\partial R_j}|\chi_{\alpha',n'}\rangle - |\chi_{\alpha',n'}\rangle\sum_j (\hbar^2/2M_j)\frac{\partial^2}{\partial R_j^2}|\alpha'\rangle. \tag{1.7}$$

The first term on the right-hand side of (1.7), the electronic kinetic energy operator $-(\hbar^2/2m)\sum_i \partial^2/\partial r_i^2$ and the interaction potential energy operator $V(r,R)$ in (1.1) are diagonal in the adiabatic basis. These terms are combined to give the zeroth-order Hamiltonian operator

$$H_0 = \sum_{\alpha,\alpha',n,n'} E_{\alpha',n'}\delta_{\alpha,\alpha'}\delta_{n,n'}|\alpha\rangle|\chi_{\alpha,n}\rangle\langle\chi_{\alpha',n'}|\langle\alpha'|$$

$$= \sum_{\alpha,n} E_{\alpha,n}|\alpha\rangle|\chi_{\alpha,n}\rangle\langle\chi_{\alpha,n}|\langle\alpha|. \tag{1.8}$$

The nondiagonal terms in (1.6), corresponding to the second and third terms in (1.7), are given by

$$H' = -\sum_{\alpha,\alpha',n,n'}\sum_j (\hbar^2/M_j)\langle\alpha|\frac{\partial}{\partial R_j}|\alpha'\rangle\langle\chi_{\alpha,n}|\frac{\partial}{\partial R_j}|\chi_{\alpha',n'}\rangle|\alpha\rangle|\chi_{\alpha,n}\rangle\langle\chi_{\alpha',n'}|\langle\alpha'| \tag{1.9}$$

and

$$H'' = -\sum_{\alpha,\alpha',n,n'}\sum_j (\hbar^2/2M_j)\langle\chi_{\alpha,n}|\chi_{\alpha',n'}\rangle\langle\alpha|\frac{\partial^2}{\partial R_j^2}|\alpha'\rangle|\alpha\rangle|\chi_{\alpha,n}\rangle\langle\chi_{\alpha',n'}|\langle\alpha'|. \tag{1.10}$$

It can be shown [1.3] that $H' \gg H''$, so that of the non-diagonal terms, we shall consider only the electronic non-adiabatic operator

$$H_{NA} = \sum_{\alpha\neq\alpha',j}(\hbar^2/M_j)^{-1}\langle\alpha|\frac{\partial V}{\partial R_j}|\alpha'\rangle[\varepsilon_{\alpha'}(R)-\varepsilon_\alpha(R)]^{-1}\frac{\partial}{\partial R_j} \tag{1.11}$$

which is a slightly re-arranged form of (1.9).

In the harmonic approximation, the expansion of $\varepsilon_\alpha(R)$ in powers of the displacements ΔR_α^j from the equilibrium positions R_0 corresponding to the electronic ground state contains terms linear in ΔR_α^j

$$\varepsilon_\alpha(R) = \varepsilon_0^0(R_0) + \sum_j [\partial\varepsilon_\alpha(R)/\partial R_j]_{R=R_0}\Delta R_\alpha^j + \tfrac{1}{2}\sum_j [\partial^2\varepsilon_\alpha(R)/\partial R_j^2]_{R=R_0}(\Delta R_\alpha^j)^2 \tag{1.12}$$

where

$$\varepsilon(R) = \tfrac{1}{2}M\omega^2 R^2$$

$$\partial\varepsilon(R)/\partial R = M\omega^2 R$$

and $\partial^2\varepsilon(R)/\partial R^2 = M\omega^2$.

Introducing the Fermion and Boson annihilation and creation operators $a_\alpha, a_\alpha^\dagger$ and $b_j, b_j^\dagger$,

$$q \equiv R_j = (\hbar/2\omega_j M_j)^{1/2}(b_j^\dagger + b_j) \tag{1.13}$$

$$p \equiv i\hbar \frac{\partial}{\partial R_j} = i(\hbar\omega_j M_j/2)^{1/2}(b_j^\dagger - b_j) \tag{1.14}$$

we rewrite the adiabatic Hamiltonian operator in second quantization

$$H_0 = \sum_\alpha \varepsilon_\alpha^0(R_0) a_\alpha^\dagger a_\alpha + \sum_{\alpha,j} \hbar\omega_\alpha^j a_\alpha^\dagger a_\alpha b_j^\dagger b_j + \sum_{\alpha,j} g_\alpha^j \hbar\omega_\alpha^j a_\alpha^\dagger a_\alpha(b_j^\dagger + b_j). \tag{1.15}$$

Similarly, the non-adiabatic perturbation operator (1.11) may be rewritten

$$H_{NA} = \sum_{\alpha \neq \alpha',j} (\hbar^2\omega_j/2M_j)^{1/2} \langle\alpha| \frac{\partial V}{\partial R_j}|\alpha'\rangle\Big|_{R=R_0} \tag{1.16}$$
$$\times [\varepsilon_\alpha(R_0) - \varepsilon_{\alpha'}(R_0)]^{-1} a_\alpha^\dagger a_{\alpha'}(b_j - b_j^\dagger)$$

where we have invoked the Condon approximation, i.e., evaluated the terms indicated at the equilibrium nuclear configuration of the ground state. In (1.15) and (1.16), ω_α^j is the frequency of the j-th normal mode corresponding to the Born-Oppenheimer state $|\alpha\rangle$, and g_α^j is a dimensionless displacement parameter which is proportional to ΔR_α^j

$$g_\alpha^j = (M_j\omega_\alpha^j/2\hbar)^{1/2} \Delta R_\alpha^j. \tag{1.17}$$

Multiplying (1.15) and (1.16) from the left by e^Q and from the right by e^{-Q} where

$$Q = \sum_{\alpha,j} g_\alpha^j a_\alpha^\dagger a_\alpha(b_j^\dagger - b_j) \tag{1.18}$$

we transform (1.15) and (1.16) to

$$\mathscr{H}_0 = \sum_\alpha \varepsilon_\alpha a_\alpha^\dagger a_\alpha + \sum_{\alpha,j} \hbar\omega_\alpha^j a_\alpha^\dagger a_\alpha b_j^\dagger b_j \tag{1.19}$$

and

$$\mathscr{H}_{NA} = \sum_{\alpha',j} (\hbar^3\omega_\alpha^j/2M_j)^{1/2} \langle\alpha| \frac{\partial V}{\partial R_j}|\alpha'\rangle [\varepsilon_\alpha(R) - \varepsilon_{\alpha'}(R)]^{-1} \tag{1.20}$$
$$\times a_\alpha^\dagger a_{\alpha'} B_\alpha^\dagger B_{\alpha'}(b_j - b_j^\dagger).$$

In (1.19) and (1.20),

$$\varepsilon_\alpha = \varepsilon_\alpha^0 - \sum_j g_\alpha^{j^2} \hbar\omega_\alpha^j \tag{1.21}$$

and

$$B_\alpha^\dagger = \exp\left[\sum_j g_\alpha^j(b_j^\dagger - b_j)\right]$$

$$B_{\alpha'} = \exp\left[\sum_j g_{\alpha'}^j(b_j - b_j^\dagger)\right]. \tag{1.22}$$

The displacement parameter g_α^j plays an important role. It arises from the fact that the equilibrium position of the adiabatic surface corresponding to the j-th normal mode of the α-th electronic state is horizontally displaced from that of the ground state. In the exponential operator

$$B_\alpha^\dagger B_{\alpha'} = \exp\left[\sum_j g^j(b_j^\dagger - b_j)\right] \tag{1.23}$$

the parameter

$$g_j = g_\alpha^j - g_{\alpha'}^j. \tag{1.24}$$

is proportional to the relative horizontal displacement of the equilibrium positions of the adiabatic surfaces for the α-th and the α'-th electronic states. Multi-phonon processes in first-order perturbation theory result from non-vanishing g_j values in the non-adiabatic Hamiltonian operator (1.20). The adiabatic Hamiltonian operator $\mathcal{H}_0$ in (1.19) is approximate in that the model does not allow for frequency distortion in the adiabatic curves. That is, we have assumed $\omega_\alpha = \omega_{\alpha'}$ for all $\alpha \neq \alpha'$. In the canonically transformed representation, the rate constant for the radiationless transition between $|\alpha\rangle$ and $|\alpha'\rangle$ can be written [1.18]

$$W_{\alpha\alpha'} = (\beta\langle \mathcal{N}_\alpha\rangle)^{-1} \int_0^\infty dt \int_0^\beta d\lambda \langle \dot{\mathcal{N}}_\alpha(-t - i\hbar\lambda)\dot{\mathcal{N}}_\alpha(0)\rangle = (\beta\langle \mathcal{N}_\alpha\rangle)^{-1} \int_0^\infty dt \int_0^\beta d\lambda$$

$$\times \langle \exp\left[-(i/\hbar)\mathcal{H}(t + i\hbar\lambda)\right] \dot{\mathcal{N}}_\alpha \exp\left[(i/\hbar)\mathcal{H}(t + i\hbar\lambda)\right] \dot{\mathcal{N}}_\alpha\rangle \tag{1.25}$$

where $\mathcal{H} = \mathcal{H}_0 + \mathcal{H}_{NA}$ is the total Hamiltonian operator. The flux operator $\dot{\mathcal{N}}_\alpha$ is given by the Heisenberg equation of motion

$$\dot{\mathcal{N}}_\alpha = (i/\hbar)[\mathcal{H}_{NA}, \mathcal{N}_\alpha]$$

$$= -(i/\hbar) \sum_{\alpha \neq \alpha', j} C_{\alpha\alpha'}^j a_\alpha^\dagger a_{\alpha'} B_\alpha^\dagger B_{\alpha'}(b_j - b_j^\dagger) + \text{Hermitian conjugate} \tag{1.26}$$

where

$$C_{\alpha\alpha'}^j = -(\hbar^3 \omega_j/2M_j)^{1/2}$$

$$\times \langle \alpha|\frac{\partial V}{\partial R_j}|\alpha'\rangle\Big|_{R=R_0} [\varepsilon_\alpha(R_0) - \varepsilon_{\alpha'}(R_0)]^{-1}. \tag{1.27}$$

The time-correlation expression (1.25) for the rate constant is valid to arbitrary order in the non-adiabatic perturbation. For the application considered here, the interaction is assumed to be small, and we evaluate $W_{\alpha\alpha'}$ to the lowest order in the perturbation by making the substitution

$$\mathscr{H} \to \mathscr{H}_0 = \sum_\alpha \varepsilon_\alpha a_\alpha^\dagger a_\alpha + \sum_{\alpha,j} \hbar \omega_\alpha^j a_\alpha^\dagger a_\alpha b_j^\dagger b_j \tag{1.28}$$

in (1.25). We obtain

$$
\begin{aligned}
W_{\alpha\alpha'} = (\hbar^2 \beta \langle N_\alpha \rangle)^{-1} &\int\limits_0^\infty dt \int\limits_0^\beta d\lambda \, \mathrm{Tr}\,\{\exp(-\beta\mathscr{H}_0)\exp[-(\mathrm{i}/\hbar)\mathscr{H}_0(t+\mathrm{i}\hbar\lambda)] \\
&\times \sum_j C_{\alpha\alpha'}^j a_\alpha^\dagger a_{\alpha'} B_\alpha^\dagger B_{\alpha'}(b_j - b_j^\dagger) \exp[(\mathrm{i}/\hbar)\mathscr{H}_0(t+\mathrm{i}\hbar\lambda)] \\
&\times \sum_k C_{\alpha\alpha'}^k a_{\alpha'}^\dagger a_\alpha B_{\alpha'}^\dagger B_\alpha (b_k^\dagger - b_k)\}/\mathrm{Tr}\,\{\exp(-\beta\mathscr{H}_0)\}
\end{aligned}
\tag{1.29}
$$

where we have assumed a Boltzmann distribution for the initial vibrational manifold. Upon evaluation of the electronic part of the ensemble average, (1.29) becomes

$$
\begin{aligned}
W_{\alpha\alpha'} = (\hbar^2 \beta)^{-1} &\int\limits_0^\infty dt \int\limits_0^\beta d\lambda \exp[-(\mathrm{i}/\hbar)(\varepsilon_\alpha - \varepsilon_{\alpha'})(t+\mathrm{i}\hbar\lambda)] \\
&\times \mathrm{Tr}\,\Big\{\exp\Big[-\beta\sum_j \hbar\omega_\alpha^j b_j^\dagger b_j - \sum_j (x_\alpha^j - x_{\alpha'}^j) b_j^\dagger b_j\Big] \exp\Big(-\sum_j x_{\alpha'}^j b_j^\dagger b_j\Big) \\
&\times \sum_j C_{\alpha\alpha'}^j \exp\Big[\sum_j g_j(b_j^\dagger - b_j)\Big] (b_j - b_j^\dagger) \exp\Big(\sum_j x_{\alpha'}^j b_j^\dagger b_j\Big) \\
&\times \sum_k C_{\alpha\alpha'}^k \exp\Big[-\sum_k g_k(b_k^\dagger - b_k)\Big] (b_k^\dagger - b_k)\Big\}/\mathrm{Tr}\,\Big\{\exp\Big(-\beta\sum_j \hbar\omega_\alpha^j b_j^\dagger b_j\Big)\Big\}
\end{aligned}
\tag{1.30}
$$

where

$$x_{\alpha'}^j = \mathrm{i}\omega_{\alpha'}^j(t+\mathrm{i}\hbar\lambda). \tag{1.31}$$

In order to evaluate the trace in (1.30), it is convenient to introduce the Hamiltonian operator

$$\mathscr{H}_{\alpha\alpha'}^{\mathrm{vib}}(t) = \sum \hbar\omega_{\alpha\alpha'}(t) b_j^\dagger b_j \tag{1.32}$$

where

$$\omega_{\alpha\alpha'}(t) = \omega_\alpha^j + (\mathrm{i}t/\hbar\beta)(\omega_\alpha^j - \omega_{\alpha'}^j). \tag{1.33}$$

The corresponding vibrational partition function is defined by

$$Z_{\alpha\alpha'}^{\mathrm{vib}}(t) = \mathrm{Tr}\,\{\exp[-\beta\mathscr{H}_{\alpha\alpha'}^{\mathrm{vib}}(t)]\} = \prod_j \{1 - \exp[-\beta\hbar\omega_{\alpha\alpha'}^j(t)]\}^{-1}. \tag{1.34}$$

By means of parameter differentiation, (1.30) may be rewritten

$$W_{\alpha\alpha'} = (\hbar^2 \beta)^{-1} \int_0^\infty dt \int_0^\beta d\lambda\, Z_{\alpha\alpha'}^{\text{vib}}(t + i\hbar\lambda)(Z_\alpha^{\text{vib}})^{-1}$$

$$\times \exp\left[-(i/\hbar)(\varepsilon_\alpha - \varepsilon_{\alpha'})(t + i\hbar\lambda)\right] \sum_j \sum_k C_{\alpha\alpha'}^j C_{\alpha\alpha'}^k g_j^{-1} \frac{\partial}{\partial\gamma_j} g_k^{-1}$$

$$\times \frac{\partial}{\partial\gamma_k} \left\langle \exp\left\{ \sum_{j,k} [g_j \lambda_j \exp(-x_\alpha^j) b_j^\dagger - g_k \gamma_k b_k^\dagger] \right. \right. \tag{1.35}$$

$$- \sum_{j,k} [g_j \lambda_j \exp(x_{\alpha'}^j) b_j - g_k \gamma_k b_k] \Big\} \exp\Big\{ \tfrac{1}{2} \sum_{j=k} g_k g_k \lambda_j \gamma_k$$

$$\left. \left. \times [\exp(x_{\alpha'}^j) - \exp(-x_\alpha^j)] \right\} \right\rangle_{\mathscr{H}_{\alpha\alpha'}^{\text{vib}}} \Big|_{\lambda_j = 1,\, \gamma_k = 1} .$$

In arriving at (1.35), we have used the operator identities [1.1]

$$\exp(\xi\, b^\dagger b)\, f(b, b^\dagger) \exp(-\xi\, b^\dagger b) = f(b\, e^{-\xi}, b^\dagger e^\xi) \tag{1.36}$$

and

$$\exp O \exp O' = \exp(O + O') \exp\tfrac{1}{2}[O, O']$$
$$[O, [O, O']] = [O', [O, O']] = O . \tag{1.37}$$

The parameter differentiation in the rate constant expression allows the ensemble average to be written in a form which can be readily evaluated in terms of the operator identity

$$\langle \exp O \rangle_{\mathscr{H}_{\alpha\alpha'}^{\text{vib}}} = \exp(\tfrac{1}{2}\langle O^2 \rangle_{\mathscr{H}_{\alpha\alpha'}^{\text{vib}}}) \tag{1.38}$$

where O is an operator linear in $b^\dagger$ and b. Using (1.38), and performing the integration over λ, we can rewrite (1.35) as

$$W_{\alpha\alpha'} = \hbar^2 \exp(-\sigma) \int_{-\infty}^{+\infty} dt \exp[f(t + i\hbar\beta)] \tag{1.39}$$

where

$$\sigma = \sum_j g_j^2 (2n_\alpha^j + 1) \tag{1.40}$$

and

$$f(t + i\hbar\beta) = -(i/\hbar)(\varepsilon_\alpha - \varepsilon_{\alpha'})(t + i\hbar\beta) + \ln[Z_{\alpha\alpha'}^{\text{vib}}(t + i\hbar\beta)] - \ln(Z_\alpha^{\text{vib}})$$

$$+ \sum_j g_j^2 \{[n_{\alpha\alpha'}^j(t + i\hbar\beta) + 1] \exp[i\omega_\alpha^j(t + i\hbar\beta)]$$

$$+ n_{\alpha\alpha'}^j(t + i\hbar\beta) \exp[-i\omega_\alpha^j(t + i\hbar\beta)]\}$$

$$-2\sum_j g_j^2 \left[n_{\alpha\alpha'}^j(t+i\hbar\beta)-n_\alpha^j\right]+\ln\left[\left(\sum_j C_{\alpha\alpha'}^j g_j\{2n_{\alpha\alpha'}^j(t+i\hbar\beta)+1\right.\right.$$

$$-\left[n_{\alpha\alpha'}^j(t+i\hbar\beta)+1\right]\exp\left[i\omega_{\alpha'}^j(t+i\hbar\beta)\right] \tag{1.41}$$

$$\left.-n_{\alpha\alpha'}^j(t+i\hbar\beta)\exp\left[-i\omega_\alpha^j(t+i\hbar\beta)\right]\}\right)^2$$

$$+\sum_j C_{\alpha\alpha'}^{j2}\{\left[n_{\alpha\alpha'}^j(t+i\hbar\beta)+1\right]\exp\left[i\omega_{\alpha'}^j(t+i\hbar\beta)\right]$$

$$\left.+n_{\alpha\alpha'}^j(t+i\hbar\beta)\exp\left[-i\omega_\alpha^j(t+i\hbar\beta)\right]\}\right]$$

where

$$n_{\alpha\alpha'}^j(t+i\hbar\beta)=\{\exp[\beta\hbar\omega_{\alpha\alpha'}^j(t+i\hbar\beta)]-1\}^{-1}. \tag{1.42}$$

We recall that $\mathcal{H}_0$ has been written in the zero-distortion limit, i.e., $\omega_\alpha^j=\omega_{\alpha'}^j$, so that

$$\omega_{\alpha\alpha'}^j(t)=\omega_\alpha^j. \tag{1.43}$$

Eq. (1.39) can be simplified if we assume that there is only one effective mode of coupling. The symbols $\sum_j g_j^2$ and $\sum_j C_{\alpha\alpha'}^j$, for example, can be replaced by $L_m g_m^2$ and $L_m C_{\alpha\alpha'}^m$, where L_m is an effective degeneracy factor, and m denotes the mode of maximum coupling responsible for the radiationless process. We further note that, in the case of radiationless transitions in rare-earth ions in crystals, for example, the horizontal displacements between the equilibrium positions of the adiabatic curves are generally small for electronic states within the $4f^n$ configuration, and the inequality in the weak coupling limit, defined as

$$\varepsilon_\alpha-\varepsilon_{\alpha'}\gg L_m g_m^2\hbar\omega_\alpha^m \tag{1.44}$$

should be valid in many cases [1.2].

Under the condition (1.43) and the assumption of one L_m-fold degenerate effective mode, (1.41) becomes

$$f(t+i\hbar\beta)=-(i/\hbar)(\varepsilon_\alpha-\varepsilon_{\alpha'})(t+i\hbar\beta)+L_m g_m^2\left[(n_m+1)\exp[i\omega_m(t+i\hbar\beta)]\right.$$

$$+n_m\exp[-i\omega_m(t+i\hbar\beta)]+\ln(\{L_m C_{\alpha\alpha'}^m g_m[(2n_m+1)-(n_m+1)$$

$$\times\exp[i\omega_m(t+i\hbar\beta)]-n_m\exp[-i\omega_m(t+i\hbar\beta)]\})^2 \tag{1.45}$$

$$+L_m|C_{\alpha\alpha'}^m|^2\{(n_m+1)\exp[i\omega_m(t+i\hbar\beta)]+n_m$$

$$\times\exp[-\omega_m(t+i\hbar\beta)]\})$$

where

$$n_m = [\exp(\hbar\omega_m\beta) - 1]^{-1} \tag{1.46}$$

is the Bose distribution number for the effective *mediating* vibrational mode. On rearrangement, (1.39) can be written as

$$W_{\alpha\alpha'} = \frac{L_m |C_{\alpha\alpha'}^m|^2}{\hbar^2} \exp[-L_m g_m^2(2n_\alpha^m + 1)] \sum_{\Delta\nu=-2}^{\Delta\nu=2} \lambda_{\Delta\nu} \int_{-\infty}^{+\infty} dt \exp[h_{\Delta\nu}(t + i\hbar\beta)] \tag{1.47}$$

where

$$h_{\Delta\nu}(t) = -(i/\hbar)(\varepsilon_\alpha - \varepsilon_{\alpha'} - \Delta\nu\hbar\omega_m)t + L_m g_m^2 [(n_m + 1)\exp(i\omega_m t) + n_m \exp(-i\omega_m t)] \tag{1.48}$$

$$\lambda_0 = L_m g_m^2 [6n_m(n_m + 1) + 1] \tag{1.49}$$

$$\lambda_1 = -2 L_m g_m^2 (2n_m + 1)(n_m + 1) + n_m + 1 \tag{1.50}$$

$$\lambda_{-1} = -2 L_m g_m^2 (2n_m + 1)n_m + n_m \tag{1.51}$$

$$\lambda_2 = L_m g_m^2 (n_m + 1)^2 \tag{1.52}$$

$$\lambda_{-2} = L_m g_m^2 n_m^2 . \tag{1.53}$$

We observe that, according to the form of (1.47) the relaxation rate of the radiationless process comprises five contributions, each corresponding to a scattering process in which the transition energy gap $\varepsilon_\alpha - \varepsilon_{\alpha'}$ is modified by the addition or subtraction of $\Delta\nu$ quanta of the mediating phonon energy.

For the cases of interest in the weak coupling limit

$$\varepsilon_\alpha - \varepsilon_{\alpha'} - \Delta\nu\hbar\omega_m \gg L_m g_m^2 \hbar\omega_m . \tag{1.54}$$

The evaluation of the time integration in (1.47) can be accomplished by the method of steepest descent [1.6, 8]

$$\int_{-\infty}^{+\infty} dt \exp[h(t + i\hbar\beta)] = (2\pi)^{1/2} [-h''(t^s)]^{-1/2} \exp[h(t^s)] \tag{1.55}$$

where the saddle point t^s is determined by the condition:

$$h'(t + i\hbar\beta)_{t + i\hbar\beta = t^s} = 0 . \tag{1.56}$$

Accordingly, we have

$$h'_{\Delta v}(t^s_{\Delta v}) = 0 = -(i/\hbar)(\varepsilon_\alpha - \varepsilon_{\alpha'} - \Delta v \hbar \omega_m) + L_m g_m^2 (n_m + 1) i \omega_m \exp(i \omega_m t^s_{\Delta v}) \tag{1.57}$$

from which we obtain

$$t^s_{\Delta v} = (i \omega_m)^{-1} \ln[(p - \Delta v)/L_m g_m^2 (n_m + 1)] \tag{1.58}$$

where

$$p \equiv (\varepsilon_\alpha - \varepsilon_{\alpha'})/\hbar \omega_m . \tag{1.59}$$

Substituting (1.58) in (1.48), we obtain

$$h_{\Delta v}(t^s_{\Delta v}) = -(p - \Delta v)\{\ln[(p - \Delta v)/L_m g_m^2 (n_m + 1)] - 1\} \tag{1.60}$$

and

$$h''_{\Delta v}(t^s_v) = -\omega_m^2(p - \Delta v) . \tag{1.61}$$

The final result for the radiationless relaxation in the weak-coupling limit is [1.2]

$$
\begin{aligned}
W_{\alpha\alpha'} &= \frac{(2\pi)^{1/2}}{\hbar^2} L_m |C^m_{\alpha\alpha'}|^2 \sum_{\Delta v = -2}^{\Delta v = 2} \lambda_{\Delta v} [-h''(t^s_{\Delta v})]^{-1/2} \\
&\quad \times \exp[-L_m g_m^2 (2 n_m + 1) + h_{\Delta v}(t^s_{\Delta v})] \\
&= \frac{(2\pi)^{1/2}}{\hbar^2} L_m |C^m_{\alpha\alpha'}|^2 \omega_m^{-1} \sum_{\Delta v = -2}^{\Delta v = 2} \lambda_{\Delta v} (p - \Delta v)^{-1/2} \\
&\quad \times \exp[-L_m g_m^2 (2 n_m + 1)] \\
&\quad \times \exp\left\{-(p - \Delta v)\left[\ln\left(\frac{(p - \Delta v)}{L_m g_m^2 (n_m + 1)}\right) - 1\right]\right\},
\end{aligned}
\tag{1.62}
$$

where the coefficients $\lambda_{\Delta v}$ are given by (1.49)–(1.53).

A prominent feature of the final expression (1.62) is that the rate constant expression is a sum of five contributions. In each of these contributions, the energy gap $\varepsilon_\alpha - \varepsilon_{\alpha'}$ of the transition appears to be modified by Δv vibrational quantum of the mediating mode. A qualitative survey of the general features of (1.62) shows that the rate constant varies sharply with the number $p - \Delta v$ of vibrational quanta that must be accepted by the vibrational mode which mediates the transition. Clearly, from the negative sign carried by $p - \Delta v$ in the exponent, we observe that the relaxation rate constant decreases exponentially with $p - \Delta v$.

1.3 The Nuclear Adiabatic Approximation

The theoretical development outlined in Section 1.2 has been specialized to multi-phonon relaxation of excited electronic states with the rate constant expression evaluated in the weak-coupling limit. In this section, we apply the relaxation theory in the strong-coupling limit in a discussion of the orientational relaxation of, say, a polar molecule (libron) between a number of equilibrium configurations. This discussion may be extended to a large number of thermally activated relaxation processes that occur in polyatomic molecules, in solids, and in liquids. It serves to illustrate how the electronic adiabatic approximation procedure of the preceding section can be modified in arriving at a formal description of nuclear adiabaticity.

We consider here the case in which configuration relaxation arises from the librational nuclear kinetic energy operator. When the libron motions are very slow compared to phonon motions (i.e., $\omega_l \ll \bar{\omega}_\rho$) and, the librational relaxation is slow compared to the lattice motions, we envisage the instantaneous adjustment of the motions of the lattice atoms to those of the libration. We invoke adiabatic libron-phonon separation and write

$$[T(r) + T(Q) + V(r, Q, R_l)]|\alpha(r, Q, R_l)\rangle = U_\alpha(R_l)|\alpha(r, Q, R_l)\rangle, \qquad (1.63)$$

where the phonon adiabatic states $|\alpha(r, Q, R_l)\rangle$ with energies $U_\alpha(R_l)$ are labeled by the configuration α, and where they vary parametrically with the librational coordinate R_l. In (1.63), $V(r, Q, R_l)$ represents all electron-electron and electron-nucleus interactions, and r and Q denote, respectively, all electronic and nuclear coordinates except for R_l. The librational nuclear eigenfunctions $\eta_{\alpha n}$ and energies E_n are accordingly evaluated from

$$[T(R_l) + U_\alpha(R_l)]|\eta_{\alpha, n}(R_l)\rangle = E_{\alpha, n}|\eta_{\alpha, n}(R_l)\rangle. \qquad (1.64)$$

In the above, the adiabatic approximation is in essence applied twice. First, the electrons are regarded as the fast subsystem with the lattice phonons regarded as the slow subsystem. Then, the lattice phonons are considered to be the fast subsystem with respect to the libration. This double separation preserves the integrity of the libration in a prescribed molecular configuration and gives physical meaning to invoking the single-minimum potential surfaces for the adiabatic configuration states $|\alpha(r, Q, R_l)\rangle$.

It follows from (1.63) and (1.64) that non-adiabatic couplings between the zeroth-order states $|\alpha(r, Q, R_l)\rangle$ occur through the off-diagonal elements of the nuclear kinetic energy operator $T(R_l)$ for the librons. The above statement can be easily verified along the lines of development given in Section 1.2. It is clear, for example, that (1.63) and (1.64) are formally equivalent to (1.2) and (1.3).

For the sake of simplicity, we consider only transitions between two orientation states of the dipole, respectively described by the nuclear adiabatic state vectors $|+\rangle$ and $|-\rangle$. In the absence of an external electric field these states are equivalent and degenerate, whereas in the presence of a field E they differ in energy by an amount

$$\Delta_\mu = 2\mu \cdot E \tag{1.65}$$

where μ is the permanent dipole moment.

Similar to the outline given in Section 1.2, the rate of transition from $|-\rangle$ to $|+\rangle$ can be written

$$W_{-+} = (\beta\langle \dot{N}_-\rangle)^{-1} \int_0^\infty dt \int_0^\beta d\lambda \langle \dot{N}_-(-t-i\hbar\lambda)\dot{N}_-(0)\rangle. \tag{1.66}$$

In (1.66), the flux operator $\dot{N}_-$ is the time derivative of the number operator $a_-^\dagger a_-$, $a_-^\dagger$, and a_- being the Fermion creation and destruction operators for the state $|-\rangle$. The flux operator is given by the Heisenberg equation of motion,

$$\dot{N}_- = (i/\hbar)[\mathcal{H}, N_-] \tag{1.67}$$

where $\mathcal{H}$, the total Hamiltonian may be written as

$$\mathcal{H} = \mathcal{H}_0 + \mathcal{H}'. \tag{1.68}$$

In (1.68), $\mathcal{H}_0$ is the zeroth-order Hamiltonian and $\mathcal{H}'$ is the perturbation responsible for the transitions.

Re-orientation transitions between the states $|-\rangle$ and $|+\rangle$ occur through the breakdown of the adiabatic approximation. The electron-"libron" coupling is caused by the kinetic energy of the librational motion of the re-orienting dipole. In the harmonic approximation, the adiabatic surfaces of the $|+\rangle$ and $|-\rangle$ states are two identical harmonic potentials displaced with respect to one another. The zero-order Hamiltonian is given in the second-quantized representation in analogy with (1.28)

$$\mathcal{H}_0 = \sum_{\alpha=-,+} \varepsilon_\alpha a_\alpha^\dagger a_\alpha + \sum_{\alpha=-,+} \sum_j \hbar\omega_\alpha^j a_\alpha^\dagger a_\alpha b_j^\dagger b_j \tag{1.69}$$

where

$$\varepsilon_\alpha = \varepsilon_\alpha^0 - \sum_j g_\alpha^{j2} \hbar\omega_\alpha^j. \tag{1.70}$$

The symbols $b_j^\dagger$ and b_j are the Boson creation and destruction operators for the j-th normal mode, including the librational mode; ε_α is the potential energy at the equilibrium configuration of the nuclei in state $|\alpha\rangle$; g_α^j, a reduced dimensionless displacement parameter associated with the j-th mode in state $|\alpha\rangle$, is given by

$$g^j_- = 0$$
$$g^j_+ = (M_j \omega^j_+/2\hbar)^{1/2} \Delta R^j_+ \tag{1.71}$$

in which M_j is the effective mass of the j-th mode, ΔR^j_+ is the displacement of the equilibrium position of the j-th mode of the adiabatic surface of state $|+\rangle$ from that of state $|-\rangle$, and ω^j_+ is the fundamental frequency of the j-th mode. Since the adiabatic surfaces of the two states are assumed to be identical,

$$\omega^j_- = \omega^j_+ = \omega^j . \tag{1.72}$$

We note that for the librational mode l, M_l becomes the moment of inertia I of the dipolar molecule and ΔR^l_α the angular displacement $\delta\theta$ of the equilibrium orientations. The librational kinetic-energy coupling operator responsible for the re-orientation transition is written after (1.20)

$$\mathcal{H}' = C^l_{-+} a^\dagger_- a_+ \exp[g^l_+(b_l - b^\dagger_l)](b_l - b^\dagger_l) \tag{1.73}$$

where the coupling parameter C^l_{-+} may be given explicitly as

$$C^l_{-+} = -(\hbar^3 \omega_l/2M_l)^{1/2}[S^l_{-+}(R_l)/\Delta(R_l)] \tag{1.74}$$

with

$$S^l_{-+}(R_l) = \langle -|\frac{\partial V(r,Q,R_l)}{\partial R_l}|+\rangle \tag{1.75}$$

and

$$\Delta(R_l) = U_-(R_l) - U_+(R_l) . \tag{1.76}$$

Now $\partial V(r,Q,R_l)/\partial R_l$ may be written approximately as

$$\partial V(r,Q,R_l)/\partial R_l = \partial V_0(r,Q,R_l)/\partial R_l + \partial \mu(r,Q,R_l)/\partial R_l \cdot E \tag{1.77}$$

where $V_0(r,Q,R_l)$ represents the interactions in the absence of a field, and $\mu(r,Q,R_l)$ is the instantaneous dipole moment. For most practical cases $|E|$ is sufficiently small, so that the second term in the above equation may be neglected.

Due to the energy denominator $\Delta(R_l)$, the coupling parameter C^l_{-+} appears to be divergent in regions where $\varepsilon_+(R_l) \simeq \varepsilon_-(R_l)$. However, it has been shown that C^l_{-+} is actually finite in this case, and (1.74) becomes [1.19]

$$C^l_{-+} = -(\hbar^3 \omega_l/2 M_l)^{1/2} \left[\frac{\partial S^l_{-+}(R_l)}{\partial R_\perp} \frac{\partial \Delta\varepsilon(R_l)}{\partial R_\perp} \right]_\Omega \tag{1.78}$$

where $R_\perp$ denotes the coordinate normal to the crossing surface Ω defined by $U_+(R_l) = U_-(R_l)$. Eq. (1.78) assumes $[\partial\Delta(R_l)/\partial R_l]_\Omega \neq 0$. (If this assumption is not true, the appropriate finite expression for C^l_{-+} can also be obtained if necessary [1.19]). From (1.74) and (1.78), we observe that the coupling parameter C^l_{-+} may have considerable variation with R_l, and this may greatly affect the theoretical rate of the relaxation process. For the weak coupling case, this effect has been studied by several workers who showed how the variation of C^l_{+-} with R can increase the predicted relaxation rate by orders of magnitude [1.20]. This variation with R is one important feature that should be incorporated in a fuller theory.

From (1.67) and (1.73), we obtain for the flux operator

$$\dot{\mathcal{N}}_- = -(i/\hbar) C^l_{-+} a^\dagger_- a_+ \exp[g^l_+ (b_l - b^\dagger_l)] + \text{h.c.} \tag{1.79}$$

so that (1.66) now yields according to (1.47)

$$W_{-+} = (|C^l_{-+}|^2/\hbar^2) \exp[-g^{l2}_+ (2n^l_- + 1)]$$
$$\times \sum_{\nu=-2}^{\nu=2} \lambda_{\Delta\nu} \int_{-\infty}^{+\infty} dt \exp[f_{\Delta\nu}(t)] \tag{1.80}$$

where

$$f_{\Delta\nu}(t) = -(i/\hbar)(\Delta_\mu - \Delta\nu\hbar\omega^l)(t + i\hbar\beta) + g^{l2}_+ \{(n^l_- + 1) \\ \times \exp[i\omega^l(t + i\hbar\beta)] + n^l_- \exp[-i\omega^l(t + i\hbar\beta)]\} \tag{1.81}$$

$$\lambda_0 = g^{l2}_+ [6 n^l_- (n^l_- + 1) + 1] \tag{1.82}$$

$$\lambda_1 = -2 g^{l2}_+ (2n^l_- + 1)(n^l_- + 1) + n^l_- + 1 \tag{1.83}$$

$$\lambda_{-1} = -2 g^{l2}_+ (2n^l_- + 1)n^l_- + n^l_- \tag{1.84}$$

$$\lambda_2 = g^{l2}_+ (n^l_- + 1)^2 \tag{1.85}$$

$$\lambda_{-2} = g^{l2}_+ n^l_- \tag{1.86}$$

and

$$n^l_- = [\exp(\hbar\omega^l\beta) - 1]^{-1} \tag{1.87}$$

is the Bose distribution function for the librational quantum levels in the initial state $|-\rangle$.

For ordinary field strengths,

$$\Delta_\mu - \Delta v\hbar\omega^l < g_+^{l2}\hbar\omega^l \tag{1.88}$$

and (1.80) can be evaluated in the strong-coupling limit by the saddle-point approximation. Eq. (1.81) can be rewritten

$$f_{\Delta v}(t+i\hbar\beta) = -(i/\hbar)(\Delta_\mu + \Delta v\hbar\omega^l)(t+i\hbar\beta) \\ + g_+^{l2}\{\cosh[\omega^l(\tfrac{1}{2}\hbar\beta + it)]/\sinh(\tfrac{1}{2}\hbar\omega^l\beta)\} . \tag{1.89}$$

The saddle point is determined by the equation,

$$f'_{\Delta v}(t^s) = 0 = -(i/\hbar)(\Delta_\mu + \Delta v\hbar\omega^l) \\ + \{i\omega^l g_+^{l2}\sinh[\omega^l(\tfrac{1}{2}\hbar\beta + it^s)]/\sinh\tfrac{1}{2}(\hbar\omega^l\beta)\} . \tag{1.90}$$

Using (1.88) and assuming that the temperature is sufficiently high so that

$$\beta^{-1} > \hbar\omega^l \tag{1.91}$$

we can expand $\sinh[\omega^l(\hbar\beta/2 + it^s)]$ to first order in $\omega^l(\hbar\beta/2 + it^s)$, and obtain

$$t^s = i[\tfrac{1}{2}\hbar\beta - (\Delta_\mu + \Delta v\hbar\omega^l)(2g_+^{l2}\omega^l/\beta)^{-1}] . \tag{1.92}$$

Accordingly,

$$f''(t^s) = -2g_+^{l2}\omega^l/\hbar\beta . \tag{1.93}$$

From (1.89) and (1.92), expanding $\cosh[\omega^l(\hbar\beta/2 + it^s)]$ to second order, we write

$$f(t^s) = [(\Delta_\mu + \Delta v\hbar\omega^l)/\hbar][\tfrac{1}{2}\hbar\beta - (\Delta_\mu + \Delta v\hbar\omega^l)(2g_+^{l2}\omega^l/\beta)^{-1}] \\ + g_+^{l2}\{1 + \tfrac{1}{2}[(\beta/2g_+^{l2})(\Delta_\mu + \Delta v\hbar\omega^l)]^2\}(\tfrac{1}{2}\hbar\omega^l\beta)^{-1} . \tag{1.94}$$

The final result for the re-orientation transition rate becomes [1.15]

$$W = [(2\pi)^{1/2}/\hbar^2]|C|^2[-f''_{\Delta v}(t^s)]^{-1/2} \\ \times \sum_{v=-2}^{v=2} \lambda_{\Delta v}\exp[-g^2(2n+1)+f_{\Delta v}(t^s)] \tag{1.95}$$

$$= [(4\pi)^{1/2}/\hbar]|C|^2(4g^2\hbar\omega kT)^{-1/2}\sum_{v=-2}^{v=2}\lambda_{\Delta v}\exp(-E_a^v/kT)$$

where

$$E_a^v = (\Delta_\mu + \Delta v\hbar\omega - g^2\hbar\omega)^2(4g^2\hbar\omega)^{-1} . \tag{1.96}$$

Here, $W = W_{-+}$, $C = C^l_{-+}$, $g = g^l_+$, $\omega = \omega^l$, and $n = n^l_-$, the subscripts and superscripts having been omitted for simplicity.

The rate equation (1.95) can be drastically simplified in view of (1.91). From (1.88) and (1.96), the activation energy can be written approximately as

$$E_a \simeq \tfrac{1}{4} g^2 \hbar \omega. \tag{1.97}$$

Expanding the coefficients λ_ν to second order in $\hbar \omega \beta$, we obtain from (1.82–86)

$$\lambda_0 = 2[3g^2 + \tfrac{1}{2} g^2 (\hbar \omega \beta)^2](\hbar \omega \beta)^{-2} \tag{1.98}$$

$$\lambda_1 + \lambda_{-1} = -\{8g^2[1 + \tfrac{1}{4}(\hbar \omega \beta)^2] + 2\hbar \omega \beta\}(\hbar \omega \beta)^{-2} \tag{1.99}$$

and

$$\lambda_2 + \lambda_{-2} = g^2[2 + (\hbar \omega \beta)^2](\hbar \omega \beta)^{-2}. \tag{1.100}$$

Using (1.97–100) in (1.95) yields the simplified expression for the transition rate [1.15]

$$W = (k T/\hbar)[(4\pi)^{1/2} |C|^2/(g^2 \hbar \omega k T)^{1/2} \hbar \omega]$$
$$\times \exp(-g^2 \hbar \omega/4 k T) \tag{1.101}$$

which is displayed in the form of the familiar Arrhenius equation. Finally, from (1.71) and (1.74) we rewrite (1.101) as

$$W = (k T/h)[(2\pi M k T/h^2)^{1/2} \Delta R]^{-1}$$
$$\times [(2h\pi/M\omega)|S|^2 \Delta \varepsilon^{-2}] \exp[-M\omega^2 (\Delta R)^2/8 k T] \tag{1.102}$$

where $S = S^l_{-+}$, $M = M_l$, and ΔR^l_+. The activation barrier

$$E_a = \tfrac{1}{8} M \omega^2 (\Delta R)^2 \tag{1.103}$$

is just the height of the intersection of the adiabatic surfaces.

1.4 Extended Applications of the Adiabatic Theory of Molecular Relaxation

The two specific applications of the adiabatic approximation detailed above illustrate the diversity of radiationless phenomena that can be treated within the framework of the adiabatic theory of molecular relaxation [1.1]. The weak coupling limit theory of electronic relaxation has been generalized to an analysis of rare-earth ion $f \rightarrow d$ (intermediate coupling) relaxation [1.3], and to the (strong coupling) description of photoisomerization [1.4] and electron transfer [1.21, 22] phenomena. With minor modifications in the form of the electronic coupling $C_{\alpha\alpha'}$ and in the choice of the adiabatic basis vectors, the weak-coupling theory of Section 1.2 can be adopted in the treatment of inter-system crossing [1.8] and non-resonant energy transfer between guest molecules in a host lattice [1.10]. The rate constant expression in Section 1.2 has been calculated to an average over a Boltzmann distribution of the initial vibronic manifold, a procedure that is particularly meaningful in condensed media. Corresponding rate constant expressions can be readily obtained [1.23–26] for single vibronic relaxation processes that are important in low gas pressure conditions.

The strong-coupling theory derived in the nuclear adiabatic approximation in Section 1.3 has been specialized to the problem of dipolar re-orientation in a condensed medium. The Arrhenius-type rate constant expression (1.102) has also been applied to a discussion [1.27] of configuration rearrangement in carbonium ions. Various coupling interactions in the nuclear adiabatic theory can give rise to alternative mechanisms for nuclear rearrangement [1.5]. In the weak-coupling limit, the nuclear adiabatic theory has provided a useful description for the relaxation of vibrationally excited molecules in condensed media [1.11–14].

We thus envisage a systematized approach to a wide variety of radiationless transitions in terms of judiciously chosen adiabatic bases of stationary states and physically meaningful non-adiabatic perturbation operators. Recent developments seem to indicate that the study of non-radiative processes has come of age in the sense that we now have within our reach a body of postulates that provide us with a general theory based on the concept of adiabaticity. This theory has in recent years imparted considerable momentum to investigations into molecular relaxation phenomena, as an increasing number of isolated experimental observations become incorporated into a unified view of the complex field.

1.5 Outline

The first recorded observation of energy transfer was described in the paper by Cario and Franck [1.28] who observed in 1923 the emission spectra of both mercury and thallium in gaseous mixtures irradiated with the light of the mercury resonance line. The thallium luminescence, sensitized by the mercury excitation, has resulted from an energy transfer transition in which the donor and the acceptor atoms are coupled not only to the electromagnetic field, but also to each other.

The history of phonon-induced relaxation traces to an early work by Hellwege [1.29] who related the probability for non-radiative decay in rare-earth salts to the energy separation between the excited state and the next lower state. Hellwege's work gave rationalization to the observation that only fluorescence from hydrated ions in the center of the lanthanide series could be observed. Ions in the center of the lanthanide series possess energy gaps that are large compared to those at the beginning or the end of the series. Non-radiative decay of these central ions would thus necessarily excite a larger number of vibrational quanta in the ligand molecules, resulting in a smaller probability for such decay.

The succeeding decades since these early investigations have seen an ever-increasing number of studies on sensitized luminescence and non-radiative relaxation phenomena. Recent investigations have been heightened by the many advances in experimental laser spectroscopy [1.30], and by a number of significant theoretical developments. The widespread interest has resulted from the breadth of research areas encompassed by radiationless relaxation effects. A partial list includes (A) non-radiative in electronically excited molecules, (B) radiationless relaxation of vibrationally excited molecules in solid matrices, (C) photodissociation, and (D) energy transfer and migration processes.

Chapter 2 written by KARL F. FREED of the University of Chicago, is entitled "Energy Dependence of Electronic Relaxation Processes in Poly-atomic Molecules". It is primarily concerned with the historical origin and current developments in the study of radiationless relaxation of electronically excited molecules. Chapter 3, authored by DENNIS J. DIESTLER of Purdue University, is titled "Vibrational Relaxation of Molecules in Condensed Media". It deals with the extension of the quantum mechanical theory of molecular relaxation to the interpretation of the observed experimental behavior of molecular vibrational relaxation in solid and liquid hosts. These two chapters are complementary in that they demonstrate the various manners in which the basically same principles of adiabatic separation of degrees of freedom can be applied leading to the theoretical description of different physical phenomena. The relaxation processes described in these two chapters are closely related to the multi-phonon decay of electronically excited ions in crystals reviewed elsewhere [1.1]. All three cases are examples of radiationless relaxation in the weak coupling limit. The separation of the normal modes of the active molecules and those of the lattice in the weak coupling theory of vibrational relaxation is generically associated with the strong coupling theory of activated rate processes outlined in Section 1.3.

The participation of phonons or molecular normal mode vibrations plays an important role in the description of radiationless relaxation phenomena outlined in the preceding sections of this introduction and discussed in greater details for specific examples in Chapters 2 and 3. The need for phonon participation has arisen in part from the energy conservation requirement in that the excitation and deexcitation processes are mediated by the corresponding processes of phonon absorption and emission.

It is necessary to point out, however, that a great many current investigations are focused on non-radiative processes in which phonons play no significant part in the relaxation mechanism. Most commonly encountered examples of such phenomena are resonant energy transfer processes of the type first discovered by Cario and Franck [1.28]. It appears that a balanced coverage of the current status of research in non-radiative relaxation should provide a review of the latest developments in the study of energy transfer. This review, given in Chapters 4 and 5, affords a contrast to the material covered in Chapters 2 and 3 and brings additional dimensions to the scope of our study.

Chapters 4 and 5, written by John C. Wright of the University of Wisconsin and Raoul Kopelman of the University of Michigan, respectively, present two rather different aspects of the energy transfer problem. Chapter 4 is mostly concerned with resonance energy transfer processes involving rare-earth ions in single crystals. In addition, certain empirical aspects of non-resonant energy transfer phenomena are described. Chapter 5 describes the onset of cooperative energy transfer effects in molecular aggregates and mixed crystals in terms of percolation theory. The energy gap law that has been derived in the weak coupling limit in Chapters 2 and 3 can be readily extended to account for the experimental behavior of off-resonance energy transfer between donor and acceptor molecules described in Chapter 4. The cooperative energy transfer effects in mixed crystals, examined in terms of percolation theory in Chapter 5, provides comparison and contrast to the theoretical descriptions [1.1] for the cooperative multi-phonon decay of excited ions in crystals and exciton interactions in the primary molecular adducts in photosynthesis. The concepts described in Chapter 5 are relatively recent developments. They offer promise for the treatment of a variety of energy transfer phenomena in condensed media.

The references cited in the reviews are extensive although by no means comprehensive. Inevitably the selection of the material reviewed reflects the scope of interest defined by the editor and the contributors. It is hoped that the present work not only offers the reader a reasonably current view of several areas of research in molecular relaxation, but also an integrated overview of the complex and multifarious research field. It is also hoped that the emphasis on theory and experiment is sufficiently well balanced so that the book is useful to the experimentalist and the theorist alike.

References

1.1   F. K. Fong: *Theory of Molecular Relaxation: Applications in Chemistry and Biology* (Wiley-Interscience, New York 1975)

1.2   F. K. Fong, S. L. Naberhuis, M. M. Miller: J. Chem. Phys. **56**, 4020 (1972);
F. K. Fong, W. A. Wassam: *ibid.* **58**, 956, 2667 (1973);
F. K. Fong, H. V. Lauer, C. R. Chilver, M. M. Miller: J. Chem. Phys. **63**, 366 (1975)

1.3   H. V. Lauer, F. K. Fong: J. Chem. Phys. **60**, 274 (1974)

1.4   W. M. Gelbart, K. F. Freed, S. A. Rice: J. Chem. Phys. **52**, 6272 (1970)

1.5 K. F. Freed, F. K. Fong: J. Chem. Phys. **63**, 2890 (1975)
1.6 S. H. Lin, R. Bersqhn: J. Chem. Phys. **48**, 2732 (1968);
     K. F. Freed, J. Jortner: J. Chem. Phys. **52**, 6272 (1970)
1.7 M. Bixon, J. Jortner: J. Chem. Phys. **48**, 715 (1968)
1.8 S. F. Fischer: J. Chem. Phys. **53**, 3195 (1970)
1.9 T. Miyakawa, D. L. Dexter: Phys. Rev. B **1**, 2961 (1970)
1.10 W. A. Wassam, F. K. Fong: J. Chem. Phys. (1976)
1.11 A. Nitzan, J. Jortner: Mol. Phys. **25**, 713 (1973);
     A. Nitzan, S. Mukamel, J. Jortner: *ibid*, **60**, 3929 (1974)
1.12 D. J. Diestler: J. Chem. Phys. **60**, 2692 (1974); Chem. Phys. **7**, 349 (1975).
1.13 S. H. Lin: J. Chem. Phys. **61**, 3810 (1974)
1.14 A. Nitzan, S. Mukamel, J. Jortner: J. Chem. Phys. **63**, 200 (1975)
1.15 F. K. Fong, D. J. Diestler: J. Chem. Phys. **57**, 4983 (1972)
1.16 C. P. Flynn, A. M. Stoneham: Phys. Rev. B **1**, 3966 (1970)
1.17 S. H. Lin, H. Eyring: Proc. Nat. Acad. Sci. (USA) **69**, 3192 (1972)
1.18 R. Kubo, M. Yokota, S. Nakajima: J. Phys. Soc. Japan **12**, 1203 (1957). For a recent review,
     see [Ref. 1.1, Chapt. 3]
1.19 K. F. Freed, S. H. Lin: Chem. Phys. **11**, 409 (1975)
1.20 B. Sharf, R. Silbey: Chem. Phys. Lett. **4**, 423 (1969); **4**, 561 (1970); **9**, 125 (1971);
     A. Nitzan, J. Jortner: J. Chem. Phys. **56**, 3360 (1972);
     V. A. Kovarskii: Fiz. Tverd. Tela. **4**, 1636 (1962) [Sov. Phys.-Solid State **4**, 1200 (1963)];
     V. A. Kovarskii, E. P. Smyavskii: *ibid*. **4**, 3202 (1962) [*ibid*. **4**, 2345 (1963)];
     M. G. Prais, D. F. Heller, K. F. Freed: Chem. Phys. **6**, 331 (1974)
1.21 R. P. Vanduyne, S. F. Fischer: Chem. Phys. **5**, 183 (1974)
1.22 N. R. Kestner, J. Logan, J. Jortner: J. Phys. Chem. **78**, 2148 (1974)
1.23 A. D. Brailsford, T. Y. Chang: J. Chem. Phys. **53**, 3108 (1970)
1.24 A. Nitzan, J. Jortner: J. Chem. Phys. **55**, 1355 (1971)
1.25 A. Nitzan, J. Jortner: J. Chem. Phys. **56**, 2079 (1972)
1.26 F. K. Fong, W. A. Wassam: J. Chem. Phys. **58**, 2667 (1973)
1.27 F. K. Fong: J. Am. Chem. Soc. **96**, 7638 (1974)
1.28 G. Cario, J. Franck: Z. Physik **17**, 202 (1923)
1.29 K. H. Hellwege: Ann. Phys. (Leipzig) **40**, 529 (1942)
1.30 H. Walther (ed.): Laser Spectroscopy of Atoms and Molecules. In: *Topics in Applied Physics*,
     Vol. 2 (Springer, Berlin, Heidelberg, New York 1976)

2. Energy Dependence of Electronic Relaxation Processes in Polyatomic Molecules

K. F. Freed

With 15 Figures

A review is presented of studies of electronic relaxation processes in poly-atomic molecules under isolated molecule conditions. The general mechanistic aspects of electronic relaxation are presented in the small, intermediate, and statistical molecule limits. A general quantum mechanical description of these phenomena is described, and an application elucidates the transition from the small to the statistical limits. An introduction is also provided to the formal projection operator-Green's function techniques which are increasingly being employed in theoretical analyses. We describe recent theoretical developments concerning the theory of the vibronic state dependence of electronic relaxation rates. First the simple Boltzmann statistics method is used for the displaced potential model, then the full many-phonon generating function methods are described. Applications are presented to cases of electronic relaxation processes with moderate and with large energy gaps between the initial and final electronic states. Detailed comparisons between experiment and theory are discussed for benzene and various deuterated benzenes. Discussions are provided of a number of other interesting aspects of electronic relaxation processes in various aromatic hydrocarbons.

2.1 Early Studies of Electronic Relaxation in Condensed Systems

There have recently been great advances in both our theoretical and experimental understanding of electronic relaxation processes in polyatomic molecules [2.1–9]. It is convenient in this review to begin by adopting a historical approach [2.6] to briefly consider those features that have contributed to the conceptual development of the subject and/or those that are necessary to the discussion of the energy dependence of these processes.

Early experimental investigations were primarily concerned with studies of the luminescence properties of large molecules in condensed media such as solutions and rigid media [2.6, 10–13]. This work could only provide measurements of the thermally averaged radiative lifetimes $\tau^{rad}(T)$ and quantum yields $\varphi(T)$ (or their equivalents). The thermally averaged pure radiative k_{rad} and non-radiative decay rates k_{nr} could then be obtained by using the customary relationship

$$\varphi(T) \equiv k_{rad}/(k_{rad} + k_{nr}) \tag{2.1}$$

$$\tau^{rad}(T) \equiv (k_{rad} + k_{nr})^{-1} . \tag{2.2}$$

In general, for the case of the aromatic molecules, which are primarily considered in this review, it was found that the observed quantum yields were less than unity (often considerably so) and that the observed radiative lifetime was shorter than the pure radiative value $(k_{rad})^{-1}$ deduced from (2.1) and (2.2) or from the integrated oscillator strength for the transition.

These observations were taken to be indicative of an electronic relaxation process. This relaxation has been explained on the basis of the simple molecular energy level scheme depicted in Fig. 2.1 which is expected to be common for

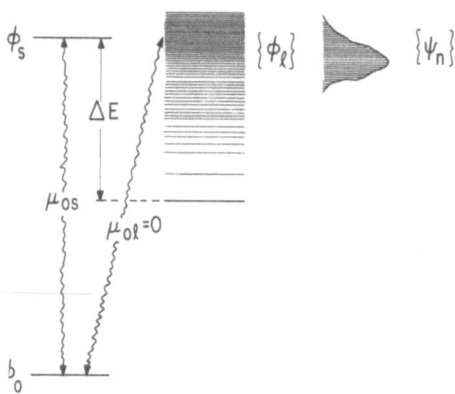

Fig. 2.1. The molecular energy level model used to discuss radiationless decay processes in polyatomic molecules. The states ϕ_0, ϕ_s, and $\{\phi_l\}$ are Born-Oppenheimer zeroth-order molecular levels. They correspond, respectively, to one of the thermally accessible ground electronic state vibronic levels, a vibronic component of an optically accessible excited electronic state, and a dense manifold of excited vibronic levels that belongs to electronic states with an origin below that of ϕ_s. μ_{0s} is the (assumed) non-zero optical transition dipole between ϕ_0 and ϕ_s, while the corresponding μ_{0l} is assumed to vanish, so optical transitions between ϕ_0 and $\{\phi_l\}$ are forbidden

all polyatomic molecules. We first assume that in zeroth-order the Born-Oppenheimer (BO) separability between the electronic and nuclear vibrational motions can be approximately invoked. Thus, the ϕ-levels in Fig. 2.1 represent these theoretical BO energy levels. There then arises the question as to which of the many particular choices of the BO approximations should be invoked. Although the purely mechanistic questions under discussion at this juncture do not depend on the choice of BO approximation, it has been the source of sufficient controversy and confusion that it is worthwhile to briefly digress to resolve this question [2.4, 14–17].

Obviously, it is best to try to make the theoretical zeroth-order states most closely resemble the states that are actually observed in reality. The crude BO approximation (CBO), however, considers all its electronic states to be determined at a single nuclear configuration $Q^{(0)}$. $Q^{(0)}$ is generally taken to be the equilibrium position of the ground electronic state. Thus, crude BO wavefunctions $\phi^{CBO}(q, Q^{(0)})$ do not allow the electrons (with coordinates q) to follow the vibrations of nuclei. It is well known that force constants derived from $\phi^{CBO}(q, Q^{(0)})$ are generally in error even by an order of magnitude [2.16]. As the nuclei vibrate, $\phi^{CBO}(q, Q^{(0)})$ forces the electrons to remain centered about Q_0. Thus, the center of charge, due to the core electrons about an atom, is described as displaced from the nuclear position, thereby generating unphysically large forces on the nuclei. In the adiabatic BO approximation (ABO) the electronic wavefunction $\phi^{ABO}(q, Q)$ are recomputed at each nuclear configuration. $\phi^{ABO}(q, Q)$, hence, describes the electrons as perfectly following the slower nuclear motion. This approximation is known to give a very accurate representation of the vibration-rotation constants of molecules [2.18] as the corrections due to electron slippage are rather small. In conclusion, the adiabatic BO approximation is the one which most faithfully represents the true molecular levels that are actually prepared experimentally. Consequently, the theoretical energy levels in Fig. 2.1 are taken as adiabatic BO ones.[1]

Returning now to Fig. 2.1, we note that it represents the polyatomic molecules as having a ground electronic state and a number of excited electronic states. One of the accessible vibronic components of the ground electronic state, ϕ_0, is explicitly shown. ϕ_0 is assumed to have a non-zero oscillator strength for an electric dipole transition to a vibronic component of an excited electronic state ϕ_s. (This oscillator strength can be intrinsic or be due to vibronic coupling with other electronic states ϕ_{vc} which are not explicitly represented in the figure. Consequently, as discussed in Section 4, the transition dipole μ_{0s} as well as all couplings v_{sl} below are taken to be effective ones, including contributions from all the other electronic states that are not displayed in Fig. 2.1). Isoenergetic with ϕ_s is a dense manifold of vibronic levels $\{\phi_l\}$ which belong to some lower electronic state(s), including high lying vibrational components of the ground electronic state. It is assumed that optical transitions between ϕ_0 and the $\{\phi_l\}$ are essentially forbidden due to unfavorable spin symmetry and/or very small Franck-Condon factors.

The luminescence properties, observed in these early experiments in condensed media, of polyatomic molecules that are characterized by the energy level scheme (Fig. 2.1) are greatly affected by the fact that any BO approximation is not exact. When studying the properties of the electronic ground states of stable molecules, it is generally rare to consider non-adiabatic corrections to the BO approximation. One situation in which these effects have a qualitative influence on the molecular energy levels arises when the molecule is a free radical and has degeneracies due to the presence of non-zero electronic spin and/or orbital angular momenta. In this case, perturbations coupling the ground and excited electronic state are introduced by some form of de-

---

[1] It could then be argued that the crude BO wavefunctions are still best for the evaluation of non-radiative decay rates and other transition rates between electronic states. However, it is well known that the evaluation of transition probabilities between different electronic states, such as oscillator strengths, is much more sensitive to errors in the wavefunction than is the calculation of properties of a single electronic state. Since an enormous admixture of other CBO electronic states is required to rectify its deficiencies in even correctly describing the force constants of molecular electronic states, this admixture should be expected to greatly affect calculated non-radiative decay rates. Therefore, as the use of adiabatic BO functions is more appropriate to individual electronic states, it is expected to likewise be far superior for the evaluation of transition rates.

generate perturbation theory. For diatomic molecules this results in the lifting of the degeneracies, the well known effects of rho-type and lambda doubling [2.19]. However, the non-adiabatic mixing of higher electronic states into the ground one is generally so small that it is neglected except insofar as it leads to a lifting of level degeneracies (and possible modification of their g-factors, etc.). Thus, in this case, the non-adiabatic corrections have a pronounced effect.

We should likewise expect that non-adiabatic interactions are of importance in situations where they can couple otherwise degenerate electronic states. Hence, in Fig. 2.1 ϕ_s and a large number of $\{\phi_l\}$ are quasi-degenerate, and they are coupled by non-adiabatic correction terms which lead to the observed non-radiative decay of ϕ_s in condensed media. This qualitative importance of the non-adiabatic interactions is often referred to as "the breakdown of the Born-Oppenheimer approximation" which is rather dramatic sounding language for the effects of couplings between degenerate zeroth-order levels.

Given the non-adiabatic interactions, the observed electronic relaxation in condensed media is easily explained in terms of simple time dependent quantum mechanics. A molecule, initially in ϕ_0, absorbs incident radiation with frequency corresponding to the energy difference between ϕ_s and ϕ_0. The molecule is taken to be excited to ϕ_s as this is the only zeroth-order level which carries oscillator strength from ϕ_0. Subsequently, the molecule in ϕ_s can either radiate to some (possibly other) vibronic component of the ground (or other) electronic state $\phi_0^{v\ 2}$ or it can undergo a non-radiative transition to the $\{\phi_l\}$ because of the non-adiabatic coupling between ϕ_s and $\{\phi_l\}$. From first-order time-dependent perturbation theory, the non-radiative decay rate is given by the "golden rule" formula ($\beta = 1/kT$ as usual)

$$k_{\mathrm{nr}}(s \to l; \beta) = \frac{2\pi}{\hbar} \sum_{i,j} p(s\,i) |\langle s\,i | V | l\,j \rangle|^2 \, \delta(E_{si} - E_{lj}). \tag{2.3}$$

In (2.3), it is assumed that the molecules in the initial electronic state are vibrationally thermally equilibrated in the dense media on time scales which are rapid compared to observed radiative decay rates from ϕ_s. Consequently, (2.3) is the thermally averaged decay rate for electronic state s, and $p(s\,i)$ is the familiar Boltzmann distribution giving the probability of finding the i-th vibrational level at vibrational thermal equilibrium. The non-adiabatic coupling is $\langle s\,i | V | l\,j \rangle$, and the Dirac delta function in (2.3) assures conservation of energy in the non-radiative decay.

---

[2] ϕ_0^v is generally a vibronic component of the ground state which is different from the initial state ϕ_0 before absorption because the fluorescence spectrum, etc., is usually red shifted from the absorption spectrum.

Because of rapid vibrational relaxation in condensed media, the initially formed vibrationally hot states $|lj\rangle$ are rapidly vibrationally deactivated by the solvent. Thus, it is more correct to also include the degrees of freedom corresponding to the host and treat $|si\rangle$ and $|lj\rangle$ as large supermolecules including both the guest molecule and host. If the host medium is "inert" in the sense that it does not modify the non-adiabatic coupling V, the molecular oscillator strengths, or the molecular energy levels E_{si} and E_{lj}, its only function is that of a heat bath. Then the summation over final levels j becomes an integration over the continuous density of states provided by the levels $|lj\rangle$ imbedded in the solvent. Rather than explicitly considering the energy levels of the "inert" host medium, it is often convenient to realize that the molecule-solvent coupling gives rise to a broadening of each of the $|lj\rangle$ levels. The summation over j for an isolated molecule now becomes

$$\sum_j \to \sum_j \int dE_{lj}\, \rho_{lj}(E_{lj}), \tag{2.4}$$

where $\rho_{lj}(E_{lj})$ is the density of guest molecule states at energy E_{lj} arising from the coupling of $|lj\rangle$ to the solvent. Substituting (2.4) into (2.3) then gives

$$k_{nr}(s \to l; \beta) = \frac{2\pi}{\hbar} \sum_{i,j} p(si)|\langle si|V|lj\rangle|^2\, \rho_{lj}(E_{si}), \tag{2.5}$$

which is essentially the thermally averaged version of the pioneering ROBINSON-FROSCH formula [2.6, 11], see Section 2.6.

The "golden rule" rate expression was then further approximated by invoking the Condon approximation which assumes that the electronic matrix elements β_e^2 can be factored out of the integration over nuclear coordinates and can be evaluated at the nuclear equilibrium position. Considering only a single vibronic level si, it was then popular to introduce the average Franck-Condon factor

$$F = \sum_j{}' |\langle i_s|j_l\rangle|^2, \tag{2.6}$$

where the primed summation in (2.6) implies that only the energy conserving vibrational overlaps are to be included. Given these approximations and the realization that all final vibrational levels $|lj\rangle$ are not as effective in the electronic relaxation process (because of the symmetry restrictions leading to small or vanishing Franck-Condon factors), the single level decay rate was written as

$$k_{nr}(si \to l) \approx \frac{2\pi}{\hbar} \beta_e^2 F \rho_{eff}, \tag{2.7}$$

where ρ_{eff} is the *effective* density of $|lj\rangle$ states. The quantity ρ_{eff} is then only defined as that density of states which assures equality between (2.7) and the rate

$$k_{\text{nr}}(si \to l) = \frac{2\pi}{\hbar} \beta_e^2 \sum_j |\langle i_s | j_l \rangle|^2 \rho_{lj}(E_{si}),$$ (2.8)

so ρ_{eff} in (2.7) plays the role in (2.7) of the "ubiquitous fudge-factor". Even given the crudeness of (2.7), at the time of its formulation it provided an enormous increase in the understanding of non-radiative decay processes in large molecules as the available experimental data were generally no more refined than (2.7) itself.

Since experiments in "inert" dense media could only provide thermally averaged rates

$$k_{\text{nr}}(s \to l; \beta) \approx \frac{2\pi}{\hbar} \beta_e^2 \langle F \rangle \rho_{\text{eff}},$$ (2.9)

where $\langle F \rangle$ is the thermal average of (2.6), it is hard to directly test the theory by using (2.9) to compare the single observed thermal decay rate $k_{\text{nr}}(s \to l; \beta)$ with the three theoretical quantities $\beta_e^2, \langle F \rangle$, and the "fudge-factor" ρ_{eff}. As noted earlier, the non-adiabatic interaction matrix elements β_e^2 are expected to be rather sensitive to the quality of approximate wavefunctions, so even wavefunctions of Hartree-Fock accuracy for a molecule like benzene could not *a priori* be expected to correctly determine β_e^2 to within an order of magnitude. Furthermore, the errors inherent in the use of the Condon approximation are probably even of greater magnitude [2.20–23]. The average Franck-Condon factor $\langle F \rangle$ depends only on the relative shapes and positions of the s and l electronic potential surfaces (and the thermal average). Consequently, given some spectroscopic data and reasonable assumptions, a direct calculation of $\langle F \rangle$ is quite possible. But even an error of 5% in the Franck-Condon factor for each individual vibrational mode in, say, benzene—a miniscule error considering the inherent approximations—would lead to an uncertainty of a factor of 4 in $\langle F \rangle$. These substantial errors are compounded by the uncertainty in the estimation of ρ_{eff} and questions of the assumption of the inertness of the host medium. Thus, a more stringent test of the theory requires first the measurement of sufficient additional experimental data to overdetermine the theoretical parameters, and the removal of some of the above questionable assumptions concerning both the theory and the interpretation of the experiments.

In the search for additional experimental information, it is natural to attempt to measure non-radiative decay rates of individual vibrational levels as in (2.8). The observed rapid vibrational relaxation in dense media [2.24] then implies the consideration of experiments in low pressure gases where vibrational deactivation of the $|si\rangle$ levels is slow compared to their radiative and non-radiative decays. In this case it then is natural to avoid the ambiguities inherent in the transition from (2.8) to (2.7), and to maintain the full density of states weighted Franck-Condon factor. Errors associated with the accurate evaluation of electronic matrix elements are minimized by the consideration of relative non-radiative decay rates [2.25–30]

$$
\frac{k_{nr}(si \to l)}{k_{nr}(si' \to l)} = \frac{\sum_j |\langle i_s | l_j \rangle|^2 \, \delta(E_{si} - E_{lj})}{\sum_j |\langle i'_s | l_j \rangle|^2 \, \delta(E_{si'} - E_{lj})},
\tag{2.10}
$$

wherein the nasty electronic factors cancel, leaving a quantity on the right in (2.10) which depends only on the relative nature of the s and l potential surfaces.

The theoretical developments associated with the understanding of the vibrational energy dependence of the non-radiative decay rates are the main focus of this review. However, since the associated experiments could only be performed in low pressure gas, there initially arose a series of conceptual problems associated with electronic relaxation in isolated gas phase poly-atomic molecules which required resolution prior to the consideration of individual vibronic level decay rates. Some of this conceptual framework is necessary in order to properly describe the energy dependence of electronic relaxation rates, so the salient features are reviewed in Sections 2.2–5. Emphasis is placed upon the major conceptual errors in the primitive theory that have not already been briefly noted. In Sections 2.6–9 we review the current status of our understanding of single vibronic level decay rates. Current work is considering the beginnings of a transition back to condensed media by a consideration of the pressure dependence of electronic relaxation [2.31–34], but this field is beyond the scope of a review on isolated molecule properties. All of these areas are in a rapid state of development, and many concepts and methods are subject to future modification. We concentrate on those features which are either pedagogically useful or which exhibit fundamental principles which will still appear in the future improved theories. A reviewer's prerogative is taken of emphasizing more strongly our own work, as the field is now too large to be covered by even this volume. The other contributions to this volume and the original literature references should provide access to the remainder of the work. In the following sections, emphasis is place on major conceptual and physical features, so an attempt is made to keep the mathe-matical analysis to a minimum in most of the sections. However, because the subsequent understanding of the original literature requires a familiarity with the techniques employed, some of them are introduced in Sections 2.4, 5, and 8, so these sections may be skipped by readers who are not interested in these details.

We should note in passing that the treatment of the energy dependence electronic relaxation processes is of relevance to the subsequent development of satisfactory theories of the energy dependence of product yield and energy distributions in photochemical reactions [2.9, 35]. In these cases, partial rates replace those in (2.10) wherein the summation is only over a subset of the possible final levels corresponding to different reaction products and/or different internal energy states of the products. The theory of these photo-dissociation rates is just beginning to be developed, but it is already clear that a number of features are in common with those considered in this review for the simpler case of electronic relaxation processes.

2.2 The Transition to Gas Phase Studies

Based on the theoretical molecular energy level scheme of Fig. 2.1, the antici-pated results of low pressure studies of electronic relaxation were readily deduced. Under "isolated" molecule conditions, the effects of collisions with other molecules in the system can safely be ignored. Thus, upon optical ex-citation the molecule is initially excited to ϕ_s since it is the only zeroth-order level in the appropriate spectral region which carries oscillator strength from ϕ_0. Because of the non-adiabatic coupling, ϕ_s is coupled to the dense manifold of discrete levels $\{\phi_l\}$. Consequently, ϕ_s is a non-stationary state of the molecular Hamiltonian (in the absence of coupling to the radiation field). So again, as in condensed media, the molecule in ϕ_s has some probability P_r of radiating a photon and a probability $1 - P_r$ of "crossing-over" to the $\{\phi_l\}$. When the ϕ_s and $\{\phi_l\}$ have the same spin symmetry, this "crossing" is termed internal conversion, while intersystem crossing corresponds to situations in which these two electronic states have different spin symmetries.

2.2.1 The Irreversibility Problem

Because of the absence of vibrationally relaxing collisions, the $\{\phi_l\}$ levels no longer rapidly lose their excess vibrational energy in the very low pressure gas. Thus, the $\{\phi_l\}$ are still coupled back to ϕ_s. Putting it another way, ϕ_s and $\{\phi_l\}$ represent a *finite* set of coupled states which are in resonance. Conse-quently, when the molecule is in $\{\phi_l\}$, it will eventually develop some probability amplitude for being in ϕ_s again. Once in ϕ_s, the molecule has probability P_r of radiating and $1 - P_r$ of crossing back to $\{\phi_l\}$. By induction, the total probability of reradiating the incident radiation is expected to be unity in an isolated molecule. This implies a unit quantum yield φ and a radiative lifetime τ^{rad} which agrees with that deduced from the integrated absorption intensity.

Electronic relaxation experiments at very low pressures were first carried out by KISTIAKOWSKY and PARMENTER and coworkers for the benzene molecule [2.36, 37]. In this case, ϕ_0 is the ground singlet S_0 state, ϕ_s is the first excited singlet electronic state S_1, while $\{\phi_l\}$ may contain high lying levels of S_0 and triplet levels T. The experimental results showed conclusively that the fluorescence quantum yield φ_f of benzene is non-zero (≈ 0.4) at low pressures, where no noticeable pressure dependence can be observed, and that the fluorescence lifetime τ_f is considerably shortened from that obtained from the oscillator strength for the $S_1 \leftarrow S_0$ transition. KISTIAKOWSKY and PARMENTER astutely noted in their classic paper [2.36, 37] that these results appeared to violate the laws of quantum mechanics—those laws which above are used to predict unit quantum yields.

2.2.2 The Small Molecule Limit

At about the same time, the situation was thrown into even more confusion by the experimental results of DOUGLAS who measured fluorescence lifetimes for small polyatomic molecules [2.38]. He found that the observed lifetimes in SO_2 and NO_2 were considerably longer than those deduced from the integrated absorption intensity. It was already well known that the absorption spectra of these molecules contain a large number of unexpected additional lines, and to date they have resisted satisfactory assignment. DOUGLAS explained both this appearance of extra lines and the lifetime lengthening on the basis of the molecular energy level scheme in Fig. 2.1 that is expected to also apply to these molecules. DOUGLAS argued that the coupling between ϕ_s and $\{\phi_l\}$ should be treated by diagonalizing the molecular Hamiltonian to yield the molecular eigenstates $\{\psi_n\}$ which are also represented schematically in Fig. 2.1. Each ψ_n contains a piece of the original ϕ_s via

$$\psi_n = C_{sn}\phi_s + \sum_l C_{ln}\phi_l . \tag{2.11}$$

If μ is the dipole operator, this implies that

$$\langle\phi_0|\mu|\psi_n\rangle = C_{sn}\langle\phi_0|\mu|\phi_s\rangle + \sum_l C_{ln}\langle\phi_0|\mu|\phi_l\rangle$$

$$= C_{sn}\langle\phi_0|\mu|\phi_s\rangle \equiv C_{sn}\mu_{0s} \tag{2.12}$$

because of the assumption that

$$\langle\phi_0|\mu|\phi_l\rangle \cong 0 \tag{2.13}$$

for the thermally accessible ground state levels ϕ_0. Since the $\phi_0 \rightarrow \psi_n$ oscillator strength $f_{0\rightarrow n}$ is proportional to $|\langle\phi_0|\mu|\psi_n\rangle|^2$,

$$f_{0\rightarrow n} \propto |C_{sn}|^2 |\mu_{0s}|^2 , \tag{2.14}$$

absorption lines are seen to those ψ_n which contain sufficient amounts of ϕ_s, i.e. have $|C_{ns}|^2$ large enough. Because of the strong $\phi_s - \{\phi_l\}$ coupling, many such extra lines appear. Similarly, since each molecular eigenstate ψ_n carries a piece of the original oscillator strength $f_{0 \to s}$, it is clear that

$$f_{0 \to n} < f_{0 \to s} \tag{2.15}$$

because of the condition that[3]

$$\sum |C_{sn}|^2 = 1, \quad \text{so} \quad |C_{sn}| < 1. \tag{2.16}$$

But (2.15) implies that

$$(\tau_{0 \to n}^{\text{rad}})^{-1} < (\tau_{0 \to s}^{\text{rad}})^{-1}, \tag{2.17}$$

so the observed fluorescence lifetime, $\tau_{n \to 0}^{\text{rad}}$, of ψ_n should be longer than that for the hypothetical zeroth-order level ϕ_s in agreement with experiment. The independence of the individual molecular eigenstates ψ_n would then imply quantum yields of fluorescence $\varphi_f(n)$ of unity in isolated molecules. Because of the rather long lifetimes of these molecules the attainment of isolated molecule conditions requires the use of extremely low pressures. The results to date are consistent with $\varphi_f(n) = 1$ at zero pressure [2.39].

2.2.3 The Molecular Eigenstates Picture

At this juncture there were two diverse sets of experiments, the large molecule benzene case and the small SO_2 and NO_2 molecules, whose behavior is expected to follow from the *same* energy level scheme Fig. 2.1. In order to resolve the question of the observation of quantum yields of less than unity for the isolated benzene molecule, radiationless transition theory was then formulated in terms of the molecular eigenstates basis that so simply explains the small molecule situation [2.2, 40–44]. Thus, provided the incident radiation is of the proper spectral and temporal properties, the non-stationary state ϕ_s is the one which is taken to be initially excited. This initial state is written as a linear superposition of the molecular eigenstates

$$\phi_s = \sum_n C_{ns} \psi_n. \tag{2.18}$$

In the absence of matter-radiation interaction, each ψ_n with energy E_n evolves separately in time, so the non-stationary state at time t is

$$\Psi(t) = \sum_n C_{ns} \exp(-i E_n t/\hbar) \psi_n. \tag{2.19}$$

---

[3] Eqs. (2.15) and (2.16) become equalities only in the absence of $s-l$ coupling.

Hence, the probability of finding the molecule in state ϕ_s at time t is

$$P_s(t) \equiv |\langle \phi_s | \Psi(t) \rangle|^2 = \left| \sum_n |C_{ns}|^2 \exp(-i\, E_n t/\hbar) \right|^2 , \tag{2.20}$$

upon using the relationship

$$\langle \phi_s | \psi_n \rangle = C_{sn} = C_{ns}^* . \tag{2.21}$$

For short enough times, $P_s(t)$ is found to generally decay exponentially with rate constant of the form

$$k_{nr}(s \rightarrow l) = \frac{2\pi}{\hbar} \sum_l |v_{sl}|^2 \rho_l(E_s) \tag{2.22}$$

where E_s is slightly shifted from the zeroth-order energy E_s^0 of ϕ_s and

$$\rho_l(E_s) = \varepsilon^{-1} \tag{2.23}$$

with ε the average effective level spacing in $\{\phi_l\}$ for those levels with appreciable v_{sl}. Eq. (2.22) is just the simple golden-rule type rate for the $s \rightarrow l$ non-radiative decay with ρ_l as the l-manifold density of states. This result is in agreement with an earlier suggestion of ROBINSON [2.6, 11] that a sufficiently large molecule provides an "effective quasi-continuum" of levels that might be able to act as its own heat bath.

Robinson's proposal, however, leaves unanswered the question of how a molecule can act as its own heat bath and of what density of final states is sufficient. The molecular eigenstates description appears to answer the former question, but does not resolve the latter. However, in the presence of coupling to the radiation field—when we admit of the possibility of spontaneous emission of radiation—the above argument in terms of molecular eigenstates runs into some considerable technical difficulties for large molecules [2.4, 41, 45–47]. If we run through arguments similar to those given above in the BO basis, the resulting conclusion is again that there should be unit quantum yield in the isolated large molecule with a fluorescence lifetime identical to that obtained from the oscillator strength for the transition. Since the molecular eigenstates and the BO bases are individually complete sets of states, the results of the calculation of physical observables, such as lifetimes and quantum yields, should be independent of the basis chosen [2.4, 45, 46]. In specific applications one basis may be preferable to another since it most nearly "diagonalizes" the problem at hand. In such a case a physical meaning is imparted to this diagonalizing basis, but any complete one could be used in order to perform the calculation. Theories which claim that only a particular basis set is correct in making a calculation should therefore be viewed with extreme skepticism[4].

The conclusion of unit quantum yields at zero pressure, etc., are, in fact valid quantum mechanical consequences of the molecular energy level model, as given in Fig. 2.1. They are manifestations of a general quantum mechanical Poincaré recurrence theorem. But then why are quantum yields of less than unity observed in benzene? Does this imply errors in the experiment? In the case of molecules in solution or in rigid media, as ROBINSON noted, the presence of the medium as a heat sink immediately allows for the possibility of irreversible electronic relaxation, and hence, fluorescence quantum yields of less than unity. However, the fact that fluorescence yields remain less than unity for many molecules, even as "isolated" conditions are approached and in a variety of inert solvents and rigid media, implies that the same intramolecular mechanism is operative under conditions of isolation and of interaction with the surroundings. These facts illuminate some of the general features of observed irreversible electronic relaxation. The broad questions raised in the above discussion and more specific questions posed by the experimental data require a complete reformulation of the theory of electronic relaxation in isolated polyatomic molecules beyond that encompassed by the simple theories outlined above. Any proper theory of radiationless processes in polyatomic molecules must be capable of discussing all of the general phenomena which are classified as involving electronic relaxation within a single unified framework.

2.2.4 Some Fundamental Questions

Although a majority of the experimental data relevant to radiationless processes have been obtained for cases of molecules in solution or in rigid media, we exhibit the fundamental phenomena in their simplest and most basic forms by considering radiationless processes in isolated molecules. We limit our focus not because of lack of interest in these processes in condensed media, but because the indications are that the dominant feature of radiationless processes are of intramolecular nature. The presence of a surrounding medium then leads to added complications in the theoretical description of the intramolecular electronic relaxation. Once we can satisfactorily describe these processes in isolated molecules, it is important to generalize the discussion

---

[4] Our previous comments on the preferability of using crude or adiabatic BO wavefunctions is based upon the fact that the latter diagonalize a significantly larger portion of the problem and thereby correspond more closely to observed states. Calculations with either basis can be made, in principle, but the lowest-order non-vanishing contribution to the golden rule rate (2.5), (2.8), etc., are not necessarily sufficient in one or both of these bases (see the discussion in Sects. 2.3 and 5).

to include the interesting effects of the external media, thereby enabling a comparison between theory and the vast majority of the experimental data.

At the outset, it is also desirable to separate those aspects of radiationless processes which are generic, i.e., independent of the specific details of a particular molecule, from those which are specific and which apply only to a certain molecule or class thereof. The general questions that we wish to pose are as follows:

1) How can the dense manifold of vibronic states $\{\phi_l\}$ lead to the observed irreversible electronic relaxation in an isolated molecule in apparent contradiction with the quantum mechanical Poincaré recurrence theorem? In other words, how should the model of Fig. 2.1 be interpreted or modified in order that this irreversible behavior be directly apparent?

2) Given that irreversible behavior can occur in isolated molecules, it is clear that Robinson's conjecture of an "effective quasi-continuum of states" must somehow be valid. This still leaves the important question of how we decide whether or not a dense manifold of states is or is not an effective quasi-continuum, i.e., whether or not it leads to dissipative behavior. Thus, criteria are required to determine when a molecule corresponds to the small, intermediate, and large molecule limits. The earliest theories of radiationless transitions which are (mostly, but not entirely) based upon the simple model of equally spaced levels $\{\phi_l\}$ with equal coupling to ϕ_s imply that irreversibility is related to the Poincaré recurrence time [2.40, 45, 46]. However, in real molecules, it is well known that some of the states in $\{\phi_l\}$, say $\{\phi_b\}$, are strongly coupled to ϕ_s because of favorable symmetry and/or Franck-Condon factors, while others, say $\{\phi_w\}$, are only weakly coupled to ϕ_s because of unfavorable symmetry and/or Franck-Condon factors. The strongly coupled states $\{\phi_b\}$ do not necessarily have a very slowly varying density of states [2.45, 46]. Thus, the requisite general criteria for irreversible electronic relaxation must reflect a dependence on the varying coupling strengths and energy level densities that are expected to be found in real molecules.

3) Are the criteria for dissipative non-radiative behavior dependent upon the nature of the radiationless process under consideration? Are the general criteria for irreversible electronic relaxation different for cases of internal conversion and intersystem crossing?

4) How can the mechanism which leads to a shortening of the decay lifetime of large polyatomic molecules also lead to a lengthening of this lifetime for small (triatomic) molecules such as SO_2, NO_2? This Douglas effect should not be separated from the phenomena observed in larger molecules. In fact, it might be expected that there also occur an interesting set of intermediate cases which reflect the union of the Douglas effect and the effects associated with the large molecule statistical limit *within the same molecule* [2.4, 8, 11, 42, 45, 46]!

5) Aside from considering processes characteristic of the isolated molecule, there are many interesting additional effects which are observed in media. How can we explain the virtual independence of the nature of the "inert"

surrounding medium of fluorescence and phosphorescence quantum yields and lifetimes? On the other hand, we also must explain, e.g., the observation that the benzene fluorescence quantum yield drops from the isolated molecule limit of $\varphi_f(0) = 0.34$ to about $\varphi_f = 0.18$ at a pressure of a few Torr.

6) Although it was noted that all physical observables are independent of the choice of the basis set, the description of the system may be simplified by the use of a particular one of the complementary pictures. Thus, a qualitative description of a particular phenomena is afforded by that basis choice which most nearly "diagonalizes" the problem. It will then be clear that the different manifestations of radiationless processes that are observed in the small, intermediate, and large molecule limits represent differences in the nature of the "diagonal" bases in these cases.

7) Lastly, because there are a large number of closely coupled excited states, it is of interest to consider whether the use of sufficiently monochromatic or coherent exciting radiation would give rise to any interference or other new effects. As the aim of the review is a description of the energy dependence of electronic relaxation processes, we can confine most of our attention to the case of excitation by conventional light sources. In order to exhibit the basic molecular phenomena, there is no need to consider the correlation functions of the exciting light. More refined treatments of the excitation process are required to describe the dependence of luminescence properties on pulse shape and coherence, and these are briefly mentioned in Section 2.5. However, the general phenomena associated with radiationless processes are observed whatever the nature of the exciting light source, be that a flash lamp, a laser, or a "pocket ultraviolet flashlight". Thus, it is clear that we reveal the basic molecular processes by limiting our discussion to an idealized representation of conventional light sources.

When different vibronic levels of ϕ_s come sufficiently close to each other, there is also the possibility of the observation of purely quantum mechanical interference phenomena such as an oscillatory radiative decay. As these interference effects are the subject of recent reviews by JORTNER and MUKAMEL [2.8], they need not be considered further herein. We merely note that a pair of interfering vibronic levels of ϕ_s (or between ϕ_s and a particular ϕ_l) represents a whole manifold of interfering rotational components of these two vibronic states in isolated molecules. Consequently, unless the rotational levels are resolved, the observed interference effects are superpositions of a large number of elementary interferences, so it still remains as an experimental and theoretical question as to whether these effects can be observed and, if so, whether they can satisfactorily be analyzed to unravel the detailed coupling mechanism[5].

---

[5] Even when the superpositions of quantum beats of disparate frequencies wash out any observable interference effects, it is possible to observe a changed quantum yield by subjecting the molecule to a coherent multiple pulse train of optical radiation [2.131]. Alternatively, when the beat superpositions produce a complicated oscillatory decay pattern, the multiple pulse experiments will alter the nature of this decay, enhancing those beat frequencies in phase with the pulses and surpressing those which are out of phase [2.131].

As noted above, our description of the general quantum mechanical theory of electronic relaxation processes is kept brief enough to supply enough conceptual framework for the subsequent treatment of the energy dependence of electronic relaxation and to enable the reader to approach the ever expanding literature in this area.

2.2.5 Irreversibility?

The first question posed by the experimental data on electronic relaxation concerns the rationalization of the apparent irreversible non-radiative decay, and this must be resolved before a detailed description of the general theory. This irreversibility paradox also arose slightly less than a hundred years ago in a different context. It will be recalled that Boltzmann, using plausible physical assumptions, developed an evolution equation which described the non-equilibrium behavior of a dilute classical fluid. This equation predicts that an initial non-equilibrium distribution of interacting structureless particles will decay to the appropriate equilibrium Maxwell-Boltzmann distribution. For finite systems, there is a recurrence theorem (due to Poincaré) which states that if we wait long enough, any finite system obeying the laws of classical mechanics will ultimately return arbitrarily close to its initial state. We know from experience that isolated systems tend to equilibrium. For instance, a system, in which all the molecules are in one-half of the container, after the removal of a partition, will eventually tend towards an equilibrium distribution wherein the particles are uniformly distributed throught the box. The recurrence theorem would imply that if we waited sufficiently long, the system would spontaneously return to its initial state with all the particles in one-half of the box. Boltzmann's equation was therefore criticized because it did not contain this recurrence which is a direct consequence of the laws of classical mechanics. Boltzmann's reply was basically: "You wait for the recurrences!" For, any reasonable estimate of the time scale involved gives a result which is many orders of magnitude longer than the age of the universe.

In a discussion of irreversibility and radiationless transitions, we again can ask about the time scales of these "theoretical recurrences" [2.4, 46]. In any real experiment there is some time limit placed upon the length of an experiment, e. g., the length of time we can spend waiting for the fluorescence. This time may be a microsecond, a second, a year, the limits of patience of a graduate student, etc. Furthermore, the simple model presented in Fig. 2.1 is only valid for a limited period of time. We have tacitly assumed the molecule to be isolated, but at any non-zero pressure the molecule will collide with another molecule after some time τ_{collis} ($\approx 10^{-4}$ s at $1\,\mu$m pressure). Even at "zero pressure", the molecule would collide with the walls of the vessel after some time τ_{wall}. Furthermore, states of $\{\phi_l\}$, which are isoenergetic with ϕ_s, are vibrationally "hot" and would spontaneously emit infrared radiation, thereby leading to a molecule with energy too low to "cross back" to ϕ_s. ("Isolated molecules" in interstellar space are persistently undergoing these processes.) Estimates of the times for infrared emission give $\tau_R \approx 10^{-4} - 10^{-3}$ s.

Because of the time limitations on the duration our experiment and the validity of the model employed, we can ask if irreversible behavior ensues for times on the order of the time limit τ_{max}. If we find relaxation for times τ_{max}, then we call the phenomena irreversible. In this case the discrete manifold $\{\phi_l\}$ is an "effective continuum" on the time scale relevant to the experiment. Because the system is considered for only a finite time τ_{max}, all energy levels, except the ground state of the system, must have some "width" in energy ε_m as required by the "uncertainty principle"

$$\varepsilon_m \gtrsim \hbar/\tau_{max} . \tag{2.24}$$

The introduction of this energy uncertainty, by associating with each excited state an additional imaginary energy $-i\varepsilon_m$, allows us to consider whether there is practical irreversibility for times on the order of τ_{max}. This procedure avoids the introduction of unattainable recurrences.

The introduction of this energy uncertainty might appear to be without rigor. However, given the interaction of the molecule with the container of the system H_{wall}, (a quantity of immense current experimental and theoretical interest) with other molecules H_{coll}, etc., we could proceed in all rigor. If we obtain an effective Hamiltonian for only the system of interest, i.e., the molecule and any radiation, this Hamiltonian has a non-Hermitian component reflecting the effects of the interaction of the molecule with the wall, etc. Thus, all energies have a small negative imaginary part $-i\varepsilon_m$ reflecting the absence of stationary states for only this subsystem of molecule and radiation. The lifetimes are then just $\tau_m = \hbar/\varepsilon_m$, and thus (2.24) corresponds to giving each state an average lifetime due to the existence of these other dissipative channels.

Now that practical irreversibility has been defined, it is necessary to obtain general criteria for the occurrence of irreversible electronic relaxation in polyatomic molecules. The system in Fig. 2.1 requires that we determine the radiative properties of a large number of closely coupled states. In order to most efficiently accomplish this task it is useful to consider some powerful analytic methods which are not necessarily well known to the photochemist. A simple analogy should, however, elucidate the basic nature of these methods. In discussing the state of a quantum mechanical system in terms of a particular basis, we could deal with the set of simultaneous equations determining the coefficients which appear in the expansion of the energy. However, it is far more convenient to employ matrix notation than work with each of the individual equations. Similarly, when we wish to choose an appropriate basis or when some general questions are posed, it is useful to employ operator notation. The description of the radiative properties of a large number of closely coupled states is conveniently treated by the use of operator techniques which in this case corresponds to Green's function or resolvent operator methods. This approach has the benefit that it explicitly determines for us that basis set which most nearly "diagonalizes" the problem under consideration, telling us, e. g., whether the BO or molecular eigenstates picture is more convenient [2.45, 46, 48].

2.3 Quantum Description of Radiationless Processes

2.3.1 Excitation and Decay

We now discuss the excitation and subsequent decay, both radiative and non-radiative, of a molecule. Because of the natural time limit τ_{max} on any real experiment (or inherent in the model) only the time dependence for times $t < \tau_{max}$ is considered. As was noted in Subsection 2.2.4 the general properties of non-radiative processes are observed independent of the specific features of the exciting light source such as whether long or short flashes are used, or whether the band pass is 10 or 100 Å, etc. We consider an idealized excitation process: Let the molecule be in the ground electronic state ϕ_0 at time $t = 0^-$ when it is subjected to a pulse of light which contains n_a photons of wave vector k_a and polarization vector e_a (the photon frequency is $c|k_a| = \omega_a$, where c is the velocity of light), n_b photons with k_b and e_b, etc. Thus, the light may be monochromatic, polychromatic, or even have a continuous spectrum. This pulse of light is taken to have a pulse length τ, τ being chosen so that

$$\tau \gg \omega^{-1}, \tag{2.25}$$

and the light can be taken to be monochromatic if desired. In the optical region (2.25) is also a practical limitation. After the time τ, all the photons are removed, and we monitor the decay of the system so prepared. We might observe the decay of the fluorescence at a single frequency, over a range of frequencies, or by measuring the quantum yield. This idealized model of excitation, therefore, corresponds to the use of a square pulse (in time) of radiation.

We assume that the light source is of the conventional incoherent variety. Classically, if $E_{opt}(\omega_i)$ denotes the electric field of the optical radiation at frequency ω_i, the field at a frequency ω_j is taken to be uncorrelated with that at ω_i. If an average is taken over the light pulse, this implies that for a chaotic pulse of infinite duration

$$\langle E_{opt}(\omega_i) E_{opt}^*(\omega_j) \rangle = \frac{4\pi}{c} \delta_{ij} P(\omega_i) \tag{2.26}$$

where $P(\omega_i)$ is the power spectrum of the light pulse. Now, if we considered probabilities of photon absorption and fluorescence, these probabilities would be bilinear in the E_{opt}. Using (2.26) we see that for conventional light sources cross terms with different frequencies would vanish, leaving a result which is a summation over the incident photon frequencies. For the case of excitation with pulses of long, but finite, duration τ, there is coherence between frequencies within an interval $\Delta\omega \approx \tau^{-1}$, and these are easily incorporated in the theory as necessary [2.8, 44], see Section 2.5.

Explicitly, let $P_{F,A}(t,\tau)$ be the probability that a photon A(i.e., k_A, e_A) is absorbed during a monochromatic flash of duration τ and then a photon F(i.e., k_F, e_F) is emitted at a time t after the flash. If $P_F(t,\tau)$ is the probability of absorption of any photon from the polychromatic flash, then (2.26) implies that

$$P_F(t,\tau) = \sum_A P_{F,A}(t,\tau). \tag{2.27}$$

(Eq. (2.27), of course, involves a summation over the discrete frequencies, wave vectors, and polarization vectors, in the pulse, and an integration over the continuous values of these parameters.) Eq. (2.27) therefore implies that we can confine our attention to monochromatic pulses, since the general result is merely obtained by summation over frequency, etc. with only minor adjustments necessary to incorporate the coherence effects that result from the finite pulse duration. Furthermore, rather than considering a description of the light pulse in terms of the electric field of the radiation, for simplicity the pulse is taken to have a fixed number of photons n_A which is the (integral number that is closest to the) most probable number of photons[6].

The Hamiltonian for the system is given by

$$\mathscr{H} \equiv \mathscr{H}_{el} + \mathscr{H}_r + \mathscr{H}_{int}, \tag{2.28}$$

where $\mathscr{H}_{el}$ is the complete molecular Hamiltonian, $\mathscr{H}_r$ describes the radiation field, and $\mathscr{H}_{int}$ describes the interaction between radiation and matter. It is convenient to treat the radiation quantum mechanically. All this means, effectively, is that we count photons. Thus, $\mathscr{H}_{int}$ can enable the absorption or emission (in this case only spontaneous emission) of a single photon to occur. With this book-keeping scheme, the elegant formal structure of the quantum theory of radiation is wholly unnecessary. The electronic Hamiltonian is divided into the BO part $\mathscr{H}_{BO}$ and the remaining interactions—vibronic, spin-orbit, etc.—$\mathscr{H}_V$:

$$\mathscr{H}_{el} = \mathscr{H}_{BO} + \mathscr{H}_V. \tag{2.29}$$

---

[6] More rigorously, for τ finite there is an energy spread $\Delta E \sim \hbar/\tau$ in the incident light. Thus, it is necessary to employ a wave packet formulation to provide the correct analysis of the system. In this formulation a pulse of light is described as being incident on the molecule from some source that is at a great distance from the molecule. The pulse duration then corresponds to the transit time of this wave packet across the molecule. Although the idealized square pulse employed here omits these energy spreads, its simplicity enables the elucidation of the simple fundamental molecular phenomena we wish to illustrate. The more general case is briefly discussed in Section 2.5.

The benefit of using the quantum treatment of radiation and a square excitation pulse is that (2.28) and (2.29) are time independent both when the pulse is on and when it is off. Thus, the initial state of the system before the pulse is

$$\Psi(0^-) = |\phi_0, n_A k_A e_A\rangle \equiv |\phi_A\rangle. \tag{2.30}$$

During the time the pulse is on, the time dependence of the system is completely given by the $\mathcal{H}$ of (2.28),

$$\Psi(t_1) = \exp(-i\mathcal{H} t_1/\hbar)\,\Psi(0^-), \quad 0 \le t_1 \le \tau. \tag{2.31}$$

At time τ, the incident light is removed, so that if we only wish to consider those molecules which have absorbed a photon during $0 \le t_1 \le \tau$ and at τ are in some electronically excited state, then this excited molecular state is

$$\psi_{\text{mol}}^{(\text{excited})}(\tau) = \langle n_A - 1, k_A e_A | \Psi(\tau)\rangle. \tag{2.32}$$

In (2.32), the bra vector $\langle n_A - 1, k_A e_A|$ merely selects, out of those states of the system at time τ, the particular states for which a single photon has been absorbed. We could, likewise, consider those molecules which have absorbed two photons by employing the bra vector $\langle n_A - 2, k_A e_A|$, etc., but we are now only interested in the dominant one photon process. At time τ, the pulse of light is shut off, so the remaining photons are removed, thereby leaving the radiation field in the vacuum state $|\text{vac}\rangle$ of no photons. The state of the whole system, molecule plus radiation field, is then

$$\Psi_{\text{excited}}(\tau) = |\text{vac}\rangle\,\psi_{\text{mol}}^{(\text{excited})}(\tau) \tag{2.33}$$

at time τ. For all times after the pulse is off, the Hamiltonian is (2.28) and is time independent, so the state evolving from (2.33) is

$$\Psi(t+\tau) = \exp(-i\mathcal{H} t/\hbar)\,\psi_{(\text{excited})}(\tau). \tag{2.34}$$

The probability $P_{F,A}(t,\tau)$ of finding the system at time t after the pulse in a state where the molecule is in a state ϕ_0^v (which may be vibrationally excited with respect to ϕ_0) and where a single photon $k_F e_F$ is present, is defined by

$$\begin{aligned} P_{F,A}(t,\tau) &\equiv |\langle \phi_F^v | \Psi(t+\tau)\rangle|^2 \\ &= |\langle \phi_F^v | \exp(-i\mathcal{H} t/\hbar)| \text{vac}\rangle \\ &\quad \times \langle n_A - 1, k_A e_A | \exp(-i\mathcal{H} \tau/\hbar)| \phi_A\rangle|^2, \end{aligned} \tag{2.35}$$

where $|\phi_F^v\rangle \equiv |\phi_0^v, k_F e_F\rangle$.

Consider some set of molecular wave functions $\{\theta_j\}$ which is complete enough to describe the most general state of the molecule after the absorption of a single photon. This completeness implies that we can introduce the states $\{\theta_j\}$ as the "intermediate" states of the molecule at time τ, or

$$\psi_{\text{mol}}^{(\text{excited})}(\tau) = \sum_j |\theta_j\rangle \langle n_A - 1, k_A e_A; \theta_j | \Psi(\tau)\rangle . \tag{2.36}$$

Let

$$A_j(\tau) = \langle \theta_j; n_A - 1, k_A e_A | \exp(-i\mathcal{H}\tau/\hbar) | \phi_A\rangle \tag{2.37}$$

be the probability amplitude that the molecule be in the state θ_j at time τ. Then (2.36) and (2.37) imply that

$$\psi_{\text{mol}}^{(\text{excited})}(\tau) = \sum_j A_j(\tau) |\theta_j\rangle . \tag{2.38}$$

Thus, the amplitudes $\{A_j(\tau)\}$ contain all the information concerning the excitation process. Substituting (2.30) into (2.33–36) gives

$$P_{F,A}(t,\tau) = \sum_{i,j} \langle \phi_F^v | \exp(-i\mathcal{H}t/\hbar) | \theta_j; \text{vac}\rangle$$
$$\times A_j(\tau) \langle \phi_F^v | \exp(-i\mathcal{H}t/\hbar) | \theta_i; \text{vac}\rangle^* A_i^*(\tau) . \tag{2.39}$$

Now, in (2.36) and (2.39), *any basis set* $\{\theta_j\}$ *could be used, provided it is complete. Therefore,* $\{\theta_j\}$ *could represent* BO *states (CBO or ABO!), the molecular eigenstates, or any other convenient basis set.* The final result is completely independent of this choice of basis set as any observable must be. Some basis choices might turn out to be more convenient because they more nearly "diagonalize" the problem, and it is usually that basis to which we ascribe "physical significance".

Since (2.39) contains a $\sum\limits_{i,j}$, and hence cross terms from θ_i and θ_j, there is the possibility of the observation of quantum mechanical interference effects when states θ_i and θ_j are coherently excited, i.e., $A_i(\tau)$, $A_j(\tau) \neq 0$. The presence of these "interference effects" could be representation dependent. For instance, if we chose a basis set $\{\mu_j\}$ so only one state μ_i is excited, $A_j(\tau) = \delta_{ij}$, there are no cross terms in (2.39), and μ_i has simple exponential decay. However, if we chose a different basis set $\{v_j\}$ which did not have $v_i = \mu_i$ for some i, then in terms of the $\{v_j\}$ basis, there would be more than one non-zero amplitude A_j, and consequently there would appear to be interference terms. We reiterate that the fluorescence probability $P_{F,A}$ is independent of this choice of basis.

2.3.2 Solution by Laplace Transformation

It is convenient to introduce the fluorescence amplitudes

$$F_j(t) \equiv \langle \phi_F^v | \exp(-i \mathcal{H} t/\hbar) | \theta_j; \text{vac} \rangle , \tag{2.40}$$

so (2.39) can be rewritten as

$$P_{F,A}(t,\tau) = \sum_{i,j} A_j(\tau) F_j(t) A_i^*(\tau) F_i^*(t) . \tag{2.41}$$

$F_j(t)$ is the probability amplitude that a molecule, initially in state θ_j with no photons present, is found in state ϕ_F^v at time t. The general form of (2.41) in terms of probability amplitudes for absorption and emission can always be obtained when there is a clear separation between the absorption and the emission processes. Apart from this assumption of separation, $F_j(t)$ is completely independent of the heuristic model of a square pulse excitation process. The model is merely introduced to show how $P_{F,A}(t,\tau)$ can be written in terms of the probability amplitudes for absorption and emission. These probability amplitudes are written in (2.37) and (2.40) in terms of the matrix elements of the evolution operator $\exp(-i \mathcal{H} t/\hbar)$ which are often difficult to evaluate directly. Instead, the probability amplitudes are obtained by the solution of the time dependent Schrödinger equation

$$i\hbar \frac{\partial}{\partial t} \Psi(t) = \mathcal{H} \Psi(t) , \tag{2.42}$$

where the time dependent wavefunction is expanded in some complete time-independent basis set $\{\psi_j\}$,

$$\Psi_k(t) = \sum_j a_{jk}(t) \psi_j , \tag{2.43}$$

thereby defining the usual probability amplitudes $a_{jk}(t)$. The extra index k has been added to $\Psi_k(t)$, $a_{jk}(t)$ in (2.43) to indicate the particular initial condition

$$a_{jk}(0) = \delta_{jk} = \begin{cases} 0 & \text{if } j \neq k , \\ 1 & \text{if } j = k , \end{cases} \tag{2.44a}$$

so

$$\Psi_k(0) = \psi_k . \tag{2.44b}$$

Given some other initial conditions

$$\Psi(0) = \sum_k \lambda_k \psi_k , \quad \lambda_k \equiv \langle \psi_k | \Psi(0) \rangle , \tag{2.44c}$$

the linearity of the Schrödinger equation (2.42) implies that the general solution is

$$\Psi(t) = \sum_{j,k} \psi_j a_{jk}(t) \lambda_k . \tag{2.45}$$

Thus, the general situation follows from a consideration of the solutions (2.43) for each k individually. In the above example we have the correspondence

$$F_j(t) \equiv a_{F;j,\text{vac}}(t) \tag{2.46a}$$

and

$$A_j(t) = a_{j,n_A-1;A}(t), \tag{2.46b}$$

where F labels the state ϕ_F^v and A denotes ϕ_0, n_A, etc.

Eq. (2.46) demonstrates the essential symmetry between absorption and emission. Because a quantum mechanical description of the radiation is employed, (2.46) properly accounts for spontaneous emission. Furthermore, during and after the pulse, the Hamiltonian for the whole system is time independent, thereby substantially simplifying the process for the evaluation of (2.42–46). Because decay processes are involved, it is far more convenient to consider the (imaginary) Laplace transform of the time dependent Schrödinger equations (2.42) with (2.43). The use of (imaginary) Laplace transforms is a very commonly used method for converting initial value problems into algebraic equations, e.g., in the theory of linear network circuit theory [2.49]. In the present case, the transformation

$$a_{jk}(t) = (2\pi i \hbar)^{-1} \int_C dE \exp(-i E t/\hbar) G_{jk}(E) \tag{2.47}$$

can be substituted into the individual equations for the $a_{jk}(t)$ (obtained by multiplying (2.42) by $\langle \psi_j |$)

$$i\hbar \frac{\partial}{\partial t} a_{jk}(t) = \sum_m \langle \psi_j | \mathcal{H} | \psi_m \rangle a_{mk}(t) \tag{2.48}$$

to obtain the transformed equations

$$\sum_m [E \delta_{jm} - \langle \psi_j | \mathcal{H} | \psi_m \rangle] G_{mk}(E) = \delta_{jk}, \tag{2.49}$$

where the initial conditions have been used in expressing the transform of $\partial a_{jk}(t)/\partial t$ as $E G_{jk}(E) - a_{jk}(0)$. The contour C in (2.47) runs from $+\infty$ to $-\infty$ above the real E axis.

Employing the notation

$$\mathcal{H}_{jm} = \langle \psi_j | \mathcal{H} | \psi_m \rangle \tag{2.50}$$

the algebraic equations can be written in matrix form as

$$E G_{jk}(E) - \sum_m \mathcal{H}_{jm} G_{mk}(E) = \delta_{jk} \tag{2.51a}$$

or

$$(E\mathbf{1} - \mathcal{H}) G(E) = \mathbf{1}, \tag{2.51b}$$

where $\mathbf{1}$ is the unit matrix $(\mathbf{1})_{jk} = \delta_{jk}$. The matrix equations (2.51) provide the transformed solution for arbitrary initial conditions as is seen by substituting the definition (2.47) into (2.45) to give

$$\Psi(t) = \sum_j \psi_j \int_C \frac{dE}{2\pi i \hbar} \exp(-i E t/\hbar) G_{jk}(E) \lambda_k . \tag{2.52}$$

Since (2.51) represents a set of matrix equations, its solution can be formally represented in terms of the matrix inverse $(E\mathbf{1} - \mathcal{H})^{-1}$ of $(E\mathbf{1} - \mathcal{H})$ via

$$G(E) = (E\mathbf{1} - \mathcal{H})^{-1}. \tag{2.53}$$

Since $\mathcal{H}$ is Hermitian, its eigenvalues are real, and (2.53) is defined along the contour C for which E is complex.

Eqs. (2.52) and (2.53) take on a particularly simple form when the basis functions $\{\chi_j\}$ are eigenfunctions of $\mathcal{H}$,

$$\mathcal{H} \chi_j = E_j \chi_j \tag{2.54}$$

with E_j as eigenvalues. In this case (2.53) and (2.52) become

$$G_{ij}(E) = \delta_{ij}(E - E_j)^{-1}, \tag{2.55}$$

and, upon performing the simple contour integral,

$$\Psi(t) = \sum_j \chi_j \exp(-i E_j t/\hbar) \langle \chi_j | \Psi(0) \rangle , \tag{2.56}$$

which is obviously the correct result.

2.3.3 The Green's Function Method

In the present case involving the simultaneous radiative and non-radiative decay of a number of closely coupled excited levels, the basis functions $|\phi_F^\nu\rangle$, for instance, form a continuum of levels, so the solution of (2.54) and the inversion of (2.53) are not trivial problems. Fortunately, general methods have been developed for the approximate evaluation of (2.53). The equations (2.51) and (2.53) refer to a particular choice of basis set, and it is often convenient in quantum mechanics to display equations in a basis set independent manner, in terms of an operator notation. Thus, we introduce the Green's operator [2.45, 46, 48, 50–52].

$$G(E) = (E - \mathcal{H})^{-1} \tag{2.57}$$

with the matrix elements

$$G_{ij}(E) = \langle \psi_i | G(E) | \psi_j \rangle \tag{2.58}$$

and the formal relationship between $G(E)$ and the evolution operator $\exp(-i\mathcal{H}t/\hbar)$

$$\exp(-i\mathcal{H}t/\hbar) = (2\pi i\hbar)^{-1} \int_C dE \exp(-iEt/\hbar) G(E). \tag{2.59}$$

The formal operator equation (2.59) is seen to be totally equivalent to the defining equation (2.47) for the $G_{jk}(E)$ upon taking the jk matrix element of (2.59) and using (2.58) and the definition

$$a_{jk}(t) \equiv \langle \psi_j | \exp(-i\mathcal{H}t/\hbar) | \psi_k \rangle \tag{2.60}$$

of the probability amplitudes [cf. (2.37), (2.40), and (2.46)] when $\mathcal{H}$ is time independent.

The Green's operator $G(E)$ is introduced, in part for mathematical convenience. Methods for approximation with $G(E)$ have been developed in scattering theory [2.50, 51]. We note that the fluorescence probability envisioned in (2.35) just corresponds to the scattering of a photon by a molecule: photon A is scattered by the molecule into photon F! Furthermore, $G(E)$ provides us with useful physical information; the poles of (2.53), (2.55), or (2.57) in the lower half E-plane are just the eigenenergies of the system, while the residues of $G(E)$ at these poles give the corresponding eigenfunctions. When considering molecular radiative and non-radiative decay, there no longer are stationary states. Consequently, $G(E)$ can be considered to have poles in the lower half E-plane, e.g., $\varepsilon_j - i\hbar\Gamma_j/2$ (see Sect. 2.5). The width Γ_j of the state j represents the decay rate of the state, or equivalently $1/\Gamma_j$ is the lifetime of this state. This conclusion easily follows by considering the decay of a single state j. Since

$$a_{jj}(t) = \langle \chi_j | \exp(-i\mathcal{H}t/\hbar) | \chi_j \rangle$$

$$\approx (2\pi i\hbar)^{-1} \int_C \frac{dE \exp(-iEt/\hbar)}{E - \varepsilon_j + i\hbar\Gamma_j/2}, \tag{2.61}$$

performing the trivial integration gives

$$a_{jj}(t) \approx \exp(-i\varepsilon_j t/\hbar - \Gamma_j t/2) \tag{2.62}$$

with the physical interpretation stated above.

In order to examine whether or not irreversible behavior ensues on time scales of relevance, we wish to examine the behavior of the system for times $t \ll \tau_{max}$. This can be done by allowing all the levels to have an imaginary energy ε_m as in (2.24). Thus, instead of using the conventional relation

$$\lim_{\text{Im}\{E\} \to 0^+} G(E) \equiv G^+(E) \tag{2.63}$$

to define the Green's operator, we have

$$G(E) \to G(E + i\varepsilon_m), \quad E \text{ real}. \tag{2.64}$$

The transform corresponding to (2.64) is the evolution operator

$$\exp(-i\mathscr{H}t/\hbar - t/\tau_{max}), \tag{2.65}$$

thereby making each state decay for $t \ll \tau_{max}$ at which time either the model of Fig. 2.1 is no longer valid or the experiment has effectively come to an end. The use of the time limit in (2.65), therefore, removes the mathematical recurrences which occur for $t \gtrsim \tau_{max}$ and which are of no physical consequence. The introduction of the imaginary component ε_m of the energy does not lead to any *ad hoc* limitation of the general theory. We could have, alternatively, introduced ε_m by including in the total Hamiltonian those terms which correspond to the physical processes which impose the natural time limit on the system. For instance, given the interaction between the molecule and the surrounding vessel, $\mathscr{H}_{wall}$, we could consider the time evolution of the system of molecule plus wall. As in the case of the use of effective Hamiltonians in, e. g., magnetic resonance [2.53], if we only focus attention on the time evolution of the molecular system, this dynamical behavior is governed by an effective Hamiltonian. This effective Hamiltonian is non-Hermitian, having a negative imaginary part, thereby leading to "energies" with negative imaginary parts and finite lifetimes of the molecular quantum states. Thus, our simple prescription (2.64) replaces the lifetimes which result, e. g., from the interaction of the molecule with the wall, by a single (or possibly more than one) typical value. This obviates the necessity for inclusion of some formal, but obviously uninteresting, details. We could always recover results which are valid for all times by taking the limit of (2.64) as $\varepsilon_m \to 0^+$.

We have so far presented the general molecular model upon which discussions of radiationless process are based, the notion of practical irreversibility, and the general time dependent formulation of the problem. We turn to the application of these fundamentals to produce a unified theory of radiationless processes in polyatomic molecules. Rather than pursuing the formal development, we summarize the basic results, while the detailed derivations are presented in Section 2.5.

2.3.4 Irreversibility Criteria

The earliest approaches to discussions of radiationless processes employ the simple model of equally spaced levels $\{\phi_l\}$ with equal coupling v_{ls} between ϕ_s and $\{\phi_l\}$, i.e., $v_{ls} = v$, all l [2.2, 40, 42, 54, 55]. As noted in Subsection 2.2.4, however, real molecules are characterized by irregular spacings between adjacent energy levels of $\{\phi_l\}$. But more importantly, the v_{ls} vary greatly with state because of symmetry restrictions and/or small Franck-Condon factors. A realistic theory must account for this fact [2.45, 46]. First, however, we quote the irreversibility conditions for isolated molecules in the simplest model [2.46].

1) When the radiative lifetimes of $\{\phi_l\}$ satisfy the inequality

$$\tau^{rad}(l) \gg \tau_{max}, \tag{2.66}$$

a situation which often corresponds to intersystem crossing from lowest excited states, the condition for irreversibility is

$$\tau_{max} \ll \tau_{rec} = \hbar \rho_l, \tag{2.67}$$

where ρ_l is the density of states in the $\{\phi_l\}$ manifold (the inverse of the spacing between levels). τ_{rec} is the Poincaré recurrence time for the system. All (2.67) says is that if we "start" in ϕ_s and "cross" to $\{\phi_l\}$, the molecule would "cross back" to ϕ_s for times of the order of τ_{rec}; however, by this time the experiment is effectively over, and this mathematical recurrence is not physically observable. Eq. (2.66) corresponds to the case, e.g., of intersystem crossing from say $\phi_s = S_1$ to $\{\phi_l\} = T_1$ when T_1 has a negligible radiative lifetime $\tau^{rad}(l)$.

2) When

$$\tau^{rad}(l) \ll \tau_{max}, \tag{2.68}$$

the $\{\phi_l\}$ decay rapidly by radiative processes. This situation often ensues in cases of internal conversion, but also occurs for the biacetyl molecule in intersystem crossing $S_1 \rightarrow T_1$ since T_1 has a relatively short radiative lifetime. In this case, if

$$\tau^{rad}(l) \ll \tau_{rec}, \tag{2.69}$$

irreversible behavior is observed provided that

$$\tau^{\text{rad}}(l) \ll \tau_{\text{max}} \, . \tag{2.70}$$

Eq. (2.69) merely states that when we "cross" from ϕ_s to $\{\phi_l\}$, if (2.69) is true, the molecule radiates from $\{\phi_l\}$ before it can "cross back" to ϕ_s. When (2.70) is also satisfied, we may allow $\tau_{\text{max}} \to \infty$, or equivalently $\varepsilon_m \to 0^+$. Therefore, in the case of (2.69), the radiative decay of $\{\phi_l\}$ leads to irreversibility of the radiationless decay.

For the real case of rapidly varying v_{ls} there are a number of possible situations which may be encountered in the small, large (statistical), or intermediate molecule cases. If attention is focused first on isolated molecules, in the general case the recurrence times are no longer necessarily simply related to the density of states. Some distribution of recurrence times $\tau_{\text{rec}}(j)$ must obviously occur. The simplest statistical limit corresponds to the case in which the $\{v_{ls}\}$ are sufficiently small and/or the level density ρ_l is sufficiently great so that all recurrence times are much larger than τ_{max},

$$\tau_{\text{rec}}(j) \gg \tau_{\text{max}} \, , \quad \text{all } j \, . \tag{2.71}$$

Eq. (2.71) is analogous to (2.67) where the experiment is over before a molecule could, in principle, "cross back" from $\{\phi_l\}$ to ϕ_s. The $\{\phi_l\}$ then act as a dissipative quasi-continuum for ϕ_s. The situation is depicted schematically in Fig. 2.2. The optically excited state of ϕ_s and $\{\phi_l\}$ is conveniently called a

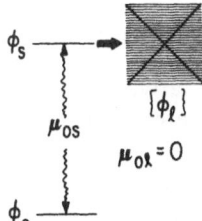

Fig. 2.2. Schematic representation of the large molecule statistical limit. The zeroth-order states are the same as those in Fig. 2.1. The $\{\phi_l\}$ form a dense manifold of states which act as a dissipative quasi-continuum on the timescales of real experiments. Hence, when molecules in ϕ_s "cross-over" to $\{\phi_l\}$, the experiment ends before they can "cross-back" to ϕ_s. ϕ_s can, therefore decay radiatively to ϕ_0^v and non-radiatively to $\{\phi_l\}$

resonance. The $\{\phi_l\}$ behaves as a continuum for ϕ_s, and therefore leads to irreversible decay just as in the cases of radiative, dissociative, and ionization continua. The approximate BO state ϕ_s is therefore coupled by two independent decay channels, radiative and non-radiative, to ϕ_0 and $\{\phi_l\}$, respectively. The quantum mechanical observables for the resonance, the quantum yield and the emission lifetime of ϕ_s, are then given by the conventional rate expressions (2.2) and (2.1), respectively. This situation, therefore, corresponds to the conventional picture of "crossing" between BO states. Although this language is heuristically convenient, since we never measure the molecule initially in ϕ_s and then in $\{\phi_l\}$, it must be emphasized that this terminology should not always be taken literally (cf. the discussion of the small molecule limit below).

2.3.5 The Radiative Damping Matrix

In the absence of the radiation field, i.e., if we ignore the possibility of spontaneous emission, the molecular eigenstates are conceptually useful in describing the radiationless decay in terms of the time development of an "initial" non-stationary state ϕ_s [2.2, 40–43]. However, when spontaneous emission is considered, because of the high degeneracy of the dense set of states $\{\psi_n\}$, these are often not useful for descriptive purposes [2.4, 45–47]. If we just consider the system of matter and radiation, the time evolution of the molecular state is governed by the effective Hamiltonian (see Sect. 2.5 for details) [2.45–47, 56]

$$\mathcal{H}_{\text{eff}} = \mathcal{H}_{\text{el}} - i\hbar\,\Gamma/2 . \tag{2.72}$$

Γ is the radiative damping matrix which is defined by

$$\Gamma_{nm} \equiv \frac{2\pi}{\hbar} \sum_{\alpha} \sum_{e} \int d\Omega_k \langle n|\mathcal{H}_{\text{int}}|\alpha, k\,e\rangle$$
$$\times \langle \alpha, k\,e|\mathcal{H}_{\text{int}}|m\rangle \, \rho_{\text{photon}}(\hbar c k) . \tag{2.73}$$

In (2.73) there is a sum over the final molecular states α ("after radiative decay") as well as a sum over the polarization e and an integration over the directions Ω_k of the emitted photon. ρ_{photon} is the density of photon states. We note that, for $n = m$, (2.73) is just the definition of the "golden rule" radiative decay rate of the state n. Thus, (2.73) is a matrix generalization of the radiative decay rate which is appropriate in the discussion of the decay of a large number of closely coupled levels. The presence of some $\Gamma_{nm} \neq 0$ can be explained physically. When $\Gamma_{nm} \neq 0$, there can be a virtual two photon process in which a molecule originally in the state n emits a photon $k\,e$ and makes a transition to some

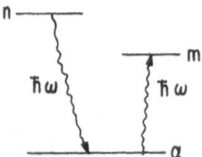

Fig. 2.3. The schematic representation of a virtual two photon process which is allowed when there is a non-zero matrix element of the radiative damping matrix connecting two states n and m. This process is negligible unless n and m are nearly degenerate

lower state α. The molecule then *re-absorbs the same photon* $k\,e$ and is excited to the state m. This *virtual* process is shown schematically in Fig. 2.3. The net effect of this virtual process is that the molecule makes a transition from state

n to state m. Such virtual processes are usually negligible *unless the states
n and m are nearly degenerate*. For

$$\hbar \Gamma_{nm} \gtrsim |E_n - E_m|, \tag{2.74}$$

where E_n and E_m are the energies of states n and m, respectively, this radiative
coupling leads to a "mixing" of the states n and m (a $50-50$ mixing if $E_n = E_m$).

For the case that the BO states $\{\phi_l\}$ have negligible radiative decay rates,
the matrices $\mathcal{H}_{el}$ and Γ in this basis are

$$\mathcal{H}_{el} = \begin{pmatrix} E_s & v_{sl_1} & v_{sl_2}\cdots \\ v_{l_1s} & E_{l_1} & \\ & & 0 \\ v_{l_2s} & 0 & E_{l_2} \\ \vdots & & & \ddots \end{pmatrix}, \quad \Gamma = \begin{pmatrix} \Gamma_s & 0 & 0 & 0\cdots \\ 0 & & & \\ 0 & & 0 & \\ 0 & & & \\ \vdots & & & \end{pmatrix}, \tag{2.75}$$

where $E_{l_1} \approx E_{l_2} \approx E_s$ and $\Gamma_s^{-1} = \tau^{\mathrm{rad}}(s)$, the radiative lifetime calculated for the
approximate state ϕ_s. The transformation to molecular states $\{\psi_n\}$, by definition,
diagonalizes $\mathcal{H}_{el}$, but it also scrambles around the matrix Γ so the latter has
a large number of non-zero off-diagonal elements. Since there are a large
number of molecular eigenstates which satisfy the condition (2.74) in the
statistical limit, these states are mixed by the radiative interaction and lose
their physical significance. Thus, as noted above, in this case it is more appropriate to consider the excited state to be a resonance which is characterized by
its observable properties. It so happens that in the statistical limit this resonance
corresponds closely to the conventional concept of a non-radiatively decaying
BO state [2.46, 57], but this result is not obtained in the general case.

2.3.6 The Statistical Limit

In the large molecule statistical limit the non-radiative decay rate is given by
the "golden-rule-like" rate expression [2.4, 14, 15, 50]

$$k_{nr}(s) = [\tau_{nr}(s)]^{-1} = \frac{2\pi}{\hbar} \sum_\alpha |v_{s\alpha}|^2 \rho_\alpha(E_\alpha). \tag{2.76}$$

In (2.76) the states α correspond to eigenfunctions χ_α of a modified effective
Hamiltonian $\mathcal{H}_{eff}$ which differs from $\mathcal{H}_{eff}$ of (2.72) in that all the interactions
between ϕ_s and $\{\phi_l\}$ are taken to vanish in $\mathcal{H}_{eff}$. Thus, $\mathcal{H}_{eff}$ is an effective
Hamiltonian which completely omits the state ϕ_s. $\rho_\alpha(E_s)$ is the density of
the energy eigenstates of $\mathcal{H}_{eff}$ at the zero-order energy E_s of ϕ_s. Because of the
presence of effective irreversibility, for time scales of relevance, $\rho_\alpha(E_s)$ is a non-

singular function of α (or E_α, the energy of $\mathcal{H}_{\text{eff}}$ corresponding to χ_α). It is often convenient to express this density of states in the usual singular form

$$\rho_\alpha(E_s) \to \delta(E_\alpha - E_s), \tag{2.77}$$

and then to consider smooth approximations to the resultant expression when (2.77) is substituted into (2.76). The use of (2.77) in (2.76), without the understanding that smooth approximations to the Dirac delta functions be taken, would just resurrect the irreversibility paradox, as discussed earlier.

The expression (2.76) is not the usual "golden-rule" rate that is obtained from second-order time dependent perturbation theory

$$k_{\text{nr}}^{(2)}(s \to l) = \frac{2\pi}{\hbar} \sum_l |v_{ls}|^2 \rho_l(E_s). \tag{2.78}$$

In (2.78) there is only a sum over the quasi-continuum $\{\phi_l\}$ which is isoenergetic with ϕ_s. $\rho_l(E_s)$ is again the density of states of this zero-order manifold. The difference between (2.76) and (2.78) comes from higher-order terms which represent the fact that in addition to the level scheme of Fig. 2.1, any real molecule has a large number of other electronic states with corresponding vibronic manifolds. If we ignore all other states besides ϕ_0, ϕ_s, and $\{\phi_l\}$ and neglect any coupling between ϕ_0 (or ϕ_0^v) and $\{\phi_l\}$, then (2.78) is exact. However, if ϕ_s and $\{\phi_l\}$ can be coupled to the same set of vibronic states $\{\phi_k\}$ by the molecular perturbation $\mathcal{H}_V$ of (2.29), then there are higher-order corrections to (2.38) which in lowest order are of the form

$$k_{\text{nr}}^{(4)}(s) = \frac{2\pi}{\hbar} \sum_{k,l}' \left| \frac{v_{sk} v_{kl}}{E_s - E_k} \right|^2 \rho_l(E_s), \tag{2.79}$$

where v_{sk} and v_{kl} are the matrix elements of $\mathcal{H}_V$. The prime upon the summation in (2.79) implies that $E_k \neq E_s$. Similarly, if the $\{\phi_l\}$ are coupled by $\mathcal{H}_V$ to a set of vibronic states $\{\phi_m\}$ which are not coupled directly to ϕ_s, there are other corrections which in lowest order are of the form

$$k_{\text{nr}}^{(3)}(s) = \frac{2\pi}{\hbar} \sum_{l,m}' \frac{v_{sl}|v_{lm}|^2}{E_s - E_m} \rho_l(E_s), \tag{2.80}$$

which merely reflect the mixing of the states $\{\phi_m\}$ into the $\{\phi_l\}$ upon consideration of the perturbation $\mathcal{H}_V$ of (2.29). The formal basis for these higher-order terms is considered in Section 2.5, cf. (2.245).

The terms (2.79), (2.80) and all higher-order effects are exactly contained in (2.76). The prescription (2.76), which is well known in the resonance theories of electron scattering, [2.50, 58], etc., is correct for any reasonable approximation to the discrete state ϕ_s. In particular, the rates calculated for a CBO or a

ABO ϕ_s are identical when (2.76) is used, whereas the approximate rates (2.78) may differ in these cases. The general expression (2.76) involves, however, terms to all orders of perturbation theory, and is useful when the series converges rapidly, preferably in second order as in (2.78). As discussed in Section 2.1, ϕ_s^{ABO} should give a better representation of this discrete state than the cruder approximation ϕ_s^{CBO}. Thus, it is more likely that the higher-order terms converge more rapidly when using ϕ_s^{ABO} than ϕ_s^{CBO}. In practice, the rates are often calculated using the simple non-vanishing lowest-order approximation. This approach is expected to be more accurate with the ABO than with the CBO formulations; however, as noted in [2.20–23, 59], caution must be exercised in not neglecting the nuclear coordinate dependence of the energy denominators, e. g., in (2.79) or (2.80), when using the ABO approach. The justification for the neglect of all higher-order terms in a given case must reside in a proof that these terms represent a minor correction; an argument that at best can only be semi-quantitative for large polyatomic molecules.

2.3.7 Small Molecule Limit

The small molecule limit is the other simple case of radiationless processes. Here the coupling v_{ls} is sufficiently large and/or the level density ρ_l is sufficiently small that typical recurrence times are very short, i.e.,

$$\tau_{rec}(j) \ll \tau_{max} \tag{2.81}$$

(and possibly $\tau^{rad}(l)$, $\tau^{rad}(s)$, etc.). In this case, using the BO picture, we would be led to say that if the molecule were initially in the state ϕ_s, then it would "resonate" many times between $\{\phi_l\}$ and ϕ_s. However, quantum mechanics deals only with observable phenomena.

If this hypothetical recurrence is sufficiently rapid that it could not be followed experimentally with the conventional experiments that have been performed and that have uncovered these phenomena, then this picture should be modified so it is expressible solely in terms of observables. For instance, we never measure benzene molecules with either of the Kekule structures nor do we observe benzene molecules undergoing transitions between these structures. A more realistic description of benzene than individual Kekule structures is obtained by taking an approximate benzene wave function to be a superposition of configurations corresponding to these two structures. As noted [2.60], when we have a perturbation V such that the time $\hbar/|V|$ is small compared with the experimental time scale, it is then necessary to combine V with $\mathcal{H}_0$ and diagonalize $\mathcal{H}_0 + V$. This is then the case in the small molecule limit.

Assume that the electronic Hamiltonian $\mathcal{H}_{el}$ of (2.29) is diagonalized by the molecular eigenstates $\{\psi_n\}$ with energies E_n. Performing the transformation, from the BO basis to these molecular eigenstates (cf. $\mathcal{H}_{eff}$ of (2.72) and (2.75)), we note that the off-diagonal matrix elements of the damping matrix are expected to be non-zero. However, in the small molecule, or resonance, limit the level density is sufficiently small that in the absence of accidental degeneracies it is expected that

$$\hbar \Gamma_{nn'} \ll |E_n - E_{n'}| \,. \tag{2.82}$$

Thus, for all practical purposes the $\{\psi_n\}$ can be taken to be the eigenfunctions of $\mathcal{H}_{\text{eff}}$. The corresponding energies are

$$\mathcal{E}_n \approx E_n - \frac{i\hbar}{2} \Gamma_s |\langle \phi_s | \psi_n \rangle|^2 \tag{2.83}$$

so that each molecular eigenstate has a lifetime (inverse fluorescence decay rate) of

$$\tau_n = 1/\Gamma_s |\langle \phi_s | \psi_n \rangle|^2 \equiv k_n^{-1} \,. \tag{2.84}$$

Since in the presence of the coupling v_{ls}

$$|\langle \phi_s | \psi_n \rangle|^2 < 1 \,, \tag{2.85}$$

(2.84) implies immediately that

$$\tau_n > 1/\Gamma_s \equiv \tau^{\text{rad}}(s) \,, \tag{2.86}$$

i.e., the observed fluorescence lifetimes $\{\tau_n\}$ are longer than the lifetime which would be obtained from the integrated absorption intensity.

In the event that (2.82) is violated for a few states, it would then be necessary to diagonalize $\mathcal{H}_{\text{eff}}$ for these states. The resulting states would then exhibit radiative anticrossing effects [2.61].

As noted above, (2.81) implies a rapid recurrence between the states ϕ_s and $\{\phi_l\}$ which effectively makes the BO picture physically meaningless in a practical sense. If we considered the type of excitation process that would be necessary in order to prepare a molecule initially in something like ϕ_s, it would turn out that we require a light source which has special coherence between a large number of different frequencies, i.e., the set of $E_n - E_0$ for $|E_n - E_s| \lesssim v_{ls}$. (For SO_2 and NO_2, $v_{ls} \gtrsim 150\,\text{cm}^{-1}$.) As this is a highly unlikely excitation process, the small molecule limit corresponds to excitation to, and radiative decay from, (essentially) molecular eigenstates. The notion of "cross over" between ϕ_s and $\{\phi_l\}$ is then misleading. On the other hand, as discussed above, in the statistical limit conventional excitation results in an initial molecular state which can often be taken as approximately ϕ_s, and therefore the notion of "cross over" leads to correct conclusion. The small molecule limit is then indicated schematically in Fig. 2.4 giving the BO and the molecular eigenstates pictures.

The experimental verification of the model of the small molecule limit requires the demonstration of unit quantum yields at zero pressure and the ability of assignment of the complex spectra in terms of the interstate mixing in Fig. 2.4. The long lifetimes of the excited electronic molecules like SO_2 and

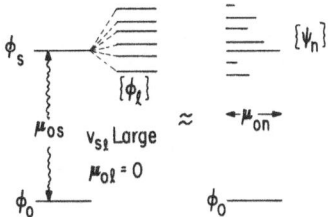

Fig. 2.4. A pictorial description of the small molecule limit. The BO states on the left are the same as those presented in Fig. 2.1. The molecular eigenstates $\{\psi_n\}$ diagonalize the molecular Hamiltonian and, in the absence of accidental degeneracies, diagonalize the effective Hamiltonian (2.72) that includes radiative processes

NO_2 [2.38] and their high cross-sections for inelastic collision processes [2.39, 62] make accurate determinations of the quantum yields extremely difficult. However, BRUS and McDONALD [2.39], and HUI and RICE [2.63] have obtained unit zero pressure quantum yields in SO_2.

The analysis of the visible absorption spectrum of NO_2 has been a long-standing, and as yet unresolved, spectroscopic problem. The spectrum of the room temperature gas consists of a very large number of lines extending throughout the visible and into the infrared, and the lack of any regularity has prevented analysis and has suggested that the excited electronic states are strongly perturbed. In a recent experiment [2.64] a technique was developed which may eventually provide some understanding of this spectrum. In this experiment a supersonic molecular beam of 3% NO_2 diluted in an argon carrier gas was crossed with a tunable visible dye laser and the fluorescence excitation spectrum (undispensed fluorescence versus exciting wavelength) was measured with a resolution of ~0.5Å over the region 5800–6100Å. Since this first experiment, further measurements have been made [2.65] at a resolution of 0.01Å over the region 5700–6700Å and further experiments to extend these measurements to 7900Å are in progress.

The rotational distribution is greatly cooled during the supersonic expansion and only a few rotational levels of the ground electronic state have any observable population. Because of this rotational cooling the fluorescence excitation spectrum is greatly simplified. The various vibronic bands are well resolved and the rotational and fine-structure assignments of the various bands are easily made. The vibrational structure is still highly irregular, but it is hoped that measurements closer to the origin of the transition will allow some interpretation of the vibrational structure.

The present rotational assignments indicate that all of the observed transitions have their transition moment parallel to the near symmetric top axis and are therefore presumably due to transitions between the $^2A_1$ ground state and a $^2B_2$ excited electronic state. No perpendicularly polarized bands have been found which indicates that most of the absorption in this region is due to a $^2B_2$ state. The effective rotational and spin-rotation constants which can be calculated for each band show substantial variation from band to band and this is further evidence that the excited state is strongly perturbed. A sample spectrum is reproduced in Fig. 2.5.

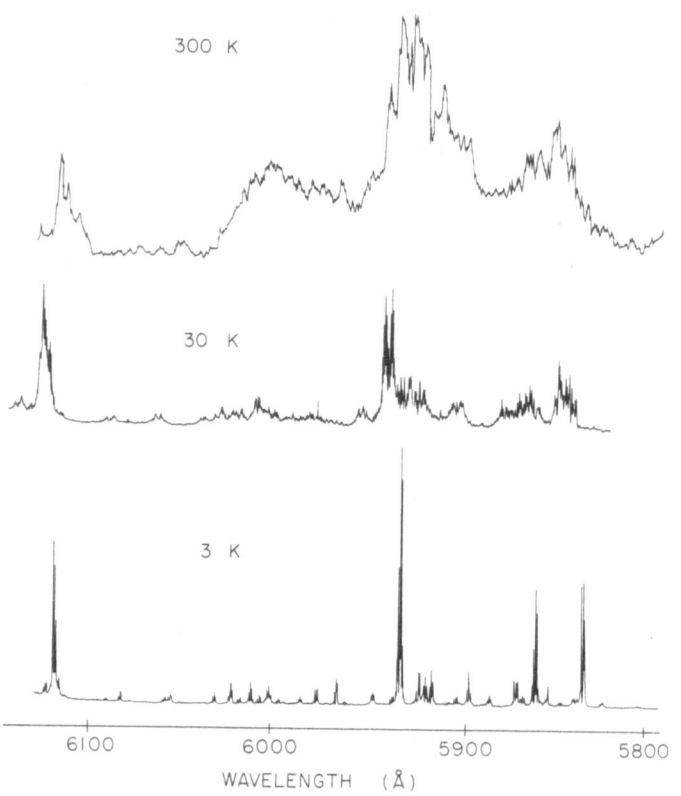

Fig. 2.5. A portion of the fluorescence excitation spectrum, taken from [2.64], obtained from NO_2 that is cooled by expansion through a supersonic nozzle. The top spectrum is for a conventional room temperature sample of pure NO_2 at 0.04 torr pressure. The middle figure is the spectrum of a supersonic beam of pure NO_2, while the bottom one is from a supersonic beam of 5% NO_2 in Ar. (The cw dye laser bandwidth was 0.5 Å for the two lower spectra.) The vibronic bands in the 3° K spectrum are clearly separated and the rotational analyses have been made [2.64, 65]

A further bit of anomalous behavior in NO_2 has recently been reported [2.66]. It has long been known that even if the exciting light is well below the dissociation limit, fluorescence from NO_2 consists not only of sharp banded structure but contains also what appears to be an underlying continuum. Previous observations [2.67] have indicated that the continuum intensity decreases relative to the banded intensity as the pressure is lowered, and this implied that bimolecular collisions were somehow involved in the production of the continuum. Recent experiments [2.66, 68] on the magnetic quenching of fluorescence in NO_2 have indicated that contrary to previous observations the banded/continuum intensity ratio is independent of pressure down to very low pressures. Further experiments which measured the banded/continuum intensity ratio as a function of the distance between the point of observation and the point of detection indicated that the continuum emitters have a much longer lifetime than the banded emitters. Therefore what had been interpreted as a decrease in continuum intensity as the pressure was decreased appears to be due to the diffusion of the long-lived continuum emitters out of the field of view of the detector. The magnetic quenching data is consistent with this interpretation and suggests either that the continuum emission comes from the isolated molecule or that the cross sections for certain bimolecular processes are extremely large [2.62]. No explanation of this phenomenon has been offered.

2.3.8 Intermediate Case

The cases of small and large molecule represent, of course, interesting and observable limits. However, there must also be an intermediate case, and this intermediate case might be expected to have certain features that resemble both the large and the small molecule limits [2.4, 8, 45, 46, 54, 69–74]. As noted already we expect that some states $\{\phi_b\}$ in $\{\phi_l\}$ are strongly coupled to ϕ_s whereas others $\{\phi_w\}$ are only weakly coupled to ϕ_s because of symmetry selection rules, etc. We could therefore envision a situation in which the $\{\phi_w\}$ are dissipative with respect to ϕ_s, forming an effective quasi-continuum which leads to irreversible decay for time scales of relevance. The $\{\phi_w\}$ therefore represent the dense manifold which occurs in the statistical limit. On the other hand, the less numerous $\{\phi_b\}$ represent a non-dissipative set of states with

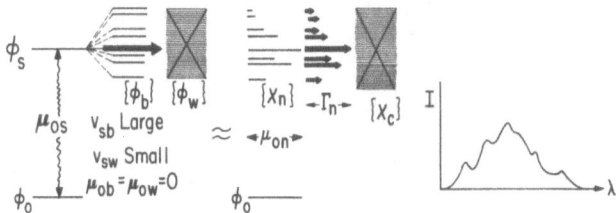

Fig. 2.6. One particular realization of the intermediate case is depicted showing the strongly and weakly coupled states, $\{\phi_b\}$ and $\{\phi_w\}$, respectively. The latter are assumed to form an effective quasi-continuum. The states $\{\chi_r\}$, which diagonalize the effective Hamiltonian (2.72), are the resonant states, each having both radiative and non-radiative decay rates. A schematic representation of the absorption spectrum is presented for this particular type of intermediate case

respect to ϕ_s. Hypothetical recurrence times between ϕ_s and $\{\phi_b\}$ are sufficiently short that these states are not independently observable in a practical sense. Thus, the set of states $\{\phi_s, \phi_b\}$ corresponds to the situation found in the small molecule limit. This intermediate case is shown schematically in Fig. 2.5a. The BO level scheme corresponds to the small molecule-limit set of states $\{\phi_s, \phi_b\}$ which are coupled to two dissipative channels, radiative decay and decay into the effective quasi-continuum $\{\phi_w\}$. The time evolution of the closely coupled "resonating" states $\{\phi_s, \phi_b\}$ is determined by an effective Hamiltonian $\mathcal{H}'_{\text{eff}}$, where

$$\mathcal{H}'_{\text{eff}} = \mathcal{H}'_{\text{el}} - i\hbar \Gamma'/2 \tag{2.87}$$

and $\mathcal{H}'_{\text{eff}}$, $\mathcal{H}'_{\text{el}}$, and Γ' only have matrix elements within the set $\{\phi_s, \phi_b\}$. Within this basis set $\mathcal{H}'_{\text{el}}$ and $\mathcal{H}_{\text{el}}$ have identical matrix elements. Γ' is a generalized damping matrix which includes both radiative and non-radiative decay. (Here we neglect other states like the $\{\phi_k, \phi_m\}$ as in (2.79) and (2.80), etc.; they can easily be incorporated.) In particular, if the $\{\phi_b\}$ have radiative lifetimes $[\Gamma^{\text{rad}}(b)]^{-1}$ the matrix Γ' in the BO basis is diagonal

$$\Gamma' = \begin{pmatrix} \Gamma^{\text{rad}}(s) + \Delta(s) & 0 & 0 & 0 \\ 0 & \Gamma^{\text{rad}}(b_1) & & 0 \\ 0 & & \Gamma^{\text{rad}}(b_2) & \\ 0 & & & \ddots \end{pmatrix} \tag{2.88}$$

where $\Delta(s)$ is the non-radiative decay rate for decay of ϕ_s into $\{\phi_w\}$, i.e., the result obtained by ignoring $\{\phi_b\}$. Thus, this implies

$$\Delta(s) = \frac{2\pi}{\hbar} \sum_w |v_{sw}|^2 \rho_w(E_s), \tag{2.89}$$

where states like the $\{\phi_k, \phi_m\}$ in (2.79) and (2.80) have been assumed to be negligible or to be incorporated into effective values of $\{v_{sw}\}$. In general, there is a set of resonant states χ_n^r which diagonalizes $\mathcal{H}'_{\text{eff}}$ with the energies

$$\mathcal{E}_n = E_n - \frac{i\hbar}{2} [\Gamma^{\text{rad}}(n) + \Delta_n] \equiv E_n - i\hbar \Gamma_n/2, \tag{2.90}$$

where we have partitioned the total decay rate of each resonant state into its radiative rate $\Gamma^{\text{rad}}(n)$ and the non-radiative rate Δ_n. The resonant state picture is presented in Fig. 2.6 for the case $\Gamma^{\text{rad}}(b) \approx 0$. In the general case the spectrum would correspond schematically to that given in Fig. 2.6. The broad background is due to the dissipative levels $\{\phi_w\}$ [assuming for the sake of this example that $\Delta(s) \gg \Gamma^{\text{rad}}(s)$] and therefore has a spectral width on the order of $\Delta(s)$. Each of the individual resonances would have a fluorescence yield and decay lifetime that is calculated in the conventional manner using (2.2) and (2.1), respectively, from their radiative and non-radiative decay rates $\Gamma^{\mathrm{rad}}(n)$ and Δ_n if these resonances were non-overlapping. In practice, many resonances are expected to be overlapping in the sense that

$$|E_n - E_{n'}| \leq \hbar |\Gamma_n + \Gamma_{n'}|/2 ,$$

and hence there are a number of quantum mechanical interference effects. The fluorescence yields and lifetimes will, however, still vary as we pass through the different parts of the spectrum of Fig. 2.5b.

If the state of the system upon excitation is initially taken to be

$$\psi_{\mathrm{mol}}(0) = \sum_n C_n \chi_n^{\mathrm{r}} , \tag{2.91}$$

after time t the excited state of the molecule evolves as

$$\psi_{\mathrm{mol}}(t) = \sum_n C_n \exp(-i E_n t/\hbar - \Gamma_n t/2) \chi_n^{\mathrm{r}} .$$

Assuming for convenience that $\Gamma^{\mathrm{rad}}(b) \approx 0$, the fluorescence probability is

$$F(t) = \left| \sum_n C_n \exp(-i E_n t/\hbar - \Gamma_n t/2) \langle \phi_0; k_F e_F | H_{\mathrm{int}} | \chi_n^{\mathrm{r}} \rangle \right|^2 ,$$

which in general contains the cross, or interference terms [2.4, 8, 45–47, 70–75]. These interference effects could only be expected to be unambiguously assigned when there are but a few overlapping resonances χ_n^{r}. However, the rotational degrees of freedom, which have been heretofore neglected, even complicate this further. Say, there are a few overlapping vibronic resonances $\chi_i^{\mathrm{r}} \ldots \chi_n^{\mathrm{r}}$ in a given energy region. Each of the zeroth-order vibronic levels, making the dominant contributions to $\{\chi_i^{\mathrm{r}}\}$, in general, have differing rotational constants. Thus, the overlapping n vibronic resonances in reality correspond to a multitude of overlapping rhovibronic levels with slightly differing energy spacings as a function of rotational quantum numbers. Thus, if interference effects, such as oscillatory decay patterns, were observed, their assignment would be problematic indeed[7].

The resonant states χ_n^{r} can be thought to be determined in a two-step process. First the set of strongly coupled states $\{\phi_s, \phi_b\}$ can be prediagonalized in a manner quite identical to that of the small molecule limit. Each of the perturbed levels, so obtained, then carries only a part of the oscillator strength for the $\phi_0 \rightarrow \phi_s$ transition and consequently a longer pure radiative lifetime than that for the zeroth-order level. The second stage of introducing the weakly

---

[7] See footnote 5.

coupled levels $\{\phi_w\}$ still leads to the resonant levels χ_n^r with radiative decay rates Γ_n^{rad} because of the dilution of ϕ_s amongst the χ_n^r. The presence of the $\{\phi_w\}$, however, implies that the χ_n^r, in addition, have non-radiative decay rates Γ_n^{nr}, and the observed decay rate

$$\Gamma_n = \Gamma_n^{rad} + \Gamma_n^{nr} \tag{2.92}$$

often considerably exceeds the pure radiative decay rate Γ_s of ϕ_s. Thus, if only a low resolution spectrum and radiative decay rate, corresponding to broad band excitation of the whole resonance arising from ϕ_s, are obtained, it might just appear that a statistical limit molecule is being investigated. However, a determination of the quantum yield, in addition, immediately signals a pure radiative lifetime lengthening [from (2.1), (2.2), and the integrated absorption intensity determined $\tau^{rad} = 1/\Gamma_s$], the hallmark of the intermediate case.

The intermediate case can be expected to be found in large molecules having very small energy gaps between the lowest-lying electronic states [2.4, 8, 69, 72, 74]. Such situations arise in the case of small splittings between the origins of the first two excited singlet states S_1 and S_2 such as naphthalene ($\Delta E_{(S_2 S_1)} \approx 3500\,cm^{-1}$) and 3,4-benzpyrene ($\Delta E_{(S_2 S_1)} \approx 3800\,cm^{-1}$). In these cases the respective observed lifetimes of S_2, $\tau_{S_2} \approx 4 \times 10^{-8}\,s$ and $\approx 7 \times 10^{-8}\,s$, are longer than those deduced from the absorption oscillator strengths, $\tau_{S_2}^{rad} \approx 10^{-8}\,s$ for both molecules. In these cases, the intersystem crossing rates are low enough that an absolute lifetime lengthening is observed as a manifestation of the intermediate case. The intermediate case also arises for small $S_1 - T_1$ (lowest excited triplet state) splittings [2.72] e.g., benzophenone where $\Delta E_{(S_1 T_1)} \approx 3000\,cm^{-1}$ and $\tau_{S_1} \approx 10^{-5}\,s > \tau_{S_1}^{rad} \approx 10^{-6}\,s$. In these cases the total density of lower vibronic states near the origin of the higher one is on the order of $10^3\,cm$. The number of strongly coupled levels $\{\phi_b\}$ is then only a small fraction of this number. Given the enormous number of rhovibronic levels, it would be very difficult to isolate quantum beats unless the molecule were excited to no more than a few well defined rhovibronic levels by the use of a high-intensity narrow-band laser[8].

The S_1 level densities of naphthalene and 3,4-benzpyrene near the S_2 origins are not high enough for the $S_2 - S_1$ coupling to lead to an irreversible internal conversion in the isolated molecule as condition (2.69) or $\tau_{rec} \gg \tau_{S_1}$ is violated. Similarly, in benzophenone the $S_1 - T_1$ coupling does not yield irreversible intersystem crossing under isolated molecule conditions. However, the presence of perturbing molecules in dense gases and condensed media lead to a rapid vibrational relaxation of the vibrationally excited $\{\phi_l\} = \{\phi_b, \phi_w\}$ that are coupled to ϕ_s. This vibrational relaxation leads, in part, to the introduction of relaxation widths γ_l and lifetimes $\hbar/\gamma_l$ which make $\{\phi_l\}$ into channels for the irreversible decay of ϕ_s because of the rapid vibrational decay of the resonances formed from $\{\phi_s, \phi_l\}$. In these cases, the structure depicted in

---

[8] See footnote 5.

Fig. 2.5 should disappear along with any variations in τ or φ across the full resonance associated with ϕ_s. Indeed, benzophenone in solution is not observed to fluoresce on a nanosecond timescale.

Another situation giving rise to the intermediate case arises in small polyatomic molecules where the occurrence of photodissociation channels converts an otherwise insufficient density of vibronic states on a lower electronic states into a true dissipative continuum. Consider the particular case of ϕ_s being coupled to a small number of zeroth-order levels $\{\phi_b\}$ which are non-dissociative. ϕ_s is assumed to not be (or negligibly be) coupled to the dissociative states $\{\phi_w\}$. It is then the weak coupling between $\{\phi_b\}$ and $\{\phi_w\}$ which leads to the eventual dissociation on the lower potential energy surface. The case of the photodissociation of chloro- and bromo-acetylene were shown by EVANS et al. [2.73] to conform to this case. In the photodissociation of formaldehyde, as studied by YEUNG and MOORE [2.76], the ϕ_s decay rates correlate extremely well with predicted $S_1 \rightarrow S_0$ internal conversion rates, so the $\phi_b - \phi_w$ coupling must be sufficiently large to give rather large Γ_b and make the zeroth-order predissociating levels ϕ_b into a true dissipative continuum, the statistical limit.

It is by now well accepted that the very diverse phenomena associated with the small, intermediate, and large molecule limits all follow from the same molecular energy level scheme Fig. 2.1. The observed differences in these cases arise from a continuous variation in the parameter $\hbar \Gamma_l \rho_l$ ($\hbar$ times ϕ_l decay rates divided by the average ϕ_l energy level spacings) for the $\{\phi_l\}$ from being much less than unity for small molecules to being much larger than unity in the statistical limit, with the intermediate case having two sets of levels $\{\phi_w\}$ and $\{\phi_b\}$ corresponding, respectively, to both these limits as

$$\hbar \Gamma_b \rho_b \ll 1 , \quad \hbar \Gamma_w \rho_w \gg 1 .$$

This transition between the small and large molecule limits can, perhaps, best be illustrated through the use of an exactly soluble model, the Bixon-Jortner model. Thus, this model, and generalizations thereof, is discussed in Section 2.4, while Section 2.5 provides a review of the underlying formal theory of the mechanistic aspects of radiationless decay processes discussed theretofore. The reader who is not interested in this more mathematical material might wish, upon first reading, to skip to Section 2.6 where a detailed discussion is given of the evaluation of non-radiative decay rates for statistical-limit molecules.

2.4 Transition Between Small and Large Molecule Limits

In general, the molecular energy level model depicted in Fig. 2.1 is not amenable to an exact quantum mechanical solution. The formal solution for the model in terms of the effective molecular Hamiltonian (2.72) is justified in more detail in Section 2.5. However, these results are based upon the application of the notions of practical irreversibility, as given in (2.63–71). In order to more thoroughly illustrate these concepts, it is useful to introduce an exactly soluble version of the model. This simple Bixon-Jortner model has been very useful in elucidating the basic features of non-radiative decay processes [2.2, 40, 42, 55, 56] so we now turn to its use in explaining the practical notions of irreversibility as well as the transition between the small, intermediate and large molecule limits.

2.4.1 Bixon-Jortner Model

The Bixon-Jortner model consists of a single state ϕ_s, carrying oscillator strength from the thermally accessible components of ϕ_0 of the ground electronic state. ϕ_s is coupled to a quasi-continuum of equally spaced levels ϕ_l with a coupling strength v that is independent of l. The level spacing is ε, and the ϕ_l are assumed to not carry any oscillator strength to ϕ_0, although this assumption may be lifted if desired. Radiative and collisional processes lead to level widths Γ_s and $\Gamma_l \equiv \Gamma$, all l, for ϕ_s and $\{\phi_l\}$, respectively. The last assumption that is necessary to yield a simple exact solution, takes the $\{\phi_l\}$ levels to run from $l = -\infty$ to $l = +\infty$, so they are not bounded from below as is normal. This last assumption has been criticized as a failing of the model, an unphysical aspect; however, it represents an assumption of mathematical convenience. It is possible to begin with a $\{\phi_l\}$ manifold that is bounded from below, and then to add and subtract a hypothetical compliment of lower levels. Because these lower levels are energetically far removed from ϕ_s, the sums in the subtracted terms can be approximated by integrals, finally leaving a slightly adjusted form of the original Bixon-Jortner model. If the bounded version of the model is employed, the lowest eigenvalue develops a large energy shift downwards from the position of the lowest zero-order level (2.7). This lowest level also contains a large admixture of ϕ_s, corresponding to a "new" separated state below the original quasi-continuum $\{\phi_l\}$ with a large oscillator strength to ϕ_0. The nature of this "new" state is perhaps best illustrated by considering couplings between two diatomic potential curves that represent a pair of electronic states [2.77]. For instance, if the higher curve is that of a bound one (ϕ_s) and the lower a repulsive curve (ϕ_l). For large enough $s-l$ coupling, the lower curve develops a small minimum that can support a bound state and give rise to the "new" state. As our interest is not in studying the properties of this new state or in introducing minor corrections to the Bixon-Jortner model, which have no significant qualitative importance, we now pass to the solution of this simple model.

  If the molecule is excited by a short pulse of incident radiation with frequencies in the general region of the expected $\phi_0 \rightarrow \phi_s$ absorption, the precise details of the observed luminescence depend on the frequency, width, and shape of the pulse as well as the molecular energy level scheme. For a short enough pulse the absorption and emission processes can be taken to be separate to a good approximation. (The general case is considered in Sect. 5.) Thus, after the pulse, the excited molecular state can be taken to be some linear combination of the zeroth-order molecular levels

$$\Psi(0) = a_s |\phi_s, \text{vac}\rangle + \sum_l b_l |\phi_l, \text{vac}\rangle . \tag{2.93}$$

The values of a_s and b_l are pulse property dependent, and, consequently, are not of interest at this juncture. It is relevant, however, to note that current technology limits the pulse width τ_p to a lower limit in the picosecond range and, hence, to an energy spread of $\hbar/\tau_p$ in this incident pulse. Thus, an idealized delta function pulse always implies $a_s = 1$, $b_l = 0$. This idealization is not one that is practically attainable for arbitrary values of the molecular parameters Γ_s, Γ_l, ε, v, so

$$\Psi(0) \doteq |\phi_s, \text{vac}\rangle \tag{2.94}$$

is often not practically attainable. A lack of realization of this fact has been a source of great confusion in parts of the literature. Because (2.93) represents the most general possible case and in order to determine when (2.94) is actually a physical possibility, we defer further discussion of the incident radiation until after the model has been solved.

2.4.2 Solution of the Model

Given the initial state (2.93), the most general form for the state at time t is again

$$
\begin{aligned}
\Psi(t) &= \exp(-i \mathcal{H} t/\hbar) \, \Psi(0) \\
&= a_s(t) |\phi_s, \text{vac}\rangle + \sum_l b_l(t) |\phi_l, \text{vac}\rangle \\
&\quad + \sum_{k,e,v} c_{kev}(t) |\phi_0^v, k\,e\rangle ,
\end{aligned} \tag{2.95}
$$

where two photon processes have been ignored. Following the development in (2.42)–(2.62), we can take the Laplace-Fourier transform of (2.95) to introduce the matrix elements of the Green's function $G(E)$. In particular we have the relationships

$$a_s(t) = (2\pi i\hbar)^{-1} \int_C dE \exp(-iEt/\hbar) \left[G_{ss}(E) a_s + \sum_l G_{sl}(E) b_l \right], \tag{2.96}$$

$$b_l(t) = (2\pi i\hbar)^{-1} \int_C dE \exp(-iEt/\hbar) \left[G_{ls}(E) a_s + \sum_{l'} G_{ll'}(E) b_{l'} \right], \tag{2.97}$$

and

$$c_F(t) = (2\pi i\hbar)^{-1} \int_C dE \exp(-iEt/\hbar) \left[G_{Fs}(E) a_s + \sum_l G_{Fl}(E) b_l \right],$$ (2.98)

where the shorthand notation

$$F = k e \phi_0^v$$ (2.99)

is employed for notational convenience.

Eqs. (2.95–99) again emphasize how the Green's function method enables the complete solution for the time dependent solution $\Psi(t)$ for arbitrary initial conditions (2.93). Using the Bixon-Jortner model, the Green's function equations (2.49) are readily found to be

$$(E - E_s) G_{ss}(E) - \sum_l v G_{ls} - \sum_F \mathcal{H}_{sF}' G_{Fs} = 1$$ (2.100a)

$$(E - E_l) G_{ls}(E) - v G_{ss}(E) = 0$$ (2.100b)

$$(E - E_F) G_{Fs}(E) - \mathcal{H}_{Fs}' G_{ss}(E) = 0$$ (2.100c)

$$(E - E_s) G_{sl}(E) - v G_{ll}(E) = 0$$ (2.100d)

$$(E - E_l) G_{ll}(E) - v G_{sl}(E) = 1,$$ (2.100e)

etc., for the $l - F$, $F - s$, $F - l$, $F - F$ matrix elements of (2.49). In (2.100), E_s, E_l, and E_F designate, respectively, the energies of ϕ_s, ϕ_l, and $|\phi_0^v, k e\rangle$, while $\mathcal{H}_{sF}'$ denotes the matrix element of the radiation-matter interaction $\mathcal{H}_{int}$ of (2.28),

$$\mathcal{H}_{sF}' = \langle \phi_s, \text{vac} | \mathcal{H}_{int} | \phi_0^v, k e \rangle = (\mathcal{H}_{Fs}')^*.$$ (2.101)

Eqs. (2.100b and c) can be rearranged to read

$$G_{ls}(E) = \frac{v}{E - E_l} G_{ss}(E)$$ (2.102a)

and

$$G_{Fs}(E) = \frac{\mathcal{H}_{Fs}' G_{ss}(E)}{E - E_l}.$$ (2.102b)

Substituting (2.102) into (2.100a) and rearrangement yields

$$G_{ss}(E) = \left[E - E_s - \sum_l \frac{v^2}{E - E_l} - \sum_F \frac{|\mathcal{H}_{sF}'|^2}{E - E_F} \right]^{-1}.$$ (2.103)

First consider the term in (2.103) involving the summation over F. As the photons have a continuous range of allowed frequencies, the summation is actually an integration,

$$\Lambda_s(E) = \sum_{v,e} \int_0^\infty dk \int d\Omega_k \frac{\rho_{\text{photon}}(\hbar c k)|\langle \phi_s, \text{vac}|H_{\text{int}}|\phi_0^v, k e\rangle|^2}{E - \hbar c k - E_0^v}, \tag{2.104}$$

where E_0^v is the energy of ϕ_0^v, c is the velocity of light $k = |k|$, and $\rho_{\text{photon}}(\hbar c k)$ is the photon density of states at photon energy $\hbar c k$, as in (2.73). As implied after (2.49), along the integration contour in (2.96–98), the integration variable has a small positive imaginary part $i\delta$, $\delta > 0$. We explicitly indicate this fact by using the substitution $E \to E' + i\delta$ with the understanding that the integration variable E' runs along the real axis. As $\delta \to 0^+$, the relation

$$\lim_{\delta \to 0^+} \frac{1}{E' + i\delta + x} = \frac{\mathscr{P}}{x} - i\pi\delta(x), \tag{2.105}$$

where $\mathscr{P}$ denotes the Cauchy principal value integral operator, is used to convert (2.104) to

$$\lim_{\delta \to 0^+} \Lambda_s(E) = D_s(E') - i\hbar\Gamma_s(E')/2 \tag{2.106}$$

where

$$\Gamma_s(E') = \frac{2\pi}{\hbar} \sum_{v,e} \int d\Omega_k \, \rho_{\text{photon}}(E - E_0^v)|\langle \phi_s, \text{vac}|\mathscr{H}_{\text{int}}|\phi_0^v, k e\rangle|^2 \tag{2.107}$$

is the radiative decay rate of the zeroth-order level ϕ_s when $E' = E_s$. The level shift is [2.50]

$$D_s(E') = \sum_{v,e} \mathscr{P} \int_0^\infty dk \int d\Omega_k \frac{\rho_{\text{photon}}(\hbar c k)|\langle \phi_s, \text{vac}|\mathscr{H}_{\text{int}}|\phi_0^v, k e\rangle|^2}{E - \hbar c k - E_0^v} \tag{2.108}$$

and is generally small and neglected after incorporating an infinite part into the definition of the electron mass [2.50].

Substituting (2.104–108) into (2.103) gives

$$G_{ss}(E' + i\delta) = \left[E' + i\delta - E_s - \sum_l \frac{v^2}{E' + i\delta - E_l} + i\hbar\Gamma_s(E')/2 \right]^{-1}, \tag{2.109}$$

where D_s has been incorporated into the definition of E_s. Because $\Gamma_s(E')$ is very slowly varying with E', it is generally permissible to replace it by the radiative decay rates

$$\Gamma_s = \Gamma_s(E_s).$$

From (2.109) the first term in $a_s(t)$ in (2.96) becomes

$$a_{ss}(t) = \lim_{\delta \to 0^+} (2\pi i\hbar)^{-1} \int_{\infty}^{-\infty} dE' \exp(-iE't/\hbar - \delta t'/\hbar)$$
$$\times \left[E' + i\delta - E_s - \sum_l \frac{v^2}{E' + i\delta - E_l} + i\hbar \Gamma_s/2 \right]^{-1}$$

(2.110)

$$= \sum_n \exp(-iE_n t/\hbar - \Gamma_n t/2) \langle s|n \rangle \langle n|s \rangle ,$$

(2.111)

where $\mathscr{E}_n = E_n - i\hbar\Gamma_n/2$ are the poles of $G_{ss}(E')$ in the lower half-plane (E_n, Γ_n real and $\Gamma_n \geq 0$) and $\langle s|n \rangle \langle n|s \rangle$ are the residues at these poles. Since the poles of $G_{ss}(E')$ are obtained as the zeros of $1/G_{ss}(E')$, the $\mathscr{E}_n$ are the solutions to the equation

$$\mathscr{E}_n = E_s - i\hbar \Gamma_s/2 + \sum_l \frac{v^2}{\mathscr{E}_n - E_l} .$$

(2.112)

In the absence of radiation-matter interaction, i.e., ignoring spontaneous emission, (2.112) becomes

$$E_n = E_s + \sum_l \frac{v^2}{E_n - E_l} ,$$

(2.113)

which is but a rewriting of the eigenvalue equation determining the molecular energy levels E_n associated with the molecular eigenstates ψ_n, as described in Section 2.3. Consequently, in this case the residues in (2.111) become

$$\langle s|n \rangle \langle n|s \rangle \to |\langle \phi_s|\psi_n \rangle|^2 ,$$

(2.114)

the probability of finding ϕ_s in the molecular eigenstates ψ_n. Eq. (2.113) can be converted into a simple algebraic equation by use of the general summation formula

$$\pi \cot \pi x = \sum_{l=-\infty}^{\infty} \frac{1}{x+l}$$

(2.115)

and the definition of the Bixon-Jortner model

$$E_l = E_s + \alpha + l\varepsilon ,$$

(2.116)

so α represents the spacing between ϕ_s and its nearest level $\phi_{l=0}$ in the $\{\phi_l\}$ manifold. Eqs. (2.113–116) yield

$$E_n = E_s + \frac{v^2}{\varepsilon} \pi \cot [\pi x(E_n)]$$

(2.117)

with the definition

$$x(E_n) = (E_n - E_s - \alpha)/\varepsilon, \quad |\alpha| < \varepsilon. \tag{2.118}$$

For the case of non-constant coupling and/or non-equal level spacing, the last term in (2.112) becomes

$$\sum_l \frac{|v_{ls}|^2}{E_n - E_l} \tag{2.119}$$

which is not generally amenable to summation as in the special case of the Bixon-Jortner model (2.117). In any event the E_n are the molecular eigenvalues and the association (2.114) persists: the zeroth-order ϕ_s has an energy which falls between a pair of l-levels (e.g. $l=0$ and 1 for $\alpha > 0$), and the final E_n values are obtained, even in the general case (2.119), by allowing ϕ_s to "repell" all the ϕ_l, shifting those for $l > 0$ up in energy and those for $l < 0$ down, with the perturbed ϕ_s itself suffering a shift that keeps it within the bounds of its neighboring eigenvalues. Thus, the Bixon-Jortner model simulates the general interleaving theorem with an exactly soluble form (2.117). Eq. (2.117) can readily be solved graphically [2.54], but this is unnecessary for our purpose. Eq. (2.114) yields the residues

$$|\langle \phi_s | \psi_n \rangle|^2 = \frac{v^2}{(E_n - E_s)^2 + (\pi v^2/\varepsilon)^2}, \tag{2.120}$$

giving a Lorentzian distribution of ϕ_s amongst the molecular eigenstates [2.2].

Of course, the neglect of spontaneous emission in (2.113–120) is merely for pedagogical purposes in order to more fully illustrate the nature of the Bixon-Jortner model. In the presence of radiation-matter interaction, the ψ_n are no longer stationary states of the complete Hamiltonian (2.28). The complex energies $\mathscr{E}_n$ of (2.112) then give the positions E_n and widths $\hbar \Gamma_n/2$ of the resonant levels χ_n^r which can differ markedly from the ψ_n, as is shown below. Even in the presence of $\mathscr{H}_{int}$, the Bixon-Jortner model enables the summation of the last term in (2.112). Before explicitly introducing this solution, we recall that there are other possible dissipative mechanisms that can lead to deactivation of the $\{\phi_s, \phi_l\}$. Firstly, the $\{\phi_l\}$ may have non-negligible radiative decay rates $\{\Gamma_l\}$ to some lower vibronic levels, and secondly there may be other non-radiative decay processes due to collisions with other molecules, with the walls of the container, etc. A treatment of the former leads to the replacement of $E_l - i\hbar \Gamma_l(E')/2$ for E_l in (2.110, 112, 113, 119), etc., while the latter in zeroth-order involves the introduction of collision induced decay rates $\{\gamma_s, \gamma_l\}$ with the substitutions $\Gamma_s \rightarrow \Gamma_s + \gamma_s$ and $\Gamma_l \rightarrow \Gamma_l + \gamma_l$ in the above [2.4, 45, 46]. Also, as discussed in Section 2.3, if the properties of $\Psi(t)$ of (2.95) are to be determined on some timescale t, the time-energy "uncertainty" relation

$$\Delta E \Delta t \geq \hbar \tag{2.121}$$

implies that an additional width $\hbar/\Delta t$ can be associated with each level to enable the determination of additional mathematical approximations that are valid on this timescale. In the following, we let Γ_s and $\Gamma_l = \Gamma$ (all l) simply denote the total decay rates for the zeroth-order levels with the understanding that any non-radiative contributions can be added when necessary.

In the general case of the Bixon-Jortner model for (2.112), with $\Gamma_l = \Gamma$ added, the analytic continuation of (2.115) yields

$$\mathscr{E}_n = E_s - i\hbar\Gamma_s/2 + (\pi v^2/\varepsilon)\cot[\pi z(\mathscr{E}_n)], \tag{2.122}$$

where

$$z(\mathscr{E}_n) = x(\mathscr{E}_n) + iy \tag{2.123}$$

with $x(\mathscr{E}_n)$ given in (2.118) and

$$y = \hbar\Gamma/2\varepsilon. \tag{2.124}$$

Substituting (2.123) into (2.122) gives

$$\mathscr{E}_n = E_s - i\hbar\Gamma_s/2 + \frac{\pi v^2}{\varepsilon}\left[\frac{1 - i\tan[\pi x(\mathscr{E}_n)]\tanh\pi y}{\tan[\pi x(\mathscr{E}_n)] + i\tanh\pi y}\right]. \tag{2.125}$$

2.4.3 Small Molecule Case

In small molecules the ratio of the average level spacing to $\hbar$ is very large compared to any radiative decay rates Γ in isolated molecules, hence we have

$$y \ll 1, \quad \text{small molecules.} \tag{2.126}$$

As an example, when $\varepsilon \approx 1$ cm^{-1} and $\Gamma \approx 10^{+8}$ s^{-1}, $\hbar\Gamma/2\varepsilon \approx 1/600$. Employing the limit (2.126) gives

$$\tanh\pi y = \pi y + \mathcal{O}[(\pi y)^3],$$

and when $\Gamma \ll \Gamma_s$, (2.125) becomes

$$\mathscr{E}_n = E_s - i\hbar\Gamma_s/2 + (\pi v^2/\varepsilon)\cot[\pi x(\mathscr{E}_n)] \tag{2.127}$$

which are just the resonant states associated with the effective Hamiltonian (2.72). To a good approximation these are given by

$$\mathscr{E}_n \approx E_n - i\hbar\Gamma_s|\langle\psi_n|\phi_s\rangle|^2/2 \equiv E_n - i\hbar\Gamma_n/2, \tag{2.128}$$

where the E_n and ψ_n are those values obtained in (2.113–120) by neglecting the radiation-matter coupling $\mathcal{H}_{\mathrm{int}}$. For Γ non-negligible, corrections to (2.128) proportional to Γ can readily be obtained as necessary. The main point is that in the small molecule limit, the resonant states correspond very closely to the molecular eigenstates, the spectroscopic view of the results of perturbations between different electronic states.

In this small molecule limit, the probability amplitude for remaining in ϕ_s is then from (2.111)

$$a_{ss}(t) = v^2 \sum_n \frac{\exp(-\mathrm{i}\,E_n t/\hbar - \Gamma_n t/2)}{(E_n - E_s)^2 + (\pi v^2/\varepsilon)^2}. \qquad (2.129)$$

If the molecule *could* somehow be prepared initially in $|\phi_s, \mathrm{vac}\rangle$, $|a_{ss}(t)^2|$ would represent the probability $P_s(t) = |\langle \phi_s, \mathrm{vac} | \Psi(t)\rangle|^2 = |a_{ss}(t)|^2$ of finding the molecule in ϕ_s at time t. Eq. (2.129) would represent $P_s(t)$ as a highly oscillatory function of time corresponding to the molecule "resonating" or "crossing" back and forth many times between ϕ_s and $\{\phi_l\}$. However, a glance at typical experimental data quickly indicates that $|\phi_s, \mathrm{vac}\rangle$ *is a most unlikely initial state after photon excitation in the small molecule limit.* For instance, in NO_2 and SO_2 if we take $\varepsilon \sim \mathcal{O}(1\ \mathrm{cm}^{-1})$ and $v \sim 150\text{–}200\,\mathrm{cm}^{-1}$, a typical value for Renner-Teller coupling, etc., (2.120) implies that a zeroth-order ϕ_s is spread over the molecular eigenstates ψ_n over a range of energies of width $\approx 150\text{–}200\,\mathrm{cm}^{-1}$. Thus, in order to initially excite ϕ_s it would be necessary to employ a coherent pulse of duration of less than 0.1 ps. In actuality v is probably larger and ε smaller, so the "minimum" pulse width is probably an upper bound on the required value.

On the nanosecond timescale of conventional experiments, barring any accidental degeneracies, the initial molecular state in $\Psi(0)$ corresponds to one of the molecular eigenstates ψ_n. In this case, it is more convenient to express (2.93, 95–99) in the molecular eigenstates basis as this leads to

$$\Psi(0) = |\psi_n, \mathrm{vac}\rangle \qquad (2.130)$$

$$\Psi(t) = d_n(t)|\psi_n, \mathrm{vac}\rangle + \sum_{k,e,v} c_{kev}(t)|\phi_0^v, k\,e\rangle \qquad (2.131)$$

with

$$d_n(t) = (2\pi \mathrm{i}\hbar)^{-1} \int_C dE \exp(-\mathrm{i}\,E t/\hbar)\, G_{nn}(E) \qquad (2.132a)$$

and

$$c_{kev} \equiv c_F = (2\pi \mathrm{i}\hbar)^{-1} \int_C dE \exp(-\mathrm{i}\,E t/\hbar)\, G_{Fn}(E). \qquad (2.132b)$$

The above analysis gives

$$G_{nn}(E) \approx (E - E_n + i\hbar\Gamma_n/2),$$ (2.133)

whereupon (2.123a) trivially yields

$$d_n(t) = \exp(-iE_n t/\hbar - \Gamma_n t/2)$$ (2.134)

corresponding to a damped pure exponential decay with radiative decay rate Γ_n. No quantum beats occur unless two of the ψ_n are accidently overlapping. Since (2.128) implies that

$$\Gamma_n \equiv \Gamma_s |\langle\psi_n|\phi_s\rangle|^2 < \Gamma_s, \quad (v \neq 0),$$ (2.135)

the observed decay rate is less than that, Γ_s, deduced from the integrated absorption intensity. Upon substitution of (2.133) into the equation for $G_{Fn}(E)$ in terms of $G_{nn}(E)$ and using the definition (2.107) and (2.135), it is readily verified that the emission quantum yield from ψ_n is unity in the absence of any other relaxation mechanisms like those arising from collisions. Thus, the Bixon-Jortner model satisfactorily summarized the essential features of the small molecule limit that survive in the general case discussed in Sections 2.3 and 2.5. We should note here in passing that the long lifetimes and the mixed state character of the ψ_n leads to the occurrence of very efficient and interesting vibrational relaxation mechanisms in these molecules [2.62].

2.4.4 Statistical Limit

The Bixon-Jortner model also undergoes an essential simplification for the large molecule statistical limit of

$$y \gg 1, \quad \text{statistical limit.}$$ (2.136)

For instance, in the $T_1 \to S_0$ intersystem crossing in naphthalene $\Delta E_{(T_1 S_0)} \approx 2 \times 10^4 \text{cm}^{-1}$ and $\varepsilon \approx 10^{-16} \text{cm}^{-1}$. Adopting a triplet decay rate of $\Gamma \approx 10^3$ s^{-1}, corresponding to infrared decay, we get $y = 10^8$ which comfortably satisfies (2.136). In the limit of (2.136), (2.125) simplifies to[9]

$$\mathscr{E}_n = E_s - i\hbar\Gamma_s/2 - i\pi v^2/\varepsilon.$$ (2.137)

Eq. (2.137) corresponds to a single solution as opposed to the infinite number in the small molecule limit. The single resonant state (2.137) in the statistical

---

[9] Actually v need not be too large for (2.137) to ensue. When $\Gamma = \varepsilon$ so $v = 1/2$, $\cot\pi z \cong -0.996i$ $\times(1 - 0.086\cos 2\pi x + 0.087 i\sin 2\pi x)$ which is already very close to $-i$.

limit corresponds identically to ϕ_s which has radiative and non-radiative decay rates Γ_s and

$$\Delta_s = 2\pi v^2/\hbar\varepsilon = (2\pi/\hbar)v^2\rho_l, \tag{2.138}$$

respectively—the kinetic view of non-radiative decay. Thus, if the initial molecular state were ϕ_s as in (2.94), (2.109) implies that

$$G_{ss}(E') = [E' - E_s + i\hbar(\Gamma_s + \Delta_s)/2]^{-1}, \tag{2.139}$$

so

$$a_{ss}(t) = \langle\phi_s|\Psi(t)\rangle = \exp[-iE_st/\hbar - (\Gamma_s + \Delta_s)t/2]. \tag{2.140}$$

Solving for $G_{sF}(E)$ by substituting (2.139) into (2.102b), then using this result and the definition (2.107) in (2.98) leads to the photon counting rate of

$$\dot{R}_s(t) = \frac{d}{dt}\sum_F|c_F(t)|^2 = \Gamma_s\exp[-(\Gamma_s + \Delta_s)t] \tag{2.141}$$

and the quantum yield of emission

$$\varphi_s = \int_0^\infty dt\,\dot{R}_s(t) = \frac{\Gamma_s}{\Gamma_s + \Delta_s} < 1. \tag{2.142}$$

It remains to be verified that (2.94) is indeed an accessible initial state. For instance, given the observed decay rate $(\Gamma_s + \Delta_s)$ of the $0-0$ level of S_1 of benzene of $10^7\,\mathrm{s}^{-1}$, it is clear that a pulse of duration ≈ 1 ns and centered about E_s automatically results in such an initial state.

2.4.5 Intermediate Case and Variable Coupling Model

The case of $y \approx 1$ is clearly within the domain of the intermediate state. TRIC has studied the detailed transition of the quasi-stationary states χ_n^r from the molecular eigenstates ψ_n in the small molecule case to the Born-Oppenheimer functions $\{\phi_s, \phi_l\}$ in the statistical limit, and hence her work included the $y \approx 1$ cases [2.54, 56]. Because of the diversity in the nature of the possible manifestations of the intermediate case, it is desirable to illustrate its behavior by extending the Bixon-Jortner model somewhat. Jortner and coworkers have used the simple model of a few discrete levels $\phi_{s_1}, \phi_{s_2}, \ldots, \phi_{s_m}$ which are coupled to a equally spaced quasi-continuum with coupling $v_{s_1}, v_{s_2}, \ldots, v_{s_m}$ which depend only on s_j and may even vanish for some of the levels. The discrete zeroth-order states may have some couplings amongst themselves, and some or all of them may carry oscillator strength from the ground state. The simplest such generalization involves the case of two levels, and the reader is referred to the original literature for a description of results [2.8, 69].

Recently MOROKUMA and FREED [2.78] have generalized the Bixon-Jortner model to provide an exactly solvable one which incorporates non-constant couplings which mimic Franck-Condon factors. This model involves the nonradiative decay of ψ_s into the $\{\phi_l\}$ where the coupling strength v_{ls} varies with l in a smooth fashion, being peaked about some $\phi_{l=0}$, which is *not* necessarily isoenergetic with ψ_s, and decreasing as $|l|$ increases (see Fig. 2.7). More explicitly the coupling strength is taken to be [2.78]

$$\langle\psi_s|\mathcal{H}_v|\phi_l\rangle = \begin{cases} v, & l=0 \\ \exp(i\,\delta_l)(v/|l|), & l\neq0. \end{cases} \tag{2.143}$$

The phases δ_l are arbitrary and may vary with l, and the l-levels are labelled by $l=0, \pm1, \pm2, \ldots$.

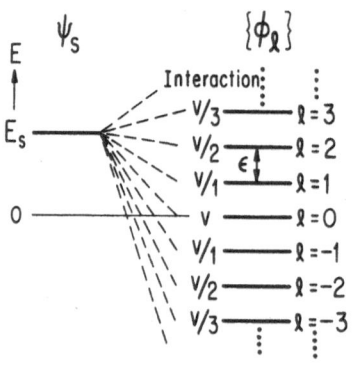

Fig. 2.7. The variable coupling model for radiationless decay. ψ_s is the only level in the figure carrying oscillator strength from the ground state. The coupling scheme of (2.143) is explicitly shown in the figure

In view of the enormous applications of the Bixon-Jortner model, a plethora of different calculations can be performed with the variable coupling model. MOROKUMA and FREED [2.78] considered the simplest case of the statistical limit where $\Gamma_l \gg \varepsilon$ and the Bixon-Jortner model simplifies considerably. In the Bixon-Jortner model, a molecule initially in ϕ_s decays with a single exponential decay rate (sum of radiative and non-radiative rates) for all times [2.4, 55]. In the variable coupling model, on the other hand, the decay of a ψ_s level is not necessarily a pure exponential, and can exhibit interference effects that would only appear in the Bixon-Jortner model for intermediate case molecules. In fact, to a reasonable approximation, the statistical limit of the variable coupling model gives results very similar to those generated by a two level Bixon-Jortner model where only one level ϕ_{s_1} is coupled to $\{\phi_l\}$ while the other ϕ_{s_2}, carries all the oscillator strength from ϕ_0 and is coupled to ϕ_{s_1}.

The variable coupling model is readily solved by using the general Green's function methods discussed above. In the statistical limit $G_{ss}(E)$ is found to have *four poles*, but two are generally negligible in intensity. Consequently, the model is well represented as involving a pair of resonances. The zeroth-order level ψ_s no longer has a Lorentzian distribution amongst the quasi-stationary states. If ψ_s were initially excited and if v is large enough, the decay would exhibit quantum beats, but, in general, ψ_s will not be an optically accessible initial state. Rather, optical excitation would yield one of the resonances which would subsequently display exponential decay.

This section presents the Bixon-Jortner model and a modification thereof in order to simply elucidate the transition between the small, intermediate and statistical limits. More realistic molecular systems are no longer exactly solvable, but they can readily be treated by the methods sketched so far. For those interested in the full theory providing the justification of some of the results of Sections 2.3 and 2.4, the next section presents a review of the full formal Green's function theory. Because of the possibility of the occurrence of quantum interference effects in intermediate cases, it is necessary to be more precise in dealing with the preparation of the initial excited molecular state. Consequently, Section 2.5 includes a brief discussion of the wave packet formalism for the description of the excitation process. As the simplified analysis in Section 2.3 does exhibit the basic molecular phenomena, those who are uninterested in the formal theory can proceed to Section 2.6 where the detailed calculation of non-radiative decay rates is considered.

2.5 The Formal Development

The use of Green's function techniques in describing both the radiative and the non-radiative molecular decays has already been motivated in Sections 2.3 and 2.4. Also, (2.64) shows how the introduction of a maximum time limit is easily managed in this formalism. Now, in the presentation of the general results of the theory special emphasis is placed upon the existence and properties of effective molecular Hamiltonians [2.72]. The use of such effective Hamiltonians is widespread in magnetic resonance theory when relaxation problems are considered [2.53]. In those cases we are interested in the time development of only a small part of the whole system. This is also characteristic of our case. In the study of radiative and non-radiative decays of polyatomic molecules our interest is centered upon the evolution in time of a set of closely coupled levels, namely ϕ_s and/or $\{\phi_l\}$, etc. The mathematical techniques, which conveniently isolate a set of molecular levels from all other levels and the radiation field, correspond to the use of projection operators [2.51]. The use of Green's functions and projection operators is already widespread in collision theory and statistical mechanics, and the necessary mathematical tools are well known. The theory of the radiative decay of unstable states, using Green's function techniques, has been reviewed by Goldberger and Watson [2.50]. Mower [2.48] has applied projection operator techniques to these Green's function methods. The following therefore provides a review of these methods as well as their generalization to include questions such as the existence of practical irreversibility.

Consider first the emission amplitudes (2.40) which are given by

$$
\begin{aligned}
F_j(t) &\equiv (2\pi i\hbar)^{-1} \int_C dE \exp(-iEt/\hbar) \langle \phi_F^v | G(E+i\varepsilon_m) | \theta_j, \text{vac} \rangle \\
&\equiv (2\pi i\hbar)^{-1} \int_C dE \exp(-iEt/\hbar) F_j(E).
\end{aligned}
\tag{2.144}
$$

Because the widths ε_m of the molecular levels are non-zero we are able to consider irreversible behavior for times $t \lesssim \hbar/\varepsilon$.

Consider the simple case of monochromatic excitation of a system in which there is intersystem crossing. Then ϕ_s is, say, the excited singlet, while $\{\phi_l\}$ corresponds to an isoenergetic quasi-continuum of triplet vibronic states. Focusing attention on an isolated molecule, for cases such as the aromatic hydrocarbons, the following set of inequalities is expected to hold

$$
\tau^{rad}(l) \gg \tau^{max} \sim \mathcal{O}(\tau_{coll}) \gg \tau^{rad}(s).
\tag{2.145}
$$

2.5.1 Projection Operator Method

Since the molecular eigenstate basis $\{\psi_n\}$ is usually expanded in terms of the BO basis $\{\phi_s, \phi_l\}$, we work in terms of the latter for convenience. The Schrödinger equation then tells us when the former "diagonalizes" the problem at hand. The state ϕ_s is isolated from the rest by considering the projection operator

$$
P_s \equiv |\phi_s, \text{vac}\rangle \langle \phi_s, \text{vac}|,
\tag{2.146}
$$

as well as the complementary projector onto all other states

$$
Q_s \equiv 1 - P_s.
\tag{2.147}
$$

The operators P_s and Q_s are Hermitian and obey the usual relations

$$
P_s^2 = P_s, \quad Q_s^2 = Q_s
\tag{2.148}
$$

satisfied by all projection operators. They also are orthogonal and hence

$$
P_s Q_s = Q_s P_s \equiv 0.
\tag{2.149}
$$

In terms of P_s and Q_s, the transform of the emission amplitude

$$F_s(E) = \langle \phi_F^v | Q_s G(E + i\varepsilon_m) P_s | \phi_s, \text{vac} \rangle \tag{2.150}$$

is written as a matrix element of the operator

$$Q_s G(E + i\varepsilon_m) P_s . \tag{2.151}$$

For notational convenience, let

$$\mathcal{E} \equiv E + i\varepsilon_m . \tag{2.152}$$

Following [2.45, 46, 48], the total Hamiltonian is separated into unperturbed

$$K = \mathcal{H}_{BO} + \mathcal{H}_r \tag{2.153}$$

and perturbation

$$V = \mathcal{H}_V + \mathcal{H}_{int} \tag{2.154}$$

parts. To determine the projection (2.151) of $G(\mathcal{E})$, we start with the definition

$$(\mathcal{E} - \mathcal{H}) G(\mathcal{E}) = 1 . \tag{2.155}$$

First this equation is multiplied on the right by P_s (inserting $1 = P_s + Q_s$)

$$(\mathcal{E} - \mathcal{H})(P_s + Q_s) G(\mathcal{E}) P_s = P_s . \tag{2.156}$$

Then take the product of (2.156) on the left with P_s and Q_s, respectively, to give

$$P_s(\mathcal{E} - \mathcal{H}) P_s(P_s G P_s) - P_s \mathcal{H} Q_s(Q_s G P_s) = P_s \tag{2.157}$$

$$-Q_s \mathcal{H} P_s(P_s G P_s) + Q_s(\mathcal{E} - \mathcal{H}) Q_s(Q_s G P_s) = 0 , \tag{2.158}$$

where (2.148) and (2.149) have been used. Since $\mathcal{E}$ is a complex variable, the operator $Q_s(\mathcal{E} - \mathcal{H}) Q_s$ has an inverse in the space spanned by Q_s; this inverse operator may have the usual singularities. Applying $[Q_s(\mathcal{E} - \mathcal{H}) Q_s]^{-1}$ on the left to (2.158) gives a formal expression for $Q_s G P_s$ in terms of $P_s G P_s$,

$$Q_s G P_s = (\mathcal{E} - Q_s \mathcal{H} Q_s)^{-1} Q_s \mathcal{H} P_s(P_s G P_s) . \tag{2.159}$$

Substitution of (2.159) into (2.158) gives

$$[(\mathcal{E} - P_s \mathcal{H} P_s) - P_s \mathcal{H} Q_s(\mathcal{E} - Q_s \mathcal{H} Q_s)^{-1} Q_s \mathcal{H} P_s] P_s G P_s = P_s , \tag{2.160}$$

or

$$P_s G(\mathscr{E}) P_s = (\mathscr{E} - P_s K P_s - P_s R_s P_s)^{-1} P_s. \qquad (2.161)$$

The operator R_s is defined by

$$R_s = V + V Q_s (\mathscr{E} - Q_s \mathscr{H} Q_s)^{-1} Q_s V \qquad (2.162)$$

and is called the level shift operator. In the BO basis $P_s G(\mathscr{E}) P_s$ has only one non-zero matrix element, namely that diagonal in $|\phi_s, \mathrm{vac}\rangle$. Thus, we get

$$\begin{aligned}
B_s(\mathscr{E}) &\equiv \langle \phi_s, \mathrm{vac} | P_s G(\mathscr{E}) P_s | \phi_s, \mathrm{vac} \rangle \\
&\equiv \langle \phi_s, \mathrm{vac} | G(\mathscr{E}) | \phi_s, \mathrm{vac} \rangle,
\end{aligned} \qquad (2.163)$$

and the inverse Laplace transform of (2.163) just gives the probability amplitude that a molecule initially in the "state" ϕ_s with no photons present remains in that state for later times. Hence, (2.163) is the quantity of interest in determining the decay rate of the "state" ϕ_s. (Here we omit any discussion as to whether ϕ_s is in fact a physically accessible state, since that will become apparent in particular cases.) Now since (2.161) describes the dynamics of the single state $|\phi_s, \mathrm{vac}\rangle$, we might consider that the quantity $P_s(K + R_s) P_s$ is the effective Hamiltonian for this level. However, this quantity is still $\mathscr{E}$-dependent, and it is only after the explicit $\mathscr{E}$-dependence has been removed (possibly by direct evaluation of the Fourier integral) that a meaningful association can be made with an effective Hamiltonian. The nature of the effective Hamiltonians is elucidated later; for the present we return to the evaluation of $Q_s G P_s$.

Substituting (2.161) into (2.159) gives, finally

$$Q_s G(\mathscr{E}) P_s = (\mathscr{E} - Q_s \mathscr{H} Q_s)^{-1} Q_s \mathscr{H} P_s (\mathscr{E} - P_s K P_s - P_s R_s P_s)^{-1}. \qquad (2.164)$$

For the case of a single level such as (2.146) the results (2.161) and (2.164) are identical to those given by GOLDBERGER and WATSON [2.50]. However, in the general case that P_s is replaced by P, a projection operator for a large number of states, the expressions (2.161) and (2.164) are the multistate generalization that is necessary to discuss the small and intermediate molecule limits.

2.5.2 Decay of a Level

In the statistical limit, only the single state projector P_s of (2.46) is required. We now need the matrix elements of the Hamiltonian (2.153) plus (2.154) to enable the calculation of (2.164) and, in turn, (2.144). For the model depicted in Fig. 2.1, where we neglect all states besides ϕ_0, ϕ_0^v, ϕ_s and $\{\phi_l\}$, the non-zero matrix elements are taken to be

$$\langle \phi_0 | \mathscr{H}_{el} | \phi_0 \rangle = 0 , \quad \text{(defining the zero of energy)} \tag{2.165a}$$

$$\langle \phi_s | \mathscr{H}_{BO} | \phi_s \rangle = E_s , \tag{2.165b}$$

$$\langle \phi_l | \mathscr{H}_{BO} | \phi_l \rangle = E_l \delta_{ll'} , \tag{2.165c}$$

$$\langle \phi_s | \mathscr{H}_V | \phi_l \rangle = v_{sl} , \tag{2.165d}$$

$$\langle nke | \mathscr{H}_r | n'k'e' \rangle = n\hbar ck \, \delta_{nn'} \delta_{kk'} \delta_{ee'} , \tag{2.165e}$$

$$\langle \phi_s, nke | \mathscr{H}_{int} | \phi_0^v, n\mathscr{H}, ke \rangle = W_{ke}^{v*}(n) . \tag{2.165f}$$

As noted in Section 2.3, it is often necessary to include other electronic manifolds, e. g., the $\{\phi_m\}$ and $\{\phi_k\}$ as in (2.79) and (2.80). These states can easily also be incorporated[10].

The diagonal matrix element of R_s of (2.162) in the state $|\phi_s, \text{vac}\rangle$ is

$$\mathscr{R}_s \equiv \langle \phi_s, \text{vac} | R_s | \phi_s, \text{vac} \rangle = \langle \phi_s, \text{vac} | V Q_s (\mathscr{E} - Q_s \mathscr{H} Q_s)^{-1} Q_s V | \phi_s, \text{vac} \rangle , \tag{2.166}$$

because by (2.165) V only has off-diagonal matrix elements. When the $\{\phi_l\}$ cannot radiate, the states $|\phi_l, \text{vac}\rangle$ and $|\phi_0^v, ke\rangle$ diagonalize the Hamiltonian submatrix $Q_s H Q_s$. Since Q_s is merely a sum over projectors onto each of these states, the expression (2.106) can be written as

$$\mathscr{R}_s(\mathscr{E}) = \sum_{v,e} \int dk | \langle \phi_s, \text{vac} | \mathscr{H}_{int} | \phi_0^v, ke \rangle |^2 (\mathscr{E} - E_0^v - \hbar ck)^{-1}$$
$$+ \sum_l |v_{ls}|^2 (\mathscr{E} - E_l)^{-1} , \tag{2.167}$$

where E_0^v is the energy of the state ϕ_0^v and a summation has been performed over all final states $|ke\rangle$, i. e., $\sum_\alpha \int dk$. $\mathscr{R}_s(\mathscr{E})$ is a sum of radiative and a non-radiative term since $\mathscr{H}_{int}$ and $\mathscr{H}_V$ do not connect the same states. Because of the $i\varepsilon_m$ in the

---

[10] In (2.165), it is also implicity assumed that we may neglect electronic and vibronic couplings $\mathscr{H}_V$ between states which have allowed radiative transitions. This restriction could be lifted if necessary.

radiative part of (2.167), this term mathematically describes irreversibility for all times, and it is permissible to set $\varepsilon_m \to 0^+$ in this term to yield

$$\mathcal{R}_s(E)_{\text{rad}} = -i\pi \sum_{v,e} \int d\Omega_k |\langle \phi_s, \text{vac}|\mathcal{H}_{\text{int}}|\phi_0^v, k\,e\rangle|^2 \rho_{\text{photon}}(E - E_0^v)$$

$$+ \sum_{v,e} \mathcal{P} \int dk |\langle \phi_s, \text{vac}|\mathcal{H}_{\text{int}}|\phi_0^v, k\,e\rangle|^2 (E - E_0^v - \hbar c k)^{-1} \qquad (2.168)$$

$$\equiv -i\hbar\, \Gamma_s(E)/2 + D_s(E).$$

The imaginary part of $\mathcal{R}_s(E)_{\text{rad}}$, when evaluated at $E = E_s$, is just the radiative decay rate of the state ϕ_s due to spontaneous emission, see (2.72). $D_s(E)$ corresponds, after the Laplace inversion, to a small radiative correction to the energy of ϕ_s. In the following, terms such as D_s are dropped as these radiative corrections are assumed to be absorbed into the zero order energies E_s.

The non-radiative part of $\mathcal{R}_s(\mathscr{E})$ contains a sum over the discrete set of states ϕ_l and appears to have a set of poles at the complex energies

$$E = E_l - i\varepsilon_m. \qquad (2.169)$$

Under certain circumstances the density of states of $\{\phi_l\}$ is sufficiently large and the variations in $\{v_{ls}\}$ are sufficiently regular that the summation over l in (2.167) can accurately be replaced by an integration

$$\sum_l \to \int dE_l \rho_l(E_l), \qquad (2.170)$$

where $\rho_l(E_l)$ is the density of states of $\{\phi_l\}$ at the energy E_l. For $\varepsilon_m \to 0^+$, the replacement (2.170) is a tautology which does not lead to an $\mathcal{R}_s(E)_{\text{nr}}$ which is a continuous non-singular function of E; however, for $\varepsilon_m \neq 0$, we can consider whether $\mathcal{R}_s(E)_{\text{nr}}$ is a non-singular function of E for $\text{Im}\{E\} \ll -\varepsilon_m$, i.e., if

$$\mathcal{R}_s(E)_{\text{nr}} = \sum_l \frac{|v_{ls}|^2 (E - E_l)}{(E - E_l)^2 + \varepsilon_m^2} - i\sum_l \frac{|v_{ls}|^2 \varepsilon_m}{(E - E_l)^2 + \varepsilon_m^2} \qquad (2.171)$$

is non-singular in this energy region. If this is in fact the case, then the approximation (2.170) with continuous $\rho_l(E_l)$ can be invoked. When $\mathcal{R}_s(E)_{\text{nr}}$ is a non-singular function of E for $\text{Im}\{E\} < -\varepsilon_a$, then it will turn out that irreversible behavior ensues for times $t < \hbar/\varepsilon_a$. The interaction $|v_{ls}|^2$ need not be a slowly varying function of l (or E_l) for $\mathcal{R}_s(E)_{\text{nr}}$ to be a smooth non-singular function of E in this part of the lower half E-plane. In fact, as notet in Section 2.3, we expect that $|v_{ls}|^2$ be large for some set of states $\{\phi_b\}$ because of, e.g. symmetry, and be small for other states $\{\phi_w\}$. In general, these two sets of states are interspersed in energy, and hence $|v_{ls}|^2$ is a rapidly varying function of E_l *even in the statistical limit.* However, if we denote symmetry species by δ, the summation in (2.171) and (2.167) may be written as

$$\sum_{\delta} \sum_{l_\delta} |v_{l_\delta s}|^2/(\mathscr{E} - E_{l_\delta}) , \tag{2.172}$$

where for a given ε_m, when the density of $\{\phi_{l_\delta}\}$ states is sufficiently large, each summation over l_δ provides a slowly varying and non-singular function of E in the energy region of interest. Thus we write

$$\mathscr{R}_s(E)_{nr} = D_s(E)_{nr} - i\hbar\,\Delta_s(E)/2 , \tag{2.173}$$

where

$$D_s(E)_{nr} = \int dE_l \rho_l(E_l)|v_{ls}(E_l)|^2 (E - E_l)[(E - E_l)^2 + \varepsilon_m^2]^{-1} \tag{2.174a}$$
$$\doteq \mathscr{P} \int dE_l \rho_l(E_l)|v_{ls}(E_l)|^2/(E - E_l)$$

and

$$\Delta_s(E) = (2/\hbar) \int dE_l \rho_l(E_l)|v_{ls}(E_l)|^2 \varepsilon_m[(E - E_l)^2 + \varepsilon_m^2]^{-1} \tag{2.174b}$$
$$\doteq (2\pi/\hbar)\rho_l(E)|v_{ls}(E)|^2 \equiv (2\pi/\hbar)\sum_l |v_{ls}|^2 \delta(E_l - E) .$$

In the second lines of (2.174a) and (2.174b) we have denoted the usual limit of $\varepsilon_m \to 0^+$. Here $\mathscr{P}$ indicates the Cauchy principle value is to be taken. In this limiting expression as in (2.174b), $\rho_l(E)$ is actually $\sum_l \delta(E - E_l)$. Thus in the limit, Δ_s is a singular function of E, thereby indicating that mathematically the non-radiative decay is not irreversible when the ϕ_l have no decay mechanisms. The expressions corresponding to a finite timescale lead to a non-singular $\Delta_s(E)$ in the statistical limit and therefore to irreversible electronic relaxation. It is often convenient to present results in the limiting form such as (2.174b) with the understanding that the expression is to be represented in terms of a smooth function.

Quantities such as $D_s(E)_{nr}$ of (2.174a) represent energy shifts due to the perturbation v_{ls}. These are absorbed into the zero-order energies E_s, by redefinition of the latter, in the same manner as is used in the case of radiative level shifts. Thus, in the statistical limit where the $\{\phi_l\}$ do not radiate, we have

$$\mathscr{R}_s = -(i\hbar/2)[\Gamma_s(E) + \Delta_s(E)] , \tag{2.175}$$

and therefore (2.163) becomes

$$B_s(E) = \left\{ E + i\varepsilon_m - E_s - \frac{i\hbar}{2}[\Gamma_s(E) + \Delta_s(E)] \right\}^{-1} . \tag{2.176}$$

Performing the Laplace inversion, as defined in Subsection 2.3.2, gives (cf. (2.61) and (2.62))[11]

$$B_s(t) = \langle \phi_s, \text{vac}|\exp(-i\mathcal{H}t/\hbar)|\phi_s, \text{vac}\rangle$$
$$= \exp\{-iE_s t/\hbar - \varepsilon_m t/\hbar - \tfrac{1}{2}[\Gamma_s(E_s) + \Delta_s(E_s)]t\}. \qquad (2.177)$$

Using the expected inequalities (2.145), we note that

$$\hbar\Gamma_s(E_s) \gg \varepsilon_m, \qquad (2.178)$$

and therefore ε_m in (2.177) is negligible. Thus, we can write

$$B_s(t) = \exp[-iE_s t/\hbar - (\Gamma_s + \Delta_s)t/2], \qquad t < \tau_{\max}, \qquad (2.179)$$

where, for convenience, the arguments (E_s) of Γ_s and Δ_s have been dropped. The probability that the molecule still be in the state ϕ_s at time t is then

$$P_s(t) = |B_s(t)|^2 = \exp[-(\Gamma_s + \Delta_s)t], \qquad t < \tau_{\max}. \qquad (2.180)$$

The lifetime in (2.180), τ, is then related to the radiative (τ^{rad}) and non-radiative (τ_{nr}) lifetimes by (2.1) and (2.2) or by

$$\frac{1}{\tau} = \frac{1}{\tau^{\text{rad}}} + \frac{1}{\tau_{\text{nr}}} \qquad (2.181)$$

with

$$\tau^{\text{rad}} = 1/\Gamma_s \quad \text{and} \quad \tau_{\text{nr}} = 1/\Delta_s. \qquad (2.182)$$

The amplitude $B_s(E)$ in (2.174) has a single pole in the lower half of the complex E-plane. This corresponds to the zero-order state ϕ_s being embedded in two different non-interfering continua, radiative and non-radiative. Thus, there is a *resonance* centered at E_s with decay rate, as given by (2.180). The emission probability is obtained from the emission amplitude (2.150). Using (2.164) and (2.176) this is easily written as

$$F_s(E) = (E + i\varepsilon_m - E_0^v - \hbar ck)^{-1} W_{ke}^v(0) B_s(E). \qquad (2.183)$$

---

[11] In arriving at (2.177), since $\Gamma_s(E)$ and $\Delta_s(E)$ are smooth functions of E, they are expanded in a Taylor series about $E = E_s$. Only the lowest-order term in this series is indicated in (2.177). The next term gives a renormalization which is uninteresting insofar as no physically new phenomena are involved.

Eq. (2.183) has two poles in the lower half E-plane[12] thereby giving (dropping the negligible ε_m)

$$F_s(t) = W_{ke}^v(0) \exp[-i(E_0^v/\hbar + ck)t]\{1 - \exp[i(E_0^v/\hbar + ck \\ - E_s/\hbar)t - (\Gamma_s + \Delta_s)t/2]\}[(E_0^v - E_s)/\hbar + ck + i(\Gamma_s + \Delta_s)/2]^{-1}. \quad (2.184)$$

The total emission probability is obtained from (2.184) by taking $|F_s(t)|^2$ and by summing over all final states v, all possible frequencies k and photon directions Ω_k, and all polarizations e

$$P_{\text{emission}}(t) \equiv \sum_{v,e} \int dk |F_s(t)|^2. \quad (2.185)$$

Using the definition of Γ_s from (2.73), the k-integration in (2.185) can be performed by trivial contour integrations to give

$$P_{\text{emission}}(t) = \frac{\Gamma_s(E_s)}{\Gamma_s(E_s) + \Delta_s(E_s)} (1 - \exp\{-[\Gamma_s(E_s) + \Delta_s(E_s)]t\}). \quad (2.186)$$

The quantum yield is just the total emission probability as $t \to \infty$,

$$\varphi \equiv P_{\text{emission}}(\infty) = \Gamma_s/(\Gamma_s + \Delta_s), \quad (2.187)$$

which upon substitution of (2.182) is just (2.1). We could take the limit $t \to \infty$ in (2.186) because ε_m has been dropped, but we note that for $t \approx \tau_{\text{max}}$, (2.178) or (2.145) imply that

$$\exp[-(\Gamma_s + \Delta_s)\tau_{\text{max}}] \ll 1,$$

so (2.187) is also $P_{\text{emission}}(\tau_{\text{max}})$ with neglible error.

2.5.3 Introduction of Decay Rates for $\{\phi_l\}$

The above example shows how the introduction of the natural time limit τ_{max} may lead to an effectively continuous density of states for $\{\phi_l\}$, thereby leading to irreversible electronic relaxation with the conventional results (2.1), (2.2), (2.174b) and (2.181).

The next case of interest is that for which the $\{\phi_l\}$ radiate rapidly, and

$$\tau_{\text{max}} \sim \mathcal{O}(\tau_{\text{coll}}) \gg \tau^{\text{rad}}(s), \quad \tau^{\text{rad}}(l) \quad (2.188)$$

---

[12] We neglect contributions from the branch cuts and close contours on the second Rieman sheet. All poles, that are considered in the lower half E-plane, refer, of course, to ones on this second sheet [2.50].

for the "isolated" molecule. The situation (2.188) often corresponds to the case of internal conversion in aromatic hydrocarbons, but it also corresponds to the case of intersystem crossing in biacetyl where the triplets can also radiate [2.79]. It is now useful to introduce the set of states $\{\phi_0^{v'}\}$ to which the $\{\phi_l\}$ radiate. We assume, for convenience, that the emission spectra from ϕ_s and $\{\phi_l\}$ are non-overlapping so $\{\phi_0^{v'}\} \neq \{\phi_0^v\}$. Furthermore, $\mathscr{H}_V$ is assumed to have no non-vanishing matrix elements between ϕ_s and $\{\phi_0^v, \phi_0^{v'}\}$. (These restrictions can easily be removed if desired.)

The net effect of this introduction of the radiative decay channel is to associate the radiative widths $\{\Gamma_l\}$ with the $\{\phi_l\}$, corresponding to the radiative lifetimes $\tau^{\mathrm{rad}}(l) = 1/\Gamma_l$ being non-zero, see Section 2.4. This result can be formally obtained from (2.187) by including the additional states $\{\phi_0^{v'}\}$. Let

$$P_l = \sum_l |\mathrm{vac}, \phi_l\rangle \langle \phi_l, \mathrm{vac}| \tag{2.189}$$

be the projector onto $\{\phi_l\}$. Then we have

$$P_l Q_s = Q_s P_l = P_l \tag{2.190}$$

and

$$Q_l = Q_s - P_l + P_s. \tag{2.191}$$

Following the development in the earlier part of this section we write

$$P_l [\mathscr{E} - Q_s \mathscr{H} Q_s]^{-1} P_l = P_l [\mathscr{E} - P_l K P_l - P_l R_l P_l]^{-1}, \tag{2.192}$$

where

$$R_l = V + V Q_l (\mathscr{E} - Q_l \mathscr{H} Q_l)^{-1} Q_l V. \tag{2.193}$$

The matrix of the inverse of (2.192) is just

$$\delta_{ll'} E_l - i\hbar \Gamma_{ll'}(E)/2, \tag{2.194}$$

when radiative level shifts are included in the zero-order energies $\{E_l\}$, and $\Gamma_{ll'}$ is defined as in (2.73) with $\alpha \to \phi_0^{v'}$. In the absence of the measurement of any coherence effects, we can assume that $\Gamma_{ll'} = \delta_{ll'} \Gamma_l$, so the damping matrix is diagonal. Alternatively, we can consider the basis $\{\chi_l\}$ which diagonalizes (2.194) instead of the states $\{\phi_l\}$.

The evaluation of (2.166) in the present case proceeds exactly as before to give[13]

$$\mathscr{R}_s(E) = -i\hbar\,\Gamma_s(E)/2 + \sum_l |v_{ls}|^2/[E + i\varepsilon_m - E_l + i\hbar\,\Gamma_l(E)/2]\,. \tag{2.195}$$

From the condition (2.183)

$$\hbar\,\Gamma_l(E) \gg \varepsilon_m, \qquad E \approx E_s\,, \tag{2.196}$$

ε_m may be neglected in (2.195). Thus, when radiative decay of $\{\phi_l\}$ is possible, we may take $\varepsilon_m \to 0^+$, and irreversible electronic relaxation is no longer precluded mathematically. In the general case of slow radiative decay of $\{\phi_l\}$, we might wish to consider retention of ε_m in (2.195). The analysis continues as before. If ε_{rec} is the maximum value of $\varepsilon_m + \hbar\,\Gamma_l(E)$ for which the summation in (2.195) is a non-singular function of energy, then $\hbar/\varepsilon_{rec} \equiv \tau_{rec}$ is a measure of the minimum recurrence time for the system. For $t < \tau_{rec}$, or $\varepsilon_m + \hbar\,\Gamma_l(E)$, ε_m, and/or $\hbar\,\Gamma_l(E) < \varepsilon_{rec}$, Eq. (2.195) leads to irreversible electronic relaxation of ϕ_s. The resulting decay expressions are again (2.180) and (2.186). In the limit $\varepsilon_m \to 0^+$ we have

$$\Delta_s(E_s) = \frac{2\pi}{\hbar} \int \frac{dE_l\,\rho_l(E)|v_{ls}(E_l)|^2\,\Gamma_l(E_s)\hbar/2}{(E_s - E_l)^2 + [\hbar\,\Gamma_l(E_s)/2]^2} \tag{2.197}$$

instead of (2.174b). Often it is convenient to formally express (2.197) as the limit $\Gamma_l \to 0^+$ and write

$$\Delta_s(E_s) \doteq (2\pi/\hbar) \sum_l |v_{ls}|^2\,\delta(E_l - E_s)\,, \tag{2.198}$$

again with the understanding that smooth representations of the Dirac delta function are to be taken. The symbolic result (2.198) is then identical to that in (2.174b) except that in the case (2.188) the radiative decay of $\{\phi_l\}$ can lead to mathematically irreversible non-radiative decay. The non-radiative decay probability can be evaluated to give

$$P_{nr}(t) = \sum_l |\langle\phi_l, \text{vac}|Q_s\,G(\mathscr{E})\,P_s|\phi_s, \text{vac}\rangle|^2$$
$$= [\Delta_s/(\Gamma_s + \Delta_s)]\{1 - \exp[-(\Gamma_s + \Delta_s)t]\}\,; \tag{2.199}$$

---

[13] Eq. (2.195), as well as a number of previous equations, assumes that the damping matrix (radiative and/or non-radiative) is diagonal for the $\{\phi_l\}$ states. This assumption is tenable, for appropriate definition of $\{\phi_l\}$, so long as we are concerned only with the decay of ϕ_s. If any subsequent decay of the $\{\phi_l\}$ is to be investigated, couplings of $\{\phi_l\}$ to the $\{\phi_m\}$ of Subsection 2.54 can lead to a damping matrix that is non-diagonal. HELLER and RICE [J. Chem. Phys. **61**, 936 (1974)] provide the most recent discussion of the problem, and earlier references can be found there.

hence, the quantum yield for non-radiative decay is just

$$\varphi_{nr} = \Delta_s/(\Gamma_s + \Delta_s) . \tag{2.200}$$

Note that our model implies that

$$\varphi + \varphi_{nr} \equiv 1 , \tag{2.201}$$

because photochemical processes, such as isomerization, have implicitly assumed to be absent. In real systems photochemistry must often be included; then (2.201) is not necessarily obeyed.

2.5.4 The Effective Coupling

As noted in Section 2.3, the above results correspond only to the model of Fig. 2.1 wherein we neglect coupling of ϕ_s and $\{\phi_l\}$ to any other vibronic states via the perturbation $\mathscr{H}_V$. Consequently, the parameters v_{ls} must be understood to correspond to effective interactions which are obtained from some sort of perturbation theory. We can, however, introduce the additional states to indicate how the effective interactions are introduced in the present formalism. Consider the decay of the state ϕ_s when these additional states $\{\phi_m\}$ are present. We assume that $|E_s - E_m| > |v_{ls}|, |v_{lm}|, |v_{sm}|$, so there is no need to worry about degeneracies. Furthermore, any or all of the interactions v_{ls}, v_{lm}, v_{sm} may be non-zero, and the states ϕ_s and $\{\phi_l\}$ are connected in some order of perturbation theory. The states $\{\phi_m\}$ might indeed correspond to more than one vibronic manifold, and it may be necessary to allow for interactions of the form $v_{m_\alpha m_\beta}$ arising from $\mathscr{H}_V$.

In order to obtain the decay rate of ϕ_s, it is sufficient to consider $B_s(E)$ of (2.163). By (2.161) this, in turn, implies the evaluation of the matrix element $\mathscr{R}_s(E)$ of (2.166). Again the real part of $\mathscr{R}_s(E)$ ultimately gives the level shift of E_s and is therefore absorbed into the latter by redefinition. Thus, only the imaginary part of $\mathscr{R}_s(E)$ is taken. The optical theorem [2.50] is

$$\text{Im}\{\mathscr{R}_s(E)\} = -\pi R_s^\dagger(E)\delta(E - Q_s K Q_s) R_s(E) , \tag{2.202}$$

where the symbolic limit $\varepsilon_m \to 0^+$ has been introduced. Thus, this implies that

$$\text{Im}\{\mathscr{R}_s(E)\} = -\pi \langle \phi_s, \text{vac}|R_s^\dagger(E)\delta(E - Q_s K Q_s) R_s(E)|\phi_s, \text{vac}\rangle . \tag{2.203}$$

Then (2.203) separates into a radiative term $-\pi\Gamma_s(E)$ and a non-radiative $-\pi\Delta_s(E)$. In the statistical limit the latter is a smooth non-singular function of E. Hence upon evaluation of the inverse (imaginary) Laplace transform it yields

$$B_s(t) = \exp\{-i E_s t/\hbar - [\Gamma_s(E_s) + \Delta_s(E_s)]t/2\} , \quad t < \tau_{max} , \tag{2.204}$$

where the uninteresting ε_m has been dropped. In the above case ϕ_s and $\{\phi_m\}$ are not isoenergetic, i.e., ϕ_s cannot "cross-over" to $\{\phi_m\}$, and then

$$k_{nr}(s \to l) = \varDelta_s(E_s) \doteq (2\pi/\hbar) \sum_l |\langle \phi_s | R_s(E_s) | \phi_l \rangle|^2 \delta(E_s - E_l) . \tag{2.205}$$

If, however, E_s can equal E_m, as would occur when ϕ_s is isoenergetic with two vibronic manifolds, then in the statistical limit both manifolds contribute, and the result is

$$\varDelta_s(E_s) = \varDelta_{s \to l}(E_s) + \varDelta_{s \to m}(E_s) , \tag{2.206}$$

where $\varDelta_{s \to l}(E_s)$ is given by (2.205) and $\varDelta_{s \to m}$ is given by an analogous rate expression for m.

The rate expression (2.205) is exact, but if $R_s(E_s)$ is expanded in powers of $\mathscr{H}_v$, Eq. (2.205) is found to contain matrix elements of all powers of $\mathscr{H}_v$ beyond the first. Thus, (2.205) includes direct, $v_{ls} \neq 0$ "crossing" as well as indirect, $v_{ls} \equiv 0$, $\infty > |v_{lm} v_{ms}/(E_s - E_m)| \neq 0$, "crossing", etc. In fact, it is well know that an exact evaluation of the right hand side of (2.205) *using either the* ABO *or the* CBO *functions* $\{\phi_s, \phi_l, \phi_m\}$ *must give identical results for the overall decay rate* $\varDelta_s(E_s)$. In practice, however, only the lowest-order terms are considered, and then different approximations need not provide identical results.

2.5.5 The Effective Molecular Hamiltonian

The small molecule limit manifests almost opposite behavior from the statistical limit. The major difference between these two limits resides in the level density ρ_l and in the coupling strength v_{ls}. Suppose we wished to evaluate the emission amplitude $F_s(E)$ in (2.150) (apart from the question discussed below of whether this state is experimentally accessible). The development (2.153) through (2.169) is identical in both limits. However, in molecules such as SO_2 and NO_2, the observed spectral density of states is $\gtrsim 0.1$ cm. Since a maximum time of 10^{-4} s corresponds to $\varepsilon_m \approx 3 \times 10^{-7}$ cm^{-1}, the level spacing is sufficiently small that the "relaxation part" of $\mathscr{R}_s(\mathscr{E})$ of (2.167) contains a set of isolated non-overlapping poles at energies $E = E_l - i\varepsilon_m$ in the complex E-plane. Thus, in the inverse (imaginary) Laplace transform of (2.144), these poles are singularities which lead to non-zero residues. If the "states" $\{\phi_l\}$ are allowed to have non-infinite radiative lifetimes as in (2.195), a lifetime as short as 10^{-8} s corresponds to $\Gamma_l \approx 3 \times 10^{-3}$ cm^{-1}. Thus, the second term of (2.195) in general corresponds to a set of isolated singularities for $E \approx E_l - i(\hbar/2)\Gamma_l(E_l)$, and this term cannot be approximated by a single smooth function of E. We could attempt to evaluate $F_s(t)$ by evaluating the residues upon substitution of (2.188) or (2.195) into (2.164) and (2.150). In effect, it turns out that in order to determine the poles and their residues, it is necessary to solve an eigenvalue equation. This result can more easily be obtained by considering, instead, the projection operator onto all the closely coupled states ϕ_s and $\{\phi_l\}$, namely

$$P = |\phi_s, \text{vac}\rangle \langle \phi_s, \text{vac}| + \sum_l |\phi_l, \text{vac}\rangle \langle \phi_l, \text{vac}| . \tag{2.207}$$

Then

$$Q = 1 - P \tag{2.208}$$

is the projector onto all other states and might contain states such as the $\{\phi_m\}$ above which are not isoenergetic with ϕ_s and $\{\phi_l\}$. For this case (2.150) is replaced by

$$F_s(E) = \langle \phi_F^v | Q \, G(E + i\varepsilon_m) \, P | \phi_s, \text{vac}\rangle , \tag{2.209}$$

and (2.159) through (2.162) follow without the subscripts s. The first thing to note is that

$$P(K + R) P = P \mathcal{H} P - i(\hbar/2) P \Gamma P - P R' P , \tag{2.210}$$

where Γ is just the radiative damping matrix (2.73) in the basis spanned by P. The latter is a non-singular slowly varying function of E in the neighborhood $E \approx E_s$. The remaining term R' is due to all other non-isoenergetic states $\{\phi_m\}$ and is a slowly varying function of E for $E \approx E_s$. It is, in fact, the effective Hamiltonian resulting from these other states once this slow E dependence is approximated by taking E to be an average energy for the nearby levels $\{\phi_s, \phi_l\}$. (More exact approaches are unnecessary for our qualitative purposes.) Rather than considering (2.163) which ultimately gives us the probability that a molecule initially in ϕ_s, *if it could ever get there*, still be there at time t, it is more instructive to consider the operator

$$B(E) = P \, G(\mathcal{E}) \, P = P [\mathcal{E} - P \mathcal{H} P - P R(\mathcal{E}) P]^{-1} , \tag{2.211a}$$

whose matrix elements give (2.163) as well as any "transition" rates between the "states" ϕ_s and $\{\phi_l\}$. As noted above the energy dependence of $R(\mathcal{E})$ is small and, as earlier, the real part of $R(\mathcal{E})$ gives rise to level shifts which are absorbed into $\mathcal{H}_{\text{el}}$ by a redefinition of the latter. Thus, the transform of (2.211a) can be written as

$$B(t) = P \exp(-i P \mathcal{H} P t/\hbar - P \Gamma P t/2) P . \tag{2.211b}$$

But (2.211b) is just an evolution operator for a quantum mechanical system of discrete levels which can undergo spontaneous emission. Removing the radiative state $|\text{vac}\rangle$, this is

$$\langle \text{vac}|B(t)|\text{vac}\rangle = P' \exp[-i P' \mathcal{H}_{\text{eff}} P' t] P' , \tag{2.212}$$

$$P' = |\phi_s\rangle\langle\phi_s| + \sum_l |\phi_l\rangle\langle\phi_l| , \tag{2.207a}$$

where $\mathcal{H}_{\text{eff}}$ is given in (2.72). As noted following (2.74) and (2.75), it is natural to consider those states χ_α which diagonalize $P' \mathcal{H}_{\text{eff}} P'$ with "energies" $E_\alpha - i\hbar\Gamma_\alpha/2$.

Since $\mathcal{H}_{\text{eff}}$ in (2.72) contains the anti-Hermitian operator $-i\hbar\Gamma/2$, $\mathcal{H}_{\text{eff}}$ is not Hermitian. It therefore has a set of right and left hand eigenfunctions χ_α^r and χ_α^l which are not identical. Furthermore, the operator which diagonalizes $\mathcal{H}_{\text{eff}}$ is not unitary. However, when a basis set is chosen such that $\mathcal{H}_{\text{el}}$ is a real matrix, $\mathcal{H}_{\text{eff}}$ is a complex symmetric matrix that can be diagonalized by an orthogonal transformation. The resonance states $\{\chi_\alpha^r\}$ are not orthogonal to each other, but the $\{\chi_\alpha^r\}$ and $\{\chi_\alpha^l\}$ form a biorthogonal set

$$\langle \chi_\alpha^l | \chi_\beta^r \rangle = \delta_{\alpha\beta} , \tag{2.213}$$

so the projector P' of (2.207a) can be represented in terms of $\{\chi_\alpha^{l,r}\}$ via

$$P' = \sum_\alpha |\chi_\alpha^r\rangle\langle\chi_\alpha^l| . \tag{2.214}$$

Using (2.213) and (2.214), (2.212) becomes

$$\sum_\alpha |\chi_\alpha^r\rangle \exp(-i E_\alpha t/\hbar - \Gamma_\alpha t/2)\langle\chi_\alpha^l| , \tag{2.215}$$

which is the evolution operator for a set of independent radiatively decaying states. When the level density is sufficiently small, matrix elements of $\mathcal{H}_{\text{eff}}$ in the molecular eigenstates basis $\{\psi_n\}$ are

$$\langle \psi_n | \mathcal{H}_{\text{eff}} | \psi_{n'} \rangle = E_n \delta_{nn'} - i\hbar\Gamma_{nn'}/2 . \tag{2.216}$$

Except for occasional accidental degeneracies, the condition (2.74) is expected to hold. Then the off diagonal terms $\Gamma_{nn'}$, $n \neq n'$, are negligible perturbations. Thus, to a good approximation, the time evolution of the excited molecule corresponds to the time development of the molecular eigenstates $\{\psi_n\}$ which may undergo spontaneous emission, as noted in (2.83) through (2.86).

2.5.6 Preparation of the Initial State

The next question concerns whether ϕ_s is a physically accessible "initial" state of the molecule. Since (2.215) indicates that the absorption spectrum in the small molecule limit corresponds to absorption to molecular eigenstates, which in the case of SO_2 and NO_2 are spread out over more than 150cm^{-1}, in order to excite ϕ_s initially it would be necessary to coherently excite all the $\{\psi_n\}$ with the appropriate phase relations such that this initial superposition is ϕ_s. This excitation is highly unlikely even with dye lasers. The notion of "crossover" is then misleading in the small molecule limit.

We have not explicitly considered the absorption amplitudes (2.137) which correspond to the absorption to an unstable molecular state. However, there is an essential symmetry between absorption and emission. This symmetry is most clearly manifested if we again use projectors onto the excited molecular states

$$P = \begin{cases} P_s = |\phi_s, \text{vac}\rangle \langle \phi_s, \text{vac}|, \\ P = |\phi_s, \text{vac}\rangle \langle \phi_s, \text{vac}| + \sum_l |\phi_l, \text{vac}\rangle \langle \phi_l, \text{vac}|. \end{cases} \tag{2.217}$$

It can be shown that the required projection of $G(\mathscr{E})$ is

$$P G(\mathscr{E}) Q = (\mathscr{E} - P \mathscr{H} P - P R P)^{-1} P \mathscr{H} Q (\mathscr{E} - Q \mathscr{H} Q)^{-1} \tag{2.218}$$

where

$$R = V + V Q (\mathscr{E} - Q \mathscr{H} Q)^{-1} Q V. \tag{2.219}$$

Eq. (2.218) appears to be the Hermitian conjugate of e. g. (2.164); however, $\mathscr{E}$ is a complex variable and therefore (2.218) cannot be simply obtained by conjugation of (2.164). Nevertheless, the evaluation of absorption amplitudes proceeds from (2.218–219) just as in the case of emission amplitudes [2.46].

Having considered the large and small molecule limits, the more complicated intermediate case becomes of interest. As noted in the previous sections, this limit has a number of features in common with the simpler limits already described. The formal development also proceeds similarly. The discussion following (2.87) naturally results from such a development. As the details have adequately been reviewed elsewhere [2.8], they need not be repeated here.

We should, perhaps, comment briefly upon a more realistic description of the preparation of the initially excited molecular state. As noted following (2.30), the idea of a monochromatic square pulse of radiation is a pure idealization that is not practically attainable unless the pulse duration is very long compared to the typical lifetimes of the molecular excited states, corresponding to ϕ_s in Fig. 2.1. As a long pulse does not correspond to the usual time-dependent experiments involving pulses of duration τ, where τ is short compared to relevant decay lifetimes

$$\tau \lesssim (\Gamma_{\text{rad}} + \Gamma_{\text{nr}})^{-1} \approx \tau_{\text{obs}}, \tag{2.220}$$

it is necessary to consider the modifications of Section 2.3 that are required for a more realistic description of the actual excitation process. The dominant effect of having τ be finite is that the optical fields $E_{opt}(\omega_i)$ and $E_{opt}(\omega_j)$ in (2.26) are no longer correlated, thereby leading to

$$\langle E_{opt}(\omega)\, E_{opt}^*(\omega')\rangle = \frac{4\pi}{c}\, P\left(\frac{\omega+\omega'}{2}, \omega-\omega'\right), \tag{2.221}$$

where P has correlations between optical frequencies over a range

$$\Delta\omega = \omega - \omega' \sim 1/\tau. \tag{2.222}$$

Thus, it is necessary to allow the pulse to at least have a frequency range given by (2.222), so (2.32) becomes modified to

$$\psi_{mol}^{(excited)}(\tau) = \sum_A \langle n_A - 1, k_A\, e_A | \Psi(\tau). \tag{2.223}$$

Labeling the individual absorption amplitudes by the particular photon, e.g., rewriting (2.37) as

$$A_j^A(\tau) = \langle \theta_j, n_A - 1, k_A\, e_A | \exp(-i\mathscr{H}\, t/\hbar) | \phi_A\rangle, \tag{2.224}$$

the final probability for the emission of photons of type F at time t is

$$P_{F,\sum_A}(t,\tau) = \sum_{i,j}\sum_{A,A'} \langle \phi_F^v | \exp(-i\mathscr{H}\, t/\hbar)|\theta_j, vac\rangle\, A_j^A(\tau)$$
$$\times \langle \phi_F^v | \exp(-i\mathscr{H}\, t/\hbar)|\theta_i, vac\rangle^* [A_i^{A'}(\tau)]^*. \tag{2.225}$$

When the summation is performed over A and A', i.e., an integration over the frequencies, etc., in the incident light pulse, Eq. (2.221) introduces the frequency correlations that arise from the finite nature of τ.

Eq. (2.225) is still a somewhat simplified picture of the effects of the excitation process as the pulse is taken to be completely abrupt. It can be utilized, however, to show the well known fact that, given a single isolated excited resonance χ, (e. g., ϕ_s in the statistical limit or a non-accidentally degenerate ψ_n in the small molecule limit) $P_{F,\sum_A}(t,\tau)$ only contains information concerning the lifetime τ_{obs} and pure radiative lifetime Γ_{rad}^{-1} of the isolated resonance along with some information concerning the shape of the incident light pulse. Thus, apart from a few quantitative details that are sometimes of considerable interest, the simplified viewpoint in Section 2.3 is adequate with regard to the basic molecular processes [2.46, 57]. However, when some $\{\psi_n\}$ are overlapping, or when the intermediate case is being studied, the whole structure of the decay pattern is strongly dependent on the nature of the exciting light. The general description of Section 2.3 in terms of interfering overlapping resonances is still qualitatively correct, but a detailed analysis of real experiments requires the inclusion of the correlations (2.221). The general case is most elegantly treated by a wave packet formalism [2.8, 80, 131]. As this approach has recently been reviewed by Jortner and Mukamel [2.8] and because our primary interest centers about the energy dependence of statistical limit electronic relaxation rates, only a brief summary of the results of this theory are given herein.

2.5.7 Wave Packet Formalism

The initial state of a real system corresponds to a situation where the molecule in its ground electronic state ϕ_0 and where a photon wave packet is incident upon the molecule. Using a discrete notation, and single photon states for simplicity, this state is

$$\Psi(0^-) = \sum_{k_A,e_A} X_{k_A e_A} |\phi_0, k_A e_A\rangle, \tag{2.226}$$

as opposed to (2.30). The general state of the system at a later time is then given by the usual result

$$\Psi(t) = \exp(-i\mathcal{H} t/\hbar)\, \Psi(0^-), \tag{2.227}$$

with $\mathcal{H}$ given in (2.28). In the absence of matter-radiation interaction the initial photon wave packet amplitudes X_{ke} introduce the directional aspects of the light beam, so (2.227) naturally propagates the photon wave packet across the molecule and off into the distance. Thus, the artificial termination of the light pulse as in (2.32–39) and implicit in (2.225) is not required (in some experiments, this separation is definitely invalid!). Thus, (2.227) describes the complete time development of the system, and the photon emission probability for photons of type F is simply

$$P_{F,X}(t) = |\langle \phi_F^v | \exp(-i\mathcal{H} t/\hbar) | \Psi(0^-)\rangle|^2. \tag{2.228}$$

Here $P_{F,X}(t)$ depends explicitly on the time interval after switching on the pulse (at $t=0$), on the light pulse X, and on nature of the photons k_F, e_F observed. Eq. (2.228) can be analyzed by the general Green's function methods presented in this section. Rather than separate excitation and decay events, (2.228) corresponds to a photon scattering event. Let P again be the projector onto the excited molecular levels as in (2.207). Upon averaging over final photon states (2.228) is

$$P_{F,X}(t) = \sum_{k_F,e_F,v} \left| \sum_{k_A,e_A} (2\pi i\hbar)^{-1} \int_C dE \exp(-iEt/\hbar) \right.$$
$$\left. \times \langle \phi_F^v | Q\, G(E)\, Q | \phi_A\rangle\, X_{k_A,e_A} \right|^2. \tag{2.229}$$

The operator $QG(E)Q$ can readily be evaluated as follows: First multiply (2.155) on the right and left by Q, then insert a factor of $1 = P + Q$ between the $\mathscr{E} - H$ and $G(\mathscr{E})$ to give

$$Q(\mathscr{E} - \mathscr{H}) Q(QGQ) - Q \mathscr{H} P(PGQ) = Q. \tag{2.230}$$

Define

$$G_Q^0 = (\mathscr{E} - Q \mathscr{H} Q)^{-1} \tag{2.231}$$

so (2.230) is converted to

$$QGQ = G_Q^0 + G_Q^0 Q \mathscr{H} P(PGQ) \tag{2.232}$$

after multiplication by G_Q^0. Substituting the expression

$$PGQ = (PGP) P \mathscr{H} Q G_Q^0, \tag{2.233}$$

which is the adjoint of (2.159), into (2.232) gives

$$QGQ = G_Q^0 + G_Q^0 Q \mathscr{H} P(PGP) P \mathscr{H} Q G_Q^0. \tag{2.234}$$

Since the emitted photons differ from the absorbed ones, say because of different frequencies or viewing direction, we have

$$\langle \phi_F^v | G_Q^0 | \phi_A \rangle \equiv \langle \phi_0^v, \mathbf{k}_F \mathbf{e}_F | G_Q^0 | \phi_0, \mathbf{k}_A \mathbf{e}_A \rangle = 0. \tag{2.235}$$

Using (2.234), and (2.235), Eq. (2.229) is rewritten as

$$P_{F,X}(t) = \sum_{\mathbf{k}_F, \mathbf{e}_F, v} \left| \sum_{\mathbf{k}_A, \mathbf{e}_A} (2\pi i \hbar)^{-1} \int_C \frac{dE \exp(-iEt/\hbar)}{(E - E_0^v - \hbar c k_F)(E - \hbar c k_A)} \right. \\ \left. \times \langle \phi_F^v | Q \mathscr{H} P[PG(E)P] P \mathscr{H} Q | \phi_A \rangle X_{\mathbf{k}_A \mathbf{e}_A} \right|^2, \tag{2.236}$$

because G_Q^0 is diagonal in the single photon states $|\phi_F^v\rangle$ and $|\phi_A\rangle$. Introducing the projector $\hat{P}$ onto those excited levels θ_j for which

$$\langle \phi_F^v | \mathscr{H}_{\mathrm{int}} | \theta_j, \mathrm{vac} \rangle \neq 0, \quad \text{or} \quad \langle \theta_j, \mathrm{vac} | \mathscr{H}_{\mathrm{int}} | \phi_A \rangle \neq 0, \tag{2.237}$$

namely

$$\hat{P} = \sum_j |\theta_j, \mathrm{vac}\rangle \langle \theta_j, \mathrm{vac}|, \tag{2.238}$$

allows (2.236) to be written as

$$P_{F,X}(t) = \sum_{k_F, e_F, v} \left| \sum_{k_A, e_A} (2\pi i\hbar)^{-1} \int_C \frac{dE \exp(-iEt/\hbar)}{(E - E_0^v - \hbar c k_F)(E - \hbar c k_A)} \right.$$
$$\left. \times \langle \phi_F | \mathscr{H}_{int} \hat{P} G(E) \hat{P} \mathscr{H}_{int} | \phi_A \rangle X_{k_A e_A} \right|^2 . \tag{2.239}$$

At this juncture, the equations become simplified if we adopt the definition of the generalized doorway states of JORTNER and MUKAMEL [2.8]

$$|N, \text{vac}\rangle = \gamma_N^{-1} \hat{P} \mathscr{H}_{int} | \phi_0, k_A e_A \rangle \tag{2.240}$$

$$|N^v, \text{vac}\rangle = (\gamma_N^v)^{-1} \hat{P} \mathscr{H}_{int} | \phi_0^v, k_A e_A \rangle \tag{2.241}$$

with

$$(\gamma_N)^2 = \sum_j |\langle \phi_0, k_A e_A | \mathscr{H}_{int} | \theta_j, \text{vac} \rangle|^2 \tag{2.242}$$

$$(\gamma_N^v)^2 = \sum_j |\langle \phi_0^v, k_F e_F | \mathscr{H}_{int} | \theta_j, \text{vac} \rangle|^2 \equiv \Gamma_N^v . \tag{2.243}$$

Substituting (2.240–243) into (2.238), then converting the E-integrations to a convolution integral, evaluating $(d/dt) P_{F,X}(t)$, and finally performing the integration over k_F leads to the photon counting rate[14]

$$\frac{d}{dt} P_{F,X}(t) = \Gamma_N \gamma_N^2 \left| \int_0^t L(t - \tau) a_{N^v N}(\tau) d\tau \right|^2 , \tag{2.244}$$

where

$$\Gamma_N = (2\pi/\hbar) \sum_v \Gamma_N^v \rho_{\text{photon}} \tag{2.245}$$

is the total radiative decay rate of the doorway state, $L(t)$ is the time dependent photon wave packet in the absence of radiation-matter interaction

$$L(t) = \exp(-i\mathscr{H}_r t/\hbar) \sum_{k_A, e_A} X_{k_A, e_A} | k_A, e_A \rangle$$
$$= \sum_{k_A, e_A} \exp(-ick_A) X_{k_A, e_A} | k_A, e_A \rangle , \tag{2.246}$$

---

[14] To properly account for the coherence properties of the pulse as in (2.221), an average must be performed over the amplitudes X_{k_A, e_A} in (2.229), (2.239), and (2.244). The final result is then equivalent to that generated by the density matrix formalism where the averaging is invoked at the beginning [2.131].

and $a_{N^vN}(\tau)$ is the probability amplitude

$$a_{N^vN}(\tau) = (2\pi i\hbar)^{-1} \int_C dE \exp(-iEt/\hbar)\langle N^v, \mathrm{vac}|G(E)|N, \mathrm{vac}\rangle, \qquad (2.247)$$

which is generally a superposition of terms involving the different resonant states χ_n.

JORTNER and MUKAMEL [2.8] discussed how no extra molecular information can be gained from a single isolated resonance by altering the nature of the incident wave packet $X_{k_Ae_A}$. They treated a number of interesting cases in detail, so the reader is referred to their work for more details.

2.6 Non-Radiative Decay Rates in the Statistical Limit: An Introduction

The foregoing theoretical description of electronic relaxation processes in polyatomic molecules focuses attention on the nature of the intramolecular dissipative process in the case of a dense manifold of (zero-order) vibronic levels, the nature of irreversibility in the electronic relaxation process, and the consequences of coupling of radiative and non-radiative decay channels. The theory in Sections 2.2–5 is mainly concerned with providing a general understanding of all of the diverse phenomena which are classified as involving radiationless transitions, and therefore does not deal with the specific details which are peculiar to a given molecule, or class of molecules. Thus, being equipped with a general qualitative understanding of the "mechanisms" leading to the different manifestations of radiationless transitions, we now investigate the quantitative factors which affect the experimentally observed non-radiative decay rates in specific classes of molecules. The simplest case, namely electronic relaxation in molecules in the statistical limit, is the one to which we specifically address ourselves.

The early experiments on the luminescence properties of molecules in condensed media accumulated a large body of data on the decay properties of aromatic molecules. The theory of ROBINSON and FROSCH [2.6, 11] mentioned in Section 2.1, deals specifically with this case, and we briefly reiterate the content of their theory in a slightly altered form.

Let us consider a molecule in an "inert" host medium. An "inert" host is one which is taken to negligibly alter the molecular energy levels of our guest molecule and to not affect the intramolecular coupling responsible for the electronic process. Upon conventional excitation (e.g. on a nanosecond time-scale) of a statistical limit molecule from the ground state ϕ_0 to an excited electronic state ϕ_s, vibrational relaxation proceeds sufficiently rapidly that we can generally take the excited molecules to be distributed amongst the vibrational components, $\{\phi_{si}\}$ of ϕ_s with a Boltzmann distribution. Provided the decay rates Γ_{si} of the individual ϕ_{si} are low enough that the states correspond to non-overlapping resonances, i.e., $\varepsilon_m + \hbar \Gamma_{si}$ is much less than spacing between adjacent ϕ_{si}, the decay of each of the ϕ_{si} can be taken to be independent.

Upon decay of ϕ_{si} into the vibronic manifold $\{\phi_{lj}\}$ (j now denotes the particular vibronic level of electronic state ϕ_l), the highly vibrationally excited $\{\phi_{lj}\}$ undergo rapid vibrational relaxation with decay rates $\Gamma_{lj}(E_{lj})$. Thus, the non-radiative decay rate Δ_{si} corresponds to the expression given in (2.197), which in discrete notation is

$$\Delta_{si}(E_{si}) = \frac{2\pi}{\hbar} \sum_{lj} \frac{|v_{si,lj}|^2 \Gamma_{lj}(E_{si}) \hbar/2}{(E_{si} - E_{lj})^2 + [\hbar \Gamma_{lj}(E_{si})/2]^2}, \tag{2.248}$$

where the indices refering to host states has been omitted for simplicity. In effect, Robinson and Frosch argued that the dominant contribution to (2.248) comes from $E_{si} \approx E_{lj}$, and in this resonant approximation (2.248) becomes

$$\Delta_{si}(E_{si}) \approx \frac{4\pi}{\hbar^2} \sideset{}{'}\sum_{lj} \frac{|v_{si,lj}|^2}{\Gamma_{lj}(E_{si})}, \tag{2.249}$$

with the primed summation indicating the restriction to

$$|E_{lj} - E_{si}| \lesssim \hbar \Gamma_{lj}(E_{si})/2. \tag{2.250}$$

As the vibrational relaxation decay rate Γ_{lj} is proportional to an "effective density" of host modes and because the summation in (2.249) also runs over these host mode states, Robinson and Frosch argued that the medium density of states essentially cancels leaving the "golden-rule" like result (2.5) that is characteristic of an isolated molecule.

The approximation (2.249) and the arguments concerning the plausibility of cancellation of the medium mode density of states are, however, quite unnecessary. Instead, under the much weaker assumption that $\Gamma_{lj}(E_{si})$, $\rho_l(E_{lj})$, and $|v_{si,lj}|^2$ are slowly varying over the range (2.250), the continuum version of (2.248) reduces to

$$\begin{aligned}
\Delta_{si}(E_{si}) &= \frac{2\pi}{\hbar} \int \frac{dE_{lj} \rho_l(E_{lj}) |v_{si,lj}|^2 \hbar \Gamma_{lj}(E_{si})/2}{(E_{si} - E_{lj})^2 + [\hbar \Gamma_{lj}(E_{si})/2]^2} \\
&= \frac{2\pi}{\hbar} \rho_l(E_{si}) |v_{si,lj}|^2 \int \frac{dE_{lj} \hbar \Gamma_{lj}(E_{si})/2}{(E_{si} - E_{lj})^2 + [\hbar \Gamma_{lj}(E_{si})/2]^2} \\
&= \frac{2\pi}{\hbar} \rho_l(E_{si}) |v_{si,lj}|^2.
\end{aligned} \tag{2.251}$$

Eq. (2.251) involves only the guest molecule density of states because the definition of an inert solvent implies that the coupling $v_{si,lj}$ is unaltered by the host medium. This means that the coupling $v_{si,lj}$ is diagonal in host states, and consequently the "inert" host undergoes no change during the electronic relaxation process. It can be shown, however, that when the host undergoes a modest change while the guest undergoes the $s \rightarrow l$ relaxation, the intramolecular result (2.251) is unchanged even though the optical absorption to ϕ_{si} shows phonon broadened lines [2.81]. Thus, the definition of an inert medium is sufficient to guarantee the purely intramolecular character of the decay rate (2.251). ROBINSON and FROSCH further stated the conviction that a large molecule could act as its owen heat bath. The justification of this conjecture had to await the more detailed theory reviewed in Sections 2.3–5.

For the purposes of the analysis of early data, the simple factorization approximation (2.7) was sufficient for many purposes. ROBINSON and FROSCH noted that the Franck-Condon factor F is approximately an exponentially decreasing function of the energy gap ΔE_{sl}, the difference between the minima of the potential surfaces for electronic states s and l. A good example of this is exhibited by $T_1 \rightarrow S_0$ non-radiative decay rates in aromatic hydrocarbons, a plot of which is adapted from SIEBRAND [2.82] in Fig. 2.8, where additional

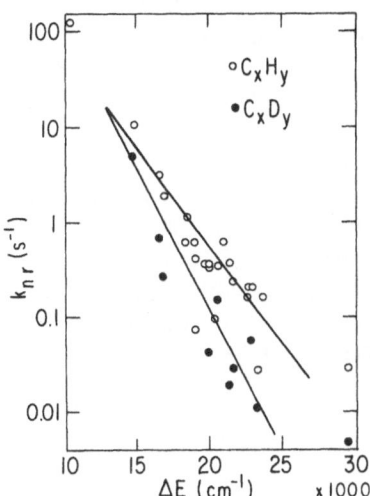

Fig. 2.8. Experimental dependence of the non-radiative decay rate for intersystem crossing ($T_1 \rightarrow S_0$) on the electronic energy gap for perhydro- and perdeutero-aromatic hydrocarbons. All data were obtained in solid solutions. The non-radiative decay rate was calculated by SIEBRAND [2.82] from (2.1) and (2.2) taking $\tau^{rad} = 30$ s. The figure was adapted from SIEBRAND's [2.82] paper with the added straight lines

straight line fits to the data are added for reasons discussed below. More recently, similar "energy-gap law" results have been observed in a series of azulene derivatives [2.83], Fig. 2.9, for relaxation between different electronic states of rare earth ions in crystals [2.84], and for $S_1 \rightarrow S_0$ internal conversion in aromatic hydrocarbons [2.132].

As noted above, the energy gap law behavior was quickly understood to be a consequence of the Franck-Condon factors. Numerical evaluation of these by BYRNE et al. [2.13] and others demonstrated the dominant role

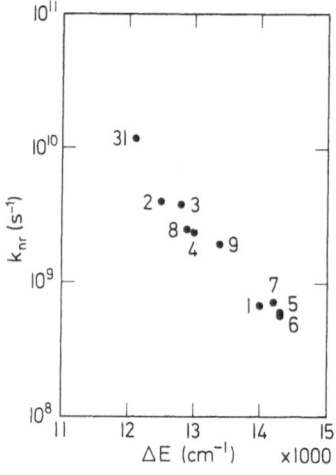

Fig. 2.9. Plot of the non-radiative $S_2 \rightarrow S_1$ internal conversion decay rate vs. the energy gap for a variety of substituted azulenes. The figure is taken from the erratum to [2.83] which includes a corrected value for the absolute fluorescence quantum yield of azulene. The compounds are explicitly given in Fig. 2.1 of [2.83]

of the C—H vibrations in the $T_1 \rightarrow S_0$ relaxation in aromatic hydrocarbons. This then provides the explanation for the dramatic deuterium isotope effect displayed in Fig. 2.8. Although the density of states are considerably larger in the perdeuterated molecules than in the perhydro ones for the large energy gaps appropriate to this $T_1 \rightarrow S_0$ relaxation, the Franck-Condon factors are much smaller in the former than in the latter case. The dominance of the Franck-Condon factors leads to the observed isotopic dependence.

In order to demonstrate the primary role of the C—H vibrations in these early studies, numerical studies of the Franck-Condon factor were employed [2.13]. Some of the more recent developments can be foreshadowed by the studies, so we now describe their basic features. We begin with the intramolecular rate expression (2.251) and then invoke a Born-Oppenheimer approximation for ϕ_{si} and ϕ_{lj} in order to write each as a product of electronic $\phi_s(q, Q^s)$, $\phi_l(q, Q^l)$ and vibrational $\chi_{si}(Q^s)$, $\chi_{lj}(Q^l)$ wavefunctions via [2.2, 4, 14, 40, 85]

$$\phi_{si}(q, Q^s) = \phi_s(q, Q^s)\chi_{si}(Q^s) \tag{2.252a}$$

$$\phi_{lj}(q, Q^l) = \phi_l(q, Q^l)\chi_{lj}(Q^l) \tag{2.252b}$$

where q denotes the electronic coordinates and $Q^s [Q^l]$ are the nuclear coordinates appropriate to $\phi_s[\phi_l]$.

With due cognizance of the fact that v is an effective interaction, as described in (2.205) ff., a generalized Condon-type (Q-centroid) approximations is invoked [2.4, 14, 23] to remove the slowly varying electronic matrix element from outside the Q integration in

$$\begin{aligned} v_{si,lj} &= \int dQ \int dq\, \phi_s^*(q, Q^s)\chi_{si}^*(Q^s)v(q, Q)\phi_l(q, Q^l)\chi_{lj}(Q^l) \\ &\approx C_{l,s}(Q^0) \int dQ\, \chi_{si}^*(Q^s)M(Q)\chi_{lj}(Q^l), \end{aligned} \tag{2.253}$$

where $C_{l,s}(Q^0)$ is the effective electronic interaction and $M(Q)$ is a Q-dependent operator whose structure is intimately related to the "mechanism" for the $s \rightarrow l$ decay. As the current discussion is independent of the structure of $C_{l,s}$ or of $M(Q)$, in this section, the latter can be replaced by unity and the former by v_{ls}, for convenience. Thus, the decay rate is that given in (2.8) with the association

$$\beta_e^2 = |C_{l,s}(Q^0)|^2 .$$

For low temperatures (e.g. 77 K) the quanta of intramolecular vibrational energy $\hbar\omega$ are much greater than kT for the aromatic hydrocarbons, so the thermally averaged rate (2.5) reduces to the decay rate of the vibrationless level of ϕ_s. Using the approximations inherent in (2.8), this is

$$\lim_{T \rightarrow 0} k_{nr}(s \rightarrow l; \beta) = k_{nr}(s0 \rightarrow l)$$
$$= (2\pi/\hbar)|v_{s,l}|^2 \sum_{j_l} |\langle 0_s | j_l \rangle|^2 \delta(E_{s0} - E_{lj}) , \qquad (2.254)$$

where $|0_s\rangle$ is shorthand for χ_{s0} and i_l represents χ_{lj}.

With the realization that j represents the collection of vibrational quantum numbers $n_1 n_2 \dots n_N$ for state χ_{lj}, Eq. (2.254) appears quite formidable. It is therefore quite natural to introduce a harmonic approximation for the vibrational modes in s and l in order to simplify the evaluation of (2.294) as much as possible. For $|0\rangle$ such an approximation is quite appropriate for the aromatic hydrocarbons. On the other hand, the $|j_l\rangle$ correspond to very highly excited vibrational states for which a harmonic approximation is quite untenable. For lack of a better approximation, we can note that a true vibrational state χ_{lj} must be a linear combination of a large number of quasi-degenerate harmonic oscillator basis states. Thus, any given harmonic oscillator state $\theta_{l\alpha}$ is distributed amongst a large number of true vibrational states χ_{lj} over a range of energies of the order of δV about $E_{l\alpha}$, where δV is a measure of the anharmonic couplings in the region of the energy $E_{l\alpha}$ of $\theta_{l\alpha}$. This implies that, by invoking the harmonic approximation for χ_{lj}, it is not necessary to be very restrictive in the satisfaction of the energy conservation condition embodied in the Dirac delta function in (2.254). This delta function should therefore be taken to be a shorthand notation for the effective continuous density of states of the θ_{lj}. This large "spread" of the harmonic θ_{lj} is an important ingredient in the evaluation of (2.254) as it enhances the accuracy of approximations that would otherwise appear to be vitiated by the seemingly strict energy restrictions in (2.254).

Since little is known about the detailed normal modes in some of the excited electronic states of, say, the aromatic hydrocarbons, it is natural at first to consider that the normal modes in s and l are identical apart from displacements in the equilibrium positions and changes in the frequencies of the modes between the two electronic states. The general case can be handled, if necessary, but this has yet to be warranted by available experimental data. (However, in the case of photochemical decompositions, this mode-mixing between initial and final states is often of prime importance [2.35].)

Given the above approximations and considerations, (2.254) is rewritten in terms of the harmonic Franck-Condon factors as

$$k_{nr}(s0 \rightarrow l) = (2\pi/\hbar)|v_{sl}|^2 \sum_{\{n_j\}} \prod_j |\langle 0_s|n_j\rangle|^2 \delta\left[\Delta E - \sum_j \hbar\omega_j^l(n_j + \tfrac{1}{2}) + \sum_j \hbar\omega_j^s/2\right],$$

(2.255)

where n_j represents the j-mode quantum number in ϕ_l, ω_j^l is the angular frequency of the mode, and the dot on the equality emphasizes the weak energy conservation interpretation.

For simplicity, we can first employ a model in which frequency shifts are ignored,

$$\omega_j^s = \omega_j^l \equiv \omega_j,$$

(2.256)

and the mode displacements are characterized by the dimensionless coupling constant

$$X_j = \tfrac{1}{2}(M_j\omega_j/\hbar)(Q_{j,0}^l - Q_{j,0}^s)^2 \equiv \tfrac{1}{2}\Delta_j^2.$$

(2.257)

In (2.257) M_j is the reduced mass for made j and $Q_{j,0}^l$ is the equilibrium position for mode j in ϕ_l (etc., for $Q_{j,0}^s$ and ϕ_s). Introducing the well known displaced oscillator Franck-Condon factors, then reduces (2.295) for this model to

$$k_{nr}(s0 \rightarrow l) \doteq (2\pi/\hbar)|v_{s,l}|^2 \exp\left(-\sum_j X_j\right)$$
$$\times \left[\prod_j \sum_{n_j=0}^{\infty} X_j^{n_j}/(n_j)!\right]\delta\left(\Delta E - \sum_j n_j\hbar\omega_j\right).$$

(2.258)

Eq. (2.258) represents the model studied by BYRNE et al. [2.13] and others in the first theoretical analyses of electronic relaxation rates. Unknown to these workers was the fact that KUBO and TOYOZAWA [2.86, 87] had previously developed the mathematical apparatus for the analysis of the decay rates (2.258), as well as the general cases of the full thermal rate, in their theory of radiative and non-radiative processes in solids. The Kubo-Toyozawa theory had also been applied to the study of Mössbauer spectra in solids and optical phonon-broadened lineshapes in solids. The application and generalization of the Kubo-Toyozawa methods to electronic relaxations processes in isolated molecules had to await the later work of LIN et al. [2.85], ENGLMAN and JORTNER [2.88], and FREED and JORTNER [2.81], that is described in Section 2.8.

Lacking the Kubo-Toyozawa theoretical apparatus, BYRNE et al. [2.13], and all earlier work, analyzed (2.258) by taking the maximum Franck-Condon factor $F_{max}(\Delta E)$ that satisfied the energy conservation condition in (2.258),

$$F_{max}(\Delta E) = \max_{\{n_j\}} \left[\prod_j X^{n_j}/(n_j)! \right] \qquad (2.259a)$$

for which

$$\sum_j n_j \hbar \omega_j = \Delta E. \qquad (2.259b)$$

The overall decay rate of ϕ_{s0} in this approximation becomes proportional to $F_{max}(\Delta E)$. A numerical analysis of (2.259) enabled the conclusion of the dominant role of the C—H vibrational motion. This numerical analysis is readily generated by evaluating

$$F(\Delta E; \{n_j\}) = \prod_j [X_j^{n_j}/(n_j)!], \qquad (2.260)$$

subject to $(2.259b)^{15}$, and then determining the maximum value, a process which is straightforward for two modes, but becomes increasingly tedious when more vibrations are present. Note benzene has 30 vibrational modes if we include the more realistic situation of non-zero frequency shifts between ϕ_l and ϕ_s.

The mode of analysis embodied in (2.259) bears a direct relationship to methods developed by Boltzmann to study equilibrium statistical thermodynamics. Such methods are described in all elementary texts on statistical thermodynamics [2.89] and physical chemistry and were, consequently, available to those presenting the earliest investigations of (2.258). Had these workers "cribbed" from the elementary texts, an analytic representation of the energy-gap law and its isotope dependence would have instantly emerged in simple transparent form [2.4, 81, 26]. Hence, we now consider these Boltzmann methods as they are actually sufficiently accurate to derive most of the information contained in (2.258). The more general case involving frequency shifts cannot adequately be treated within the framework of the Boltzmann method, but it can be handled by a generalization of the Kubo-Toyozawa theory that are presented in Section 2.8.

2.7 Boltzmann Statistics

It will be recalled [2.89] that Boltzmann considered a set of particles which could be placed into N states of energies ε_j,

$$\varepsilon_j \equiv \hbar \omega_j. \qquad (2.261)$$

---

[15] The discussion following (2.254) implies that non-integral values of n_j are permissible in (2.259a).

If there is no restriction as to the number of particles that can be placed into the j-th state, we might place n_1 objects into ε_j, n_2 into $\varepsilon_2, \ldots, n_N$ into ε_N. If the particles are distinguishable and furthermore the j-th state has degeneracy X_j, Boltzmann statistics states that the total number of ways of distributing the particles according to $\{n_1, n_2, \ldots, n_N\}$ is

$$F(\{n_i\}) = \prod_{j=1}^{N} X_j^{n_j}/(n_j)! . \tag{2.262}$$

The particles may distribute themselves with any set of $\{n_j\}$ subject to the constraints on the system. Let us take the only constraint to be that there be a fixed total energy $\varDelta E$,

$$\varDelta E = \sum_{j=1}^{N} n_j \hbar \omega_j . \tag{2.263}$$

Since all distributions $\{n_i\}$ satisfying (2.263) are equally *a priori* probable, the total number of ways of realizing the system is the microcanonical partition function

$$W(\varDelta E) = \sum_{\{n_j\}} \left[\prod_{j=1}^{N} X_j^{n_j}/(n_j)! \right] \delta \left(\varDelta E - \sum_{j=1}^{N} n_j \hbar \omega_j \right) . \tag{2.264}$$

The correspondence between (2.264) and (2.258) should be striking. Note that the total number

$$V = \sum_{j=1}^{N} n_j \tag{2.265}$$

of particles has not been fixed in order to generate this correspondence. Although the degeneracy X_j in Boltzmann statistics is a positive integer (often unity), there is nothing to prevent the use of non-integral values of X_j as required in (2.258). The elementary texts rarely express the partition function with the delta function as in (2.264). Rather, for large total energies $\varDelta E$, they employ the canonical ensemble which is obtained by finding the maximum value of the summand $F(\{n_j\})$ in (2.264) subject to the constraint (2.263) as introduced by the method of Lagrange multipliers. This is just identical to the numerical method (2.259) used for the non-radiative decay rates.

In order to repeat this analysis, we maximize $\log F(\{n_j\})$ in the customary fashion. Using Stirling's approximation [cf. (2.271)] for $\ln(n!)$ and (2.262) gives

$$\ln F(\{n_j\}) = \sum_{j=1}^{N} \left[n_j \ln(X_j/n_j) + n_j \right] . \tag{2.266}$$

Thus, if b is the Lagrange multiplier to insure (2.263), the maximization process to determine $F_{max}(\Delta E)$ is

$$\frac{\partial}{\partial n_i}\left[\ln F(\{n_j\})+b\sum_j n_j\hbar\omega_j\right]\Bigg|_{\{n_i^0\}}=0, \quad \text{all} \quad i=1,...,N. \tag{2.267}$$

This use of a Lagrange multiplier to satisfy (2.259b) and (2.263) is permissible for the electronic relaxation rates because of the implied weak energy conservation constraint in (2.258). Eq. (2.267) implies that

$$-\ln n_i^0 + \ln X_i + b\hbar\omega_i=0$$

or

$$n_i^0 = X_i\exp(b\hbar\omega_i), \quad \text{all } i, \tag{2.268}$$

and consequently

$$F_{max}(\Delta E, \{n_j^0\})=\prod_{j=1}^N \frac{X_j^{n_j^0}}{\Gamma(n_j^0+1)}, \tag{2.269}$$

where the algebraic equation

$$\Delta E=\sum_{j=1}^N n_j^0\hbar\omega_j=\sum_{j=1}^N \hbar\omega_j X_j\exp(b\hbar\omega_j) \tag{2.270}$$

fixes the value of the Lagrange multiplier b. Note that the $\{n_j^0\}$ need not turn out to be all integers; non-integral values are consistent with the weak energy constraint implicit in (2.258). Thus, (2.269) contains $\Gamma(n_j^0+1)$ instead of $(n_j^0)!$; alternatively we should use Stirling's approximation

$$\Gamma(n_j^0+1)=(n_j^0)^{n_j^0}\exp(-n_j^0) \tag{2.271}$$

in (2.269) for consistency with (2.266).

2.7.1 Illustrative Example: $T_1 \rightarrow S_0$ Decay in Benzene

At this juncture it is convenient to introduce typical values of X_j, ΔE and ω_j, and the benzene molecule is chosen as an example. Only totally symmetric vibrations may have $X_j=0$, and in benzene only the $\nu_1\,a_{1g}$ C—C skeletal mode and the $\nu_2\,a_{1g}$ C—H stretch qualify. For the $T_1 \rightarrow S_0$ transition, $\omega_{\nu_1}=990\,\text{cm}^{-1}$, $\omega_{\nu_2}=3130\,\text{cm}^{-1}$ (S_0 values), $X_{\nu_1}=1.1$, $X_{\nu_2}=0.037$, and $\Delta E(T_1 S_0)=29\,847$ cm^{-1}. Thus, although X_1 is an order of magnitude greater than X_2, ω_{ν_2} is three times ω_{ν_1}. Because n_i^0 in (2.268) depends exponentially on ω_i and, as shown below, $b\hbar\omega_{v_1}=6.0$, $\hbar\omega_{v_2}n_{v_2}^0\gg\hbar\omega_{v_1}n_{v_1}^0$. To verify this statement, let us for the moment ignore v_1. In this case (2.270) is trivially solved to give

$$b'=(\hbar\omega_{v_2})^{-1}\ln(\Delta E/X_{v_2}\hbar\omega_{v_2})=5.6/\hbar\omega_{v_2}.\tag{2.272}$$

Reintroducing v_1 and writing

$$b=b'+(\hbar\omega_{v_2})^{-1}\ln f\tag{2.272a}$$

converts (2.271) to (dividing through by ΔE)

$$1=f+\Omega_{v_1}f^{p_{v_1}}\tag{2.273}$$

where

$$\Omega_{v_1}=\frac{X_{v_1}\hbar\omega_{v_1}}{\Delta E}\left(\frac{\Delta E}{X_{v_2}\hbar\omega_{v_2}}\right)^{p_{v_1}}=0.21\tag{2.274}$$

and

$$p_{v_1}=\omega_{v_1}/\omega_{v_2}=0.32.\tag{2.275}$$

The numerical values in (2.272), (2.274), and (2.275), of course, only pertain to $T_1\to S_0$ decay in benzene. Solution of (2.273) does verify the approximate value of b quoted above, and this gives $n_{v_2}^0=7.6$ and $n_{v_1}^0=6.0$.

Benzene is singular among the aromatic hydrocarbons in having a larger X for the C—C bond for the $S_0\to S_1$ transition and presumably also for the $T_1\to S_0$ transition. Thus, in these other molecules $n_{v_1}^0\hbar\omega_{v_1}$ represents a smaller fraction of ΔE than in benzene. Consequently, as a first approximation, we retain the highest frequency totally symmetric C—H stretches (v_2 for benzene) j_M. Writing

$$d_M X_M=\sum_{j_M=1}^{d_M}X_{jM}\tag{2.276a}$$

$$\omega_M=\sum_{j_M=1}^{d_M}\omega_{jM}/d_M,\tag{2.276b}$$

using the one-frequency approximation of (2.272), and employing (2.271), converts (2.269) into

$$F_{\max}(\Delta E)=\exp(-\gamma\Delta E/\hbar\omega_M),\tag{2.277}$$

where

$$\gamma=\ln(\Delta E/d_M X_M\hbar\omega_M)-1\tag{2.278}$$

is the analytic representation of the energy gap law for aromatic hydrocarbons as first derived by ENGLMAN and JORTNER [2.88] by using the more sophisticated generating function methods of KUBO and TOYOZAWA. The above derivation, on the other hand, employs analyses commonly found in undergraduate treatments of statistical thermodynamics. Note how (2.278) automatically explains the observed deuterium isotope effect since ω_M drops from $\sim 3100 \, cm^{-1}$ to $\sim 2300 \, cm^{-1}$ upon deuteration while γ changes only slightly.

2.7.2 Inclusion of Other Modes

As FREED and JORTNER showed [2.8], the energy gap law persists when due cognizance is taken of the presence of other modes, possibly with frequency shifts, and the propensity rules for the promoting modes (see Sect. 2.8 for a discussion of the coupling operator M of (2.253)). In this general case γ is affected somewhat by the presence of the other modes and ΔE is shifted, but the general structure of (2.277) is maintained as described in Section 2.8. In fact, the Boltzmann statistics enables the treatment of the former. For example, we now consider the modification of γ that arises from the remaining modes in (2.269) and (2.270). As a byproduct of this analysis, general criteria are obtained which determine those vibrations which act as good accepting modes [2.29], i.e., for which $n_j^0 \hbar \omega_j^l$ is appreciable.

From the illustrative case of the benzene $T_1 \rightarrow S_0$ decay, it is clear that the higher frequency totally symmetric modes are most likely to be the best acceptors. Thus, these modes are taken to be the comparison modes. If some other vibrations are then found to be the dominant ones, they may be substituted for the comparison modes. It is convenient, although by no means necessary, to collect together those modes with nearly the same frequencies. For the modes of maximum frequency, Eq. (2.276) defines the total vibrational coupling strength $d_M X_M$ for these modes, and ω_M is the average frequency. This procedure may be continued for the remaining modes to define d_α, X_α, and ω_α for other groups of closely related vibrations.

The energy conservation condition (2.270) is then rewritten as

$$\Delta E = d_M X_M \hbar \omega_M \exp(b \hbar \omega_M)$$
$$+ \sum_{\alpha \neq M} d_\alpha X_\alpha \hbar \omega_\alpha \exp(b \hbar \omega_\alpha). \tag{2.279}$$

Similar to (2.273), the change in variables

$$b = (\hbar \omega_M)^{-1} \ln(\Delta E f / d_M X_M \hbar \omega_M) \tag{2.280}$$

and division of (2.279) by ΔE converts it to

$$1 = f + \sum_{\alpha \neq M} \Omega_\alpha f^{p_\alpha} \tag{2.281}$$

with

$$\Omega_\alpha = \frac{d_\alpha X_\alpha \hbar \omega_\alpha}{\Delta E} \left(\frac{\Delta E}{d_M X_M \hbar \omega_M} \right)^{p_\alpha} \tag{2.282}$$

and

$$p_\alpha = \omega_\alpha / \omega_M \tag{2.283}$$

just as in (2.274) and (2.275).

Consider first the situation where

$$\Omega_\alpha \ll 1, \quad p_\alpha \lesssim 1, \quad \text{all } \alpha, \tag{2.284}$$

so in (2.281) the $\sum_{\alpha \neq M} \Omega_\alpha f^{p_\alpha}$ is negligible compared to f, leading to the zeroth-order approximation

$$f^{(0)} \approx 1. \tag{2.285}$$

Substitution of (2.285) into (2.280) and that result into (2.268) gives

$$n_\alpha^0 = \Delta E \Omega_\alpha / \hbar \omega_\alpha, \tag{2.286}$$

whereupon (2.284) implies

$$\hbar \omega_\alpha n_\alpha^0 / \Delta E \ll 1, \tag{2.287}$$

which is just the condition that the α-modes be poor acceptors. Condition (2.287) is therefore the criterion for α being a poor acceptor. When the modes $\alpha \neq M$ satisfy (2.284), a better approximation to (2.281) than (2.285) is obtained by itering the zeroth-order result (2.285) to give the first-order solution

$$f^{(1)} = 1 - \sum_{\alpha \neq M} \Omega_\alpha, \tag{2.288}$$

and the second-order one

$$f^{(2)} = 1 - \sum_{\alpha \neq M} \Omega_\alpha \left(1 - \sum_{\beta \neq M} \Omega_\beta \right)^{p_\alpha}, \tag{2.289}$$

etc.

Substituting (2.270) into (2.269) with (2.271) yields

$$F_{\max}(\Delta E, \{n_j^0\}) = \exp \left[-b \Delta E + \sum_\alpha n_\alpha^0 \right] \tag{2.290}$$

$$\equiv \exp[-\gamma' \Delta E / \hbar \omega_M], \tag{2.291}$$

where

$$\gamma' = \gamma + \ln f + \sum_{\alpha \neq M} (1 - p_\alpha^{-1}) \Omega_\alpha f^{p_\alpha} \tag{2.292a}$$

and

$$\gamma = \ln(\Delta E/d_M X_M \hbar \omega_M) - 1, \tag{2.292b}$$

and f is still the solution to (2.281). Eq. (2.292) implies that the sole effect, within the Boltzmann statistics method, of the remaining modes is just the addition of $\gamma' - \gamma$ in (2.291). When modes α are poor acceptors, so (2.284) and (2.287) are satisfied, to lowest order in Ω_α, it is readily found that

$$\gamma' = \gamma - \sum_{\alpha \neq M} p_\alpha^{-1} \Omega_\alpha + \mathcal{O}(\Omega_\alpha^2). \tag{2.293}$$

For the $T_1 \rightarrow S_0$ decay in benzene, the percentage change in γ from (2.293) is

$$100(\gamma' - \gamma)/\gamma \approx 12\%, \qquad T_1 \rightarrow S_0 \quad \text{in benzene}, \tag{2.294}$$

which is clearly non-negligible, thereby exhibiting the contribution of the C—C vibrations to the full energy gap law (2.291). On the other hand, in the case of the $S_1 \rightarrow T_1$ decay in benzene, the relevant molecular parameters are $\Delta E \approx 8\,200\,\text{cm}^{-1}$, $X_M = 0.002$, $X_{v_1} = 0.025$, giving $\Omega_{v_1} = 0.03$, and

$$100(\gamma' - \gamma)/\gamma = 1.5\%, \qquad S_1 \rightarrow T_1 \quad \text{in benzene}, \tag{2.295}$$

which is getting small.

It is, of course, obvious that when $\Omega_\beta \gtrsim 1$ for some mode β, that vibration is a good accepting mode and numerical solution of the two-mode equation

$$1 = f + \Omega_\beta f^{p_\beta}$$

may be necessary. The important point, however, is the fact that when (2.284) is satisfied, a simple algebraic solution to (2.281) yields an accurate value for the shift $\gamma' - \gamma$ in (2.292a), and numerical analysis is unnecessary.

It is perhaps worthwhile at this juncture to reconsider the deuterium isotope effect in aromatic hydrocarbons. Experimental determinations of X_{v_1} and X_{v_2} are not available for $C_6 D_6$. Since the v_1 mode involves primarily C—C vibrations,—$\omega_{v_1}^{S_0}(C_6 H_6) = 990\,\text{cm}^{-1}$ while $\omega_{v_1}^{S_0}(C_6 D_6) = 945\,\text{cm}^{-1}$—it is reasonable to take [2.26, 25]

$$X_{v_1}(C_6 D_6) \approx X_{v_1}(C_6 H_6). \tag{2.296}$$

Likewise, as the v_2 mode is mostly a C—H stretch, we may take the C—H and C—D bond length changes to be the same [2.26, 25], or

$$\omega_{v_2}(C_6 D_6) X_{v_2}(C_6 D_6) \approx \omega_{v_2}(C_6 H_6) X_{v_2}(C_6 H_6) . \tag{2.297}$$

Eq. (2.297) implies that the dominant part of (2.292–293), namely γ, is unchanged upon deuteration. However, for the $T_1 \to S_0$ decay $\Omega_{v_1} = 0.33$, signaling the fact that v_1 is becoming a more efficient acceptor, but the overall shift $(\gamma' - \gamma)/\gamma$ is now $= 18\%$. Thus, the major effect of deuteration on the exponential part of the rate (2.291) is the reduction of ω_{v_2}, $\omega_{v_2}^{S_0}(C_6 H_6) = 3130 \, cm^{-1}$ and $\omega_{v_2}^{S_0}(C_6 D_6) = 2340 \, cm^{-1}$.

2.7.3 Analysis of the Method

It should again be emphasized that the above analysis is only approximate. Firstly, any alterations of the Franck-Condon factors due to the anharmonicies in ϕ_{1j} are neglected, but these anharmonicities enable us to consider non-integral values of n_α^0. Any numerical errors arising from anharmonicities are partially offset by the consideration of relative rates (see below) and the use of effective frequencies and coupling strengths. The Boltzmann method does not provide the pre-exponential factor A in the full rate expression

$$k_{nr}(s0 \to l) = (2\pi/\hbar) |v_{s,l}|^2 A \exp\left(-\sum_\alpha X_\alpha d_\alpha\right) \exp(-\gamma' \Delta E/\hbar \omega_M) . \tag{2.298}$$

In some cases the most serious error in the above analysis arises from the neglect of the frequency shifts which the vibrational modes undergo in the $s \to l$ transition [2.25]. These frequency changes can, however, readily be included by the more general methods described in the following section. Before turning to the more involved algebra, we can sketch out the basic ideas surrounding the emphasis on the consideration of relative non-radiative decay rates [2.4, 25–30].

The evaluation of accurate absolute non-radiative decay are still, perhaps, beyond the capabilities of present theory. To begin with, the evaluation of the electronic matrix elements requires the availability of rather accurate electronic wavefunctions since transition matrix elements generally have percentage errors of considerable magnitude greater than the corresponding transition energies. Even if we were provided with accurate enough wavefunction for, say, benzene, it is still not known whether truncation of the full rate expression to the lowest non-vanishing order is, in fact, an accurate procedure [2.4, 14], so such truncation can introduce considerable errors. Lastly, the use of the popular Condon approximation, to evaluate the electronic factor in (2.253) at the equilibrium configuration of one of the states, grossly underestimates this electronic matrix element often by as much as a couple of orders of magnitude [2.20–23]. An accurate treatment of (2.253) with the Q-centroid method [2.23] requires electronic wavefunctions ϕ_s and ϕ_l for large variations in some of the nuclear positions, and this is certainly beyond the capabilities of present technology. We are, consequently, inexorably tied to rather large uncertainties in predicting the values of the electronic matrix elements.

The remaining Franck-Condon factors may, at first sight, seem simple by comparison. However, they too involve a large degree of errors. BURLAND and ROBINSON [2.60] have shown how anharmonicities in the dominant C—H vibrations can increase the Franck-Condon factors again by a couple of orders of magnitude, so their neglect can lead to gross underestimates of the rates. Furthermore, in a molecule like benzene there are 30 vibrations whose Franck-Condon factors are to be incorporated. Lacking the relevant spectroscopic data concerning, say, triplet states, there are enough adjustable parameters associated with these modes to fit any observed single non-radiative decay rate.

2.7.4 Relative Decay Rates

It should be clear from the above discussion that the evaluation of absolute non-radiative decay rates is, at best, an uncertain task. It may also appear that the ability to surmise anything at all about these rates is vitiated all by the above-noted errors. Fortunately, this is not at all the case. When the effective electronic matrix elements in (2.253) are independent of the initial vibronic level χ_{si}, as is often expected to be the case, the relative non-radiative decay rates (ϕ_{s0} is the vibrationless level) $k_{nr}(si\rightarrow l)/k_{nr}(s0\rightarrow l)$ [cf. (2.10)] have these ellusive electronic matrix elements cancel out as a proportionality factor. The resulting relative non-radiative decay rates from individual vibronic levels are then only a function of the s and l potential surfaces.

Consider those highly anharmonic vibrations which are of significance for the absolute rates. If the anharmonicities have the net effect of multiplying the Franck-Condon factors by an average factor δ, δ then cancels in evaluating the relative decay rates, thereby minimizing the errors incurred in the use of the harmonic approximation. (Note again that some of the effects of anharmonicity are employed in the use of the weak energy conservation constraints (2.258).)

We can likewise show that the evaluation of relative decay rates also enables the minimization of errors to be incurred as a result of lack of information regarding the myriad of poor accepting modes that are not excited initially in χ_{si} [2.26–29]. To illustrate this point, it is necessary to discuss a method for the calculation of the relative decay rates. Denote those vibrations, that are excited in the initial state χ_{si}, as the optical modes. Since one prerequisite for the consideration of single vibronic level decay rates is the assignability of that level, i.e., it cannot be overlapped too much by neighboring vibronic bands, the optical modes cannot perforce involve more than a few modes. The initial energy

$$E(m_a) = \Delta E + \sum_a \hbar \omega_a^s (m_a + \tfrac{1}{2})$$

that is converted to vibrational energy in ϕ_l can ultimately reside in the optical modes $\{a\}$ or in the remaining modes. The individual decay rates can be displayed in a fashion that explicitly exhibits this partitioning of the energy between the optical and non-optical modes [2.25–29]

$$
\begin{aligned}
k_{nr}(si \rightarrow l) = (2\pi/\hbar)|v_{sl}|^2 \sum_{\{n_a\}} & \left[\prod_a |\langle m_a|n_a\rangle|^2 \right] \\
& \times \sum_{\{n_j\}} \prod_{j \neq a} |\langle 0_s|n_j\rangle|^2 \delta\left[\Delta E(m_a, n_a) - \sum_{j \neq a} n_j \hbar \omega_j^l \right]
\end{aligned}
\tag{2.299}
$$

with

$$
\Delta E(m_a, n_a) = \Delta E + \sum_a \hbar(\omega_a^s m_a - \omega_a^l n_a) + \sum_{\text{all } j} \tfrac{1}{2}\hbar(\omega_a^s - \omega_a^l). \tag{2.300}
$$

Eq. (2.299) just notes the obvious fact that if the optical modes have $\{n_a\}$ quanta in ϕ_l, they have taken up $\sum_a \hbar\omega_a^l(n_a + \tfrac{1}{2})$ of the available energy, so the remaining energy $\Delta E(m_a, n_a)$ given by (2.300) is left for the non-optical modes. The rate expression (2.299) can be recast in a suggestive form by the definition of a partial non-radiative decay rate

$$
\begin{aligned}
k'_{nr}(s m_a \rightarrow l n_a) = (2\pi/\hbar)|v_{sl}|^2 \sum_{\{n_j\}} & \prod_{j \neq a} |\langle 0_s|n_j\rangle|^2 \\
& \times \delta\left[\Delta E(m_a, n_a) - \sum_{j \neq a} \hbar \omega_j^l n_j \right],
\end{aligned}
\tag{2.301}
$$

which apart from an overall factor of $\prod_a |\langle m_a|n_a\rangle|^2$ corresponds to the non-radiative decay rate from χ_{si} to the set of final χ_{lj} where the optical modes $\{a\}$ have $\{n_a\}$ quanta. Eq. (2.301) in fact resembles the non-radiative decay rate of a vibrationless level in a molecule having vibrations $j \neq a$ and an energy gap of $\Delta E(m_a, n_a)$. We simply note that (2.299) and (2.301) can be combined to give

$$
k_{nr}(si \rightarrow l) = \sum_{\{n_a\}} \prod_a |\langle m_a|n_a\rangle|^2 k'_{nr}(s m_a \rightarrow l n_a), \tag{2.302}
$$

which weights all possible unobserved final optical mode states with the appropriate Franck-Condon factors.

From the empirical successes of the energy gap law, it is clear that the approximate evaluation of (2.301) by the use of the Boltzmann method or the more general Kubo-Toyozawa theory is sufficient to represent this partial decay rate. The only approximation employed in calculating the harmonic decay rate (2.299) then involves the use of an approximate partial non-radiative decay rate (2.301) for a vibrationless molecular fragment. This is an approximation that need only be invoked once as opposed to the necessity of determining a new effective energy-gap law for each initial χ_{si}.

For the case of no frequency changes, the Boltzmann method, described above, gives a good representation of the exponential part of the partial decay rate as in (2.298) with a superscript prime replacing the zero. The above analysis shows that a poor accepting mode has a negligible effect on γ', so the exponential part of the rate involving $\gamma' \Delta E(m_a, n_a)'/\hbar \omega_M$ is independent of the poor acceptors. It is shown in Section 2.8 how this remains in effect when frequency changes in the poor acceptors are also incorporated. Thus, the only remaining influence of the poor accepting modes lies in the factor $\exp\left(-\sum_{j \neq a} d_j X_j\right)$ and the pre-exponential A of (2.298). The former obviously cancels in the evaluation of relative decay rates, while the latter is shown in Section 2.8 to be relatively insensitive to m_a and n_a, so it too cancels out when relative rates are taken.

In conclusion, only the good accepting modes and the optical modes contribute to the relative non-radiative decay rates. The other modes help to provide the quasi-continuous density of states that make the decay irreversible, and they considerably affect absolute rates and absolute isotope effects, but their efficient cancellation in the relative decay rates is an important simplification in view of our lack of precise knowledge concerning all the modes. The good acceptors, apart from those contained in the optical modes, can be combined into some effective theoretical or empirical energy gap law for the remaining vibrationless level. Thus, the approach rests on our ability to accurately represent the energy gap law for the decay of the vibrationless level of an electronic state. If one of the optical modes has considerable anharmonicity, there is no difficulty in using the anharmonic Franck-Condon factors in (2.302) [2.28]. When one of the optical modes is a good accepting mode, the summation in (2.302) involves situations where the effective energy gap $\Delta E(m_a, n_a)$ tends towards zero—it can, of course, never be negative. It is, therefore, necessary to obtain accurate representations of the energy gap law for k'_{nr} which are valid in the limit of zero energy gap. This generalized energy gap law has indeed been derived by asymptotic analysis [2.28, 90]. For the case in which frequency changes are ignored, the final result is just [2.90]

$$k'_{nr} \propto X_M^{n_M^0}/\Gamma(n_M^0 + 1) \tag{2.303}$$

with

$$n_M^0 = \Delta E(M_a, n_a)/\hbar \omega_M . \tag{2.304}$$

It is gratifying to see that the result of the lengthy analysis just produces the Franck-Condon factor for non-integral n_M^0 that would be guessed by intuition from the Boltzmann statistics method. Eqs. (2.303) and (2.304) have the correct energy gap law behavior (2.277) and (2.278) when n_M^0 is large, but they also reproduce the correct behavior as $n_M^0 \to 0$. For cases involving frequency changes, the asymptotic analysis also leads to the analytically continuation of the original Franck-Condon factor to non-integral final quantum numbers [2.28].

The above discussion centers around an elementary derivation of the fundamental energy gap law and its subsequent use to generate accurate means for evaluating the relative non-radiative decay rates from individual vibronic levels. The next subsection contains the treatment of this energy gap law by the general Kubo-Toyozawa methods which are necessary to adequately incorporate frequency changes in the modes and the pre-exponential factors. The reader who is uninterested in the mathematical details involved may skip to Section 2.9 where a comparison between theoretical and experimental results of single vibronic level decay rates is given. For these readers, we can now quote the results, analogous to (2.284), determining criteria for when modes with frequency changes are poor (or good) acceptors [2.29].

2.7.5 Criteria for Good Acceptors: Frequency Shifts

Let the comparison mode be the mode M of maximum frequency for which $d_M X_M$ is non-negligible. We determine the accepting ability of a non-totally symmetric mode α by observing how it competes with M for the available energy ΔE. The vibration α is characterized by the parameter

$$\xi_\alpha = (\omega_\alpha^s - \omega_\alpha^l)/(\omega_\alpha^s + \omega_\alpha^l) . \tag{2.305}$$

Assuming the zero-point energy shift is included in ΔE, the conservation equations can be obtained from Section 2.8 to read

$$\Delta E = \hbar \omega_M n_M^0 + \hbar \omega_\alpha^l n_\alpha^0 \tag{2.306}$$

with

$$n_M^0 = d_M X_M \exp(b\hbar\omega_M) \tag{2.307}$$

as before, and the theory in Section 2.8 yields the definition

$$n_\alpha^0 = \xi_\alpha^2 \exp(2b\hbar\omega_\alpha^l)[1 - \xi_\alpha^2 \exp(2b\hbar\omega_\alpha^l)]^{-1} . \tag{2.308}$$

Use of (2.280) in (2.306) leads to the equation

$$1 = f + (\hbar\omega_\alpha^l/\Delta E)\Omega_\alpha' f^{2p_\alpha}[1 - \Omega_\alpha' f^{2p_\alpha}]^{-1} , \tag{2.309}$$

with

$$\Omega_\alpha' = \xi_\alpha^2 (\Delta E/X_M d_M \hbar \omega_M)^{2p_\alpha} . \tag{2.310}$$

The condition that α be a poor accepting mode is

$$\Lambda_\alpha' = (\hbar\omega_\alpha^l/\Delta E)\Omega_\alpha' \ll 1 . \tag{2.311}$$

The condition (2.311) can also be shown to be equivalent to

$$\hbar \omega_\alpha^l n_\alpha^0 / \Delta E \ll 1 , \tag{2.312}$$

a more explicit representation of the poor accepting power of mode α. When (2.311) is violated, α becomes a good acceptor.

When a totally symmetric vibration has both geometry and frequency shifts, it is natural to expect that the criterion governing the accepting abilities of the mode must somehow resemble the individual cases (2.284) and (2.311) for modes with just geometry and frequency shifts, respectively. Indeed, the following criteria for a poor accepting mode,

$$\Lambda_\alpha', \Omega_\alpha (1 - \xi_\alpha)^2 \ll 1 , \tag{2.313}$$

can likewise be shown to hold [2.29]. Again violation of (2.313) is indicative of a good acceptor. For these cases of frequency shifts, accurate representations of the full energy gap law in forms analogous to (2.291–292), but with the pre-exponential, can be generated by the methods of Section 2.8.

2.8 The Generating Function Method

The general thermally averaged non-radiative decay rate (2.3) involves a double summation over all possible initial and final vibronic components of s and l. As in Sections 2.6 and 2.7, we continue to assume that a) The electronic coupling operator is an effective one, as in (2.205), which includes the effects of other electronic levels. b) Only inert media and non-overlapping resonances are treated. c) The zeroth-order wavefunctions are the Born-Oppenheimer function (2.252). Earlier sections discuss the crude factorization approximations to (2.3) and the more sophisticated Boltzmann method for the decay of the vibrationless level.

In this section we proceed in an entirely different fashion by recognizing that (2.3), in the harmonic approximation, is completely analogous to the formal expressions for the probability of nuclear recoil in Mössbauer spectroscopy[16] in solids [2.91] and to the transition probabilities in the theory of line shapes and zero-phonon lines in the optical spectra of solids [2.86, 87, 92]. In an extensive study of radiative and non-radiative processes in solids, KUBO and TOYOZAWA [2.86] have demonstrated that the full "golden rule" expression (2.3), containing the double sum over initial and final states, and involving any arbitrary operator, can be written exactly as a single definite (Fourier or Laplace) integral. In the context of non-radiative phenomena, the theory

---

[16] For a recent review, see U. GONSER (ed.) in: *Topics in Applied Physics*, Vol. 5: Mössbauer Spectroscopy (Springer, Berlin, Heidelberg, New York 1975).

has been applied to multi-phonon ionization processes in solids and also to absorption by *F*-centers (trapped electrons) and localized impurities [2.92]. Since a solid is just a large molecule, Lin and Bersohn [2.85b], and later Englman and Jortner [2.88], proposed that molecular non-radiative decay processes in the statistical limit be considered to be (intramolecular) multi-phonon processes in a large molecule. Lin and Bersohn analyzed the complete cumbersome rate expression and have discussed some of the implications of the "golden rule" rate as to the nature of the promoting and accepting modes and the temperature dependence and isotope effects in radiationless transitions. Lin has also demonstrated how the inclusion of the nuclear momentum operator gives rise to the propensity rule for the promoting modes [2.85a]. Englman and Jortner [2.88] use the many-phonon approach to the full rate expression (2.3) to derive approximate expressions for the non-radiative decay probability (2.3) which could be reduced to simple analytic forms for two limiting cases that are determined by the magnitude of a reduced vibrational coupling strength function (not to be confused with the electronic coupling strength) [2.93–95]. The strong coupling limit (which corresponds to the existence of substantial horizontal displacement of the multi-dimensional potential energy surfaces of the two electronic states) may be appropriate to some photochemical rearrangement processes and some radiationless transitions between charge transfer states (e.g. $A^+B^- \leftrightarrow AB$ type), while the weak coupling limit (for which the relative horizontal displacement of the two potential energy surfaces is small) is appropriate to the description of electronic relaxation processes in aromatic hydrocarbons and *possibly* some hetero-aromatics. Englman and Jortner [2.88] were able to show that the full rate expression (2.3) does in fact imply both the experimentally observed energy gap law for radiationless transition rates in aromatic hydrocarbons and the nature of the accepting modes. The studies of Lin and Bersohn [2.85b], and of Englman and Jortner [2.88] involved several approximations which were introduced to simplify the rather cumbersome theoretical expressions for the non-radiative decay probability in the harmonic approximation. Lin's formal expression recast the non-radiative decay rate in terms of a Fourier integral, and even these expressions involve some approximations concerning the role of frequency changes between the two electronic states. The equations of Englman and Jortner are physically transparent, but only the simple model system (involving just shifted identical potential surfaces and a single dominant accepting mode) was studied, leading to the over-simplified energy gap law (2.277–278). These authors also neglected the dependence of the rate(s) on the nuclear kinetic energy operator.

Freed and Jortner [2.81] use the generating function method of Kubo and Toyozawa [2.86] to derive more general expressions for the non-radiative rate in the harmonic approximation. As is shown in the following, these expressions can be considerably simplified for the displaced potentials model and also for a more general case of physical interest which involves both displacement of potential surfaces and changes in the vibrational frequencies between the two electronic states. In the strong coupling case the decay probability can be recast in terms of a traditional rate equation which involves a generalization of the Eyring pre-exponential factor. In the weak coupling limit the effect of the promoting modes appears explicitly in the energy gap law. The role of different intramolecular vibrational modes (characterized either by displacement of the origin or modification of the force constants for the potential surfaces) on the non-radiative decay probability is explicitly assessed.

Although the effects of the external host medium can readily be included by considering the guest molecule and the surrounding medium modes to form a single "super-molecule" [2.81], this topic is not treated further as it is covered in other chapters in this volume. Similarly, a detailed treatment of the examples of the strong coupling limit associated with activated rate processes is more appropriate to the chapters that specifically deal with condensed media[17]. Finally, the effects of vibrational relaxation processes on decay rates, in situations where the Boltzmann thermal distribution $p(si)$ are not maintained, is a very interesting one [2.31–34], but it is beyond our limited goal of treating the electronic relaxation processes in isolated polyatomic molecules. In fact, the thermal distribution $p(si)$ of (2.3) is maintained only for the sake of generality; applications pertain to the decay of single vibronic levels.

The generating function method is designed to compute the generalized density of states function of the form (2.3) (e. g., the density of states weighted by the matrix elements of an arbitrary operator). Thus, this technique can be used to calculate the line shape of optically allowed or forbidden transitions in large molecules, where the coupling operator v is a constant or is proportional to one of the oscillator displacements, respectively. Although we focus attention on the non-radiative decay probability, it should be borne in mind that these methods are quite general [2.96]. We give a brief summary of the method so as to introduce the notation used and to specify the approximations employed, separating those approximations which are necessary from those which are merely convenient. We attempt to keep the number of cumbersome equations to a minimum.

2.8.1 Generalized Lineshape Function and Generating Function

Consider the generalized line shape function

$$F(E) = Z^{-1} \sum_{i,j} |v_{si,lj}|^2 \exp(-\beta E_{si}) \delta(\Delta E + E_{lj} - E_{si} - E),$$

$$Z = \sum_i \exp(-\beta E_{si}),$$

(2.314)

---

[17] Furthermore, the strong coupling limit rate expressions can be shown to be a special limit of the general saddle point methods described below. It is, therefore, not separately discussed here.

where it is now convenient to have the energies of the vibronic levels in each electronic manifold (E_{si} and E_{lj}) be measured from the potential minimum of that manifold. Thus, the energy gap ΔE between the minima of the two electronic states is written as

$$\Delta E = E_{s0} - E_{l0} - \text{(zero point energy shift)}. \tag{2.315}$$

The transition probability (2.315) is obviously

$$k_{nr}(s \to l; \beta) = (2\pi/\hbar) F(0). \tag{2.316}$$

Rather than use (2.314), we consider its Fourier transform

$$f(t) = \int_{-\infty}^{\infty} \exp[i(E - \Delta E)t/\hbar] F(E) dE, \tag{2.317}$$

which can be expressed in the form

$$f(t) = Z^{-1} \sum_{i,j} v_{si,lj} \exp(i E_{lj} t/\hbar) v_{lj,si} \exp(-i E_{si} \tau/\hbar), \tag{2.318}$$

where

$$\tau = t - i\hbar\beta. \tag{2.319}$$

As is shown below, it is considerably easier to evaluate $f(t)$ than $F(E)$. From $f(t)$ the line shape function is obtained by the inverse transformation

$$F(E) = (2\pi\hbar)^{-1} \int_{-\infty}^{\infty} f(t) \exp[-i(E - \Delta E)t/\hbar] dt. \tag{2.320}$$

Thus the transition probability is

$$k_{nr}(s \to l; \beta) = \hbar^{-2} \int_{-\infty}^{\infty} f(t) \exp(-i\Delta E t/\hbar) dt. \tag{2.321}$$

In order to derive the desired closed form for the generating function (2.318), it is recast in terms of the coupling operators

$$v_{sl}(Q) = \int dq\, \phi_s(q, Q) v(q, Q) \phi_l(q, Q) \tag{2.322}$$

(where Q implies *all the nuclear coordinates*) and the time dependent sum-over-states-functions for nuclear motion in the electronic states s and l, which are

$$G_s(Q^s, \tau; \bar{Q}^s, 0) = \sum_i \chi_{si}(Q^s) \chi_{si}^*(\bar{Q}^s) \exp(-i E_{si} \tau/\hbar) \tag{2.323a}$$

and

$$G_l(\bar{\boldsymbol{Q}}^l, -t; \boldsymbol{Q}^l, 0) \equiv G_l^*(\boldsymbol{Q}^l, t; \bar{\boldsymbol{Q}}^l, 0)$$
$$= \sum_j \chi_{lj}(\bar{\boldsymbol{Q}}^l) \chi_{lj}^*(\boldsymbol{Q}^l) \exp(\mathrm{i}\, E_{lj} t/\hbar)\,, \tag{2.323b}$$

respectively. The G_s and G_l are the time-dependent Green's functions for nuclear motion in s and l, respectively. Eqs. (2.322) and (2.323) enable (2.317) to be written as a set of definite integrals over the nuclear coordinates, viz,

$$f(t) = Z^{-1} \int\int d\boldsymbol{Q}\, d\bar{\boldsymbol{Q}} [v_{sl}(\boldsymbol{Q})\, G_l(\bar{\boldsymbol{Q}}^l, -t; \boldsymbol{Q}^l, 0)]$$
$$\times [v_{ls}(\bar{\boldsymbol{Q}})\, G_s(\boldsymbol{Q}^s, \tau; \bar{\boldsymbol{Q}}^s, 0)]\,. \tag{2.324}$$

The result (2.324) effectively gets rid of the nasty sum of (2.317) provided a) the sums-over-states (2.323) can be expressed in closed form, and b) the Q-integrals in (2.324) can exactly be evaluated. The important point is that useful models satisfying these two criteria are available, and for these cases the rate (2.321) can be written as a single definite integral with a known function $f(t)$ from (2.324). We now consider in more detail the evaluation of the closed form expression for $f(t)$ from (2.324).

To our knowledge, the only *useful* case for which a *time-dependent* oscillator Green's function (2.323) is known in closed form is for the harmonic oscillator. As discussed in Section 2.7, this harmonic model should be adequate for treatments of relative decay rates. However, it would be desirable if anharmonic oscillator Green's functions could likewise be employed. MIKAMI et al. [2.133] have introduced anharmonicity by the use of harmonic oscillator Green's functions with anharmonicity included through first order of perturbation theory. Diagonal anharmonicities are considered, and these are found to increase calculated rates by an average of about two orders of magnitude in conformity with calculations of BURLAND and ROBINSON [2.60].

Again, as in Sections 2.6 and 2.7, the formal result (2.324) is greatly simplified in the case that the vibrational motion in both electronic states can be expressed in terms of independent oscillators; these need not be harmonic, or have the same frequencies in both states. In this case the vibrational wave functions and energies of (2.323) are separable into products and sums, respectively, of the modes, so the Green's functions for each of the vibronic states are separable into products of Green's functions g_j^α for each of the vibrational modes j in both of the electronic states $\alpha = l$ or s

$$G_\alpha(\boldsymbol{Q}^\alpha, T; \bar{\boldsymbol{Q}}^\alpha, 0) = \prod_{j=1}^{N} g_j^\alpha(Q_j^\alpha, T; \bar{Q}_j^\alpha, 0)\,. \tag{2.325}$$

In the case of dissociation processes, the above assumption of parallel modes in s and l is quite untenable; a proper analysis requires the use of the appropriate, and generally very different, normal modes of both the initial electronic state s and the electronic state l of the fragments [2.35]. The necessary generalized theory was given by BAND and FREED, and the interested reader is referred to their works for more details [2.35].

For the parallel mode approximation, the harmonic potential surfaces can be written in terms of the dimensionless coordinates q_j^s and q_j^l, the effective masses M_j, the normal mode angular frequencies ω_j^s and ω_j^l, and the dimensionless shift Δ_j of (2.257),

$$U_s = \tfrac{1}{2} \sum_{j=1}^{N} \hbar \omega_j^l \beta_j (q_j + \Delta_j)^2$$

$$U_l = \tfrac{1}{2} \sum_{j=1}^{N} \hbar \omega_j^l q_j^2 - \Delta E .$$

(2.326)

As is usual, q_j is expressed in terms of the original displacements Q_j^s and Q_j^l by

$$q_j^s = (M_j \omega_j^s / \hbar)^{1/2} Q_j^s = \beta_j^{1/2} (q_j^l + \Delta_j^l)$$

(2.327a)

$$q_j \equiv q_j^l = (M_j \omega_j^l / \hbar)^{1/2} Q_j^l$$

(2.327b)

where

$$\beta_j = \omega_j^s / \omega_j^l .$$

(2.328)

The well known one-dimensional harmonic oscillator Green's function appropriate to (2.365) is

$$g_j^\alpha(Q_j^\alpha, T; \bar{Q}_j^\alpha, 0) = [M_j \omega_j^\alpha / 2 \pi i \hbar \sin(\omega_j^\alpha T)]^{1/2}$$
$$\times \exp\{(i M_j \omega_j^\alpha / 4 \hbar) [(\bar{Q}_j^\alpha - Q_j^\alpha)^2 \cot(\omega_j^\alpha T/2)$$
$$- (\bar{Q}_j^\alpha + Q_j^\alpha)^2 \tan(\omega_j^\alpha T/2)]\} .$$

(2.329)

2.8.2 Vibronic Coupling Operators

It is still necessary to determine the vibronic coupling operators (2.322) in order to evaluate (2.324). As discussed already, the precise form of this coupling operator is not terribly important to the study of relative decays. For completeness, we briefly sketch the customary treatment. Even though higher-order terms in the expansion of (2.205) are not necessarily negligible, the lowest-order non-vanishing contributions are the only ones generally retained.

Following LIN [2.84], and BIXON and JORTNER [2.40] the coupling operator v_{sl} is written as

$$v_{sl}(\boldsymbol{Q}) = \sum_{k=1}^{p} C_{sl}^{k} i \hbar (M_k)^{-1/2} \partial/\partial Q_k^l . \tag{2.330}$$

For internal conversion the dominant term

$$C_{sl}^{k} \equiv J_{sl}^{k} = \hbar (M_k)^{-1/2} \langle \phi_s(\boldsymbol{q}, \boldsymbol{Q}^s) | i \, \partial/\partial Q_k^l | \phi_l(\boldsymbol{q}, \boldsymbol{Q}^l) \rangle \tag{2.331}$$

is retained. For the case of intersystem crossing in aromatic hydrocarbons, the coupling is generally taken as

$$C_{sl}^{k} = \sum_{m \neq l, s} \left[\frac{\langle \phi_s | H_{SO} | \phi_m \rangle J_{ml}^{k}}{U_s(\boldsymbol{Q}) - U_m(\boldsymbol{Q})} + \frac{J_{sm}^{k} \langle \phi_m | H_{SO} | \phi_l \rangle}{U_m(\boldsymbol{Q}) - U_l(\boldsymbol{Q})} \right], \tag{2.332}$$

where H_{SO} is the spin orbit coupling operator, and $U_s(\boldsymbol{Q}) - U_m(\boldsymbol{Q})$ or $U_l(\boldsymbol{Q}) - U_m(\boldsymbol{Q})$ is the difference between the potential surfaces of the spin-orbit coupled states. The summation over k in (2.330) is taken over the small number of promoting modes ($p \ll N$), for which the J^k terms are appreciable.

The coupling operator $C_{sl}^{k}(\boldsymbol{Q})$ is explicitly a function of the nuclear co-ordinates, but, as described in (2.253), it is generally more slowly varying than the products of vibrational functions $\chi_{si}(\boldsymbol{Q}) \chi_{lj}(\boldsymbol{Q})$ in the region where the latter are non-negligible. Consequently, $C_{sl}^{k}(\boldsymbol{Q})$ is removed from inside the integral and evaluated at the $\boldsymbol{Q}$-centroid $\boldsymbol{Q}^0$. Here $\boldsymbol{Q}^0$ specifies a nuclear configuration analogous to the familiar r-centroid of diatomic spectroscopy. However, in contrast to cases of radiative transitions, where both the initial and final vibronic states are observed (i.e., can be deduced), the final vibronic level in l is not singled out. Thus, it is much too cumbersome to define a different $\boldsymbol{Q}$-centroid for each possible final lj individually. Instead, the $\boldsymbol{Q}$-centroid is determined by the initial state, or distribution thereof, of the non-radiative decay process [2.23]. The $\boldsymbol{Q}$-centroid also accounts for the density of states weighted Franck-Condon character of the decay rate. In many cases $C_{sl}^{k}(\boldsymbol{Q})$ is very slowly varying with initial state si, so it can be taken to be a constant. For simplicity, we assume that there is only a single promoting mode, where-upon the constancy of $C_{sl}^{k}(\boldsymbol{Q}^0)$ assures that the electronic factor cancels in relative rates[18].

---

[18] When two or more promoting modes are present, the situation can be somewhat more compli-cated. However, when the promoting modes are similar and are not among the optical modes, the electronic factors still cancel to a good approximation. Since experimental work has yet to unambiguously assign a promoting mode, the complexities of multiple promoting modes are presently somewhat academic. They can, however, be readily incorporated by including pheno-menological parameters to describe the ratios of electronic factors for different promoting modes. Consequently only the single promoting mode rate expressions need be explicitly given here.

2.8.3 Generating Function Expressions for Decay Rates

Using the simplified single promoting mode assumption, substituting (2.330) and (2.325) into (2.324), and invoking the Q-centroid approximation leads to

$$f(t) = |C_{sl}^k| \, \tilde{f}_k(t) \prod_{j \neq k} f_j(t) \,, \tag{2.333}$$

where the $f_j(t)$ terms correspond to the single mode generating functions for optical absorption. These terms, which arise from Franck-Condon vibrational overlap factors, are given by

$$f_j(t) = Z_j^{-1} \iint dQ_j \, d\bar{Q}_j \, g_j^s(Q_j^s, \tau; \bar{Q}_j^s, 0) \, g_j^l(\bar{Q}_j^l, -t; Q_j^l, 0) \,, \tag{2.334}$$

where

$$Z_j = \sum_{n_j} \exp[-\beta \hbar \omega_j^s (n_j + \tfrac{1}{2})] = [2\sinh(\beta \hbar \omega_j^s / 2)]^{-1} \tag{2.335}$$

is the partition function. The $\tilde{f}_k(t)$ correspond to single mode generating functions which involve the nuclear momentum operator for the promoting mode Q_k.

$$\tilde{f}_k(t) = \hbar^2 (Z_k M_k)^{-1} \iint dQ_k \, d\bar{Q}_k \left[i \frac{\partial}{\partial Q_k} g_k^s(Q_k^s, \tau; \bar{Q}_k^s, 0) \right]$$
$$\times \left[i \frac{\partial}{\partial \bar{Q}_k} g_k^s(\bar{Q}_k^l, -t; Q_k^l, 0) \right]. \tag{2.336}$$

We have completed the formal representation of the non-radiative transition probability in the harmonic approximation by expressing it as the Fourier transform of a product of generating functions. Using (2.334–336) and the explicit form (2.325) of the Green's function, then, the calculation of the generating function is reduced to a straightforward although somewhat cumbersome evaluation of elementary integrals. The general results are only required here for particular limiting cases, and these limiting forms are quoted as necessary. We can now make contact with the results of Section 2.7 by considering the simple case of the displaced potential surface model.

We first invoke the simplifying assumption that the normal modes and their frequencies are the same in the two electronic states except for displacements in the origins of the normal coordinates. Thus, we can set $\beta_j = 1$ or $\omega_j^s = \omega_j^l = \omega_j$ for all the normal modes j. The single mode generating functions now take the simple form for $\Delta_k = 0$,

$$f_j(t) = \exp\{-X_j^2 [\coth y_j (1 - \cos \omega_j t) - i \sin \omega_j t]\} \tag{2.337}$$

$$\tilde{f}_k(t) = (\hbar \omega_k / 2) [\cos(\omega_k t) \coth y_k + i \sin \omega_k t] \,, \tag{2.338}$$

with $y_j \equiv \beta \hbar \omega_j^s/2$. The transition probability is then obtained by writing (2.333) as products of terms from (2.337) and (2.338) and then substituting this into the Fourier integral (2.321). The resulting equation for the transition probability is the appropriate generalization of the formula for the optical line shape, where the transition moment is now the nuclear momentum operator.

In order to introduce the weak and strong coupling limits it is convenient to collect the positive and negative exponentials that come from products of $f_j(t)$. We define

$$G_+(t) = \sum_{j=1}^{N} X_j(\bar{n}_j + 1)\exp(i\omega_j t) \tag{2.339a}$$

$$G_-(t) = \sum_{j=1}^{N} X_j \bar{n}_j \exp(-i\omega_j t), \tag{2.339b}$$

where n_j corresponds to the number of excited vibrations with frequency ω_j at thermal equilibrium

$$\bar{n}_j = [\exp(\beta \hbar \omega_j) - 1]^{-1}, \tag{2.340}$$

and $G_\pm$ are often called phonon generating functions. The dimensionless quantity

$$G = G_+(0) + G_-(0) = \sum_{j=1}^{N} X_j^2(2\bar{n}_j + 1) \tag{2.341}$$

is often referred to as the phonon coupling strength. The final expression for the non-radiative decay rate in the simple displaced potential surface model is then

$$k_{nr}(s \to l; \beta) = |C_{sl}^k|^2(\omega_k/4\hbar)\exp(-G)$$

$$\times \left\{ (\coth y_k + 1) \int_{-\infty}^{\infty} dt \exp[-i(\Delta E/\hbar - \omega_k)t + G_+(t) + G_-(t)] \right.$$

$$\left. + (\coth y_k - 1) \int_{-\infty}^{\infty} dt \exp[-i(\Delta E/\hbar + \omega_k)t + G_+(t) + G_-(t)] \right\}, \tag{2.342}$$

where we see that the net effect of the promoting mode is to decrease or increase the effective energy gap to $\Delta E \pm \hbar \omega_k$ with the temperature dependence of $\coth y_k \pm 1$. This propensity rule for the promoting mode follows naturally from the assumption that $\Delta_k = 0$ [2.29].

In the limit of zero temperature (e. g., $\beta\hbar\omega_j \gg 1$ for all j) this result is reduced to the much simpler form

$$
\begin{aligned}
k_{nr}(s0 \rightarrow l) = |C_{sl}^k|^2 \, (\omega_k/2\hbar) \exp\left(-\sum_{j=1}^{N} X_j\right) \\
\times \int_{-\infty}^{\infty} dt \exp\left[-i(\Delta E/\hbar - \omega_k)t + \sum_{j=1}^{N} X_j \exp(i\omega_j t)\right],
\end{aligned}
\tag{2.343}
$$

where the effective energy gap is now only $E - \hbar\omega_k$, since the promoting mode which has no quanta at zero temperature must gain one quantum. Eq. (2.343) contains only the positive frequency parts $\exp(i\omega_j t)$ since the vibrationless, zero temperature state has no quanta which must be distributed in the electronic relaxation process.

Eqs. (2.342) and (2.343) involve an extension of the treatment of ENGLMAN and JORTNER [2.88] since the effect of the promoting modes is now explicitly included in conformity with the work of LIN [2.85a].

In order to get some further insight into the nature of the different features of electronic relaxation processes and possibly some photochemical rearrangement reactions, it is useful to consider some limiting cases which are determined by the magnitude of the coupling strength, (2.341). Define the molecular rearrangement energy in the excited state

$$
E_M = \sum_{j=1}^{N} X_j \hbar\omega_j,
\tag{2.344}
$$

which corresponds just to half of the Stokes shift for the two electronic states under consideration. For the model, which just involves displaced potential surfaces, the various coupling limits can be defined as follows:

a) In the strong coupling limit

$$
G \gg 1,
\tag{2.345}
$$

and therefore the displacements of the origins between the two electronic states considerably exceed the root mean square vibrational displacements, $(\hbar/M_j\omega_j)^{1/2}$, for at least some of the normal modes. In other words, in this case $E_M \gg \hbar\langle\omega\rangle$ (where $\langle\omega\rangle$ is the mean vibrational frequency), so that the molecular rearrangement energy considerably exceeds the mean vibrational energy and the relative displacement of the potential energy surfaces is large.

b) The weak coupling limit, which is generally found for internal conversion and intersystem crossing in aromatic hydrocarbons, is encountered when $G \leq 1$ or $E_M \approx \hbar\langle\omega\rangle$, so the relative displacement for each normal mode is small.

2.8.4 Displaced Potential Surface Model at Low Temperatures

We now focus attention on the weak coupling limit and concentrate on the low temperature case, (2.343). The case of frequency shifts is considered afterwards, and the general saddle point method is also applicable to the strong coupling limit, so both weak and strong coupling cases are particular limits of the general results derived below. The simple model of the displaced potential surfaces is of considerable interest as it enables us to make contact with the Boltzmann method and to develop some general theoretical methods which are also useful for the more complicated case of the complete treatment of the "harmonic molecule". Furthermore, this simple model is useful in gaining some insight concerning the pertinent molecular parameters which determine the non-radiative transition probability.

The reduced displacements Δ_j between different $\pi \to \pi^*$ excited states of aromatic hydrocarbons have been estimated by BYRNE et al. [2.13], and it is now well established that electronic relaxation processes in these molecules correspond to the weak coupling situation. It is useful to adopt the procedure suggested by BYRNE et al. [2.13] and to combine the normal molecular vibrations into several ($\alpha = 1, \ldots, n$) groups which are:

1) C—H (or C—D) stretches, 2) skeletal stretches, 3) skeletal bends, 4) C—H (or C—D) bends, and 5) out of plane modes. This classification was recently extended by BURLAND and ROBINSON [2.60] who provided extensive spectroscopic data obtained for the first excited singlet and triplet states and for the ground state of the benzene molecule. The vibrational frequencies within each group are taken to be equal (say ω_α for the α-th group). Let d_α be the number of modes within the α-th group (i.e., the degeneracy number), so $\sum_{\alpha=1}^{n} d_\alpha = N$. Obviously the reduced displacements within each group can be different, as each contains vibrational normal modes of different symmetry. It will be convenient to define a mean reduced displacement Δ_α for the α-th group,

$$2 d_\alpha X_\alpha \equiv d_\alpha \Delta_\alpha^2 = \sum_{j \in \alpha} \Delta_j^2 , \qquad (2.346)$$

where the summation of (2.346) is taken over the reduced displacements of the nearly degenerate modes in this group and

$$\sum_{j=1}^{N} \hbar \omega_j X_j = \sum_{\alpha=1}^{n} d_\alpha \hbar \omega_\alpha X_\alpha . \qquad (2.347)$$

The integrals appearing in (2.343), determining the transition probability, have the form

$$I_k = \int_{-\infty}^{\infty} dt \exp\left[-i\Delta E_k t/\hbar + \sum_{\alpha=1}^{n} X_\alpha \exp(i\omega_\alpha t) \right]$$

$$\equiv \int_{-\infty}^{\infty} dt \exp[A(t)] , \tag{2.348}$$

where the effective energy gap is

$$\Delta E_k = \Delta E - \hbar\omega_k . \tag{2.349}$$

2.8.5 Saddle Point Approximation and General Rate Expressions

The integral I_k can be evaluated for large values of ΔE_k by saddle point integration (see below). The saddle point t_0 in the complex t-plane is obtained from the relation

$$-\Delta E_k + \sum_{\alpha=1}^{n} d_\alpha X_\alpha \hbar \omega_\alpha \exp(i\omega_\alpha t_0) = 0 , \tag{2.350}$$

where X_α was defined in (2.257). Note that (2.350) is identical to (2.279) if we make the association

$$it_0 \leftrightarrow b \tag{2.351}$$

and realize that the promoting mode has reduced the available energy from ΔE to ΔE_k. Thus, the generating function method for the displaced potential surface model reduces to the same energy partitioning relationship as the Boltzmann method. However, as (2.350) involves complex variables, there are, in general, an infinite number of solutions of (2.350) for t_0. It should be anticipated, however, that only the single solution to (2.279) is relevant as the Boltzmann method must give the proper energy partitioning in the limit of large ΔE. NITZAN and JORTNER [2.97] have examined the contributions from all of these solutions for t_0, and they have verified that retention of the one, corresponding to the solution of the Boltzmann method, does, indeed, yield the correct result. It should also be noted that the counterpart of the weak energy conservation constraint in the Boltzmann method is an implicit damping factor in all t-integrands (2.321, 342, 343, 348) which account for the effective continuous density of states in the l-manifold for timescales of relevance to the decay of ϕ_s.

The generating function method differs from the Boltzmann method because the former also leads to an explicit representation for the preexponential factor. From (2.348) and (2.350) or (2.279), it can readily be shown that

$$\left. \frac{d^2 A(t)}{dt^2} \right|_{t=t_0} < 0 , \tag{2.352}$$

and consequently the saddle-point approximation to (2.348) is

$$I_k \cong \left[-2\pi \Big/ \frac{d^2 A(t)}{dt^2} \Big|_{t=t^0} \right]^{1/2} \exp[A(t_0)].$$

(2.353)

Substituting (2.350), (2.351), and (2.279) into (2.353) gives

$$I_k = \left[\omega_M \Delta E_k f_k \left(1 + \sum_{\alpha \neq M} \Omega_{\alpha,k} P_\alpha f_\alpha^{p_\alpha - 1} \right) / 2\pi \right]^{1/2} \exp(-\gamma_k' \Delta E_k / \hbar \omega_M),$$

(2.354)

where γ_k' is simply obtained from (2.292) by just appending the subscript k to ΔE and Ω_α to convert them to ΔE_k and $\Omega_{\alpha,k}$, the latter being defined by performing the same transformation on (2.282), i.e.,

$$\Omega_{\alpha,k} = \frac{d_\alpha X_\alpha \hbar \omega_\alpha}{\Delta E_k} \left(\frac{\Delta E_k}{d_M X_M \hbar \omega_M} \right)^{p_\alpha},$$

(2.355)

etc. For small $\Omega_{\alpha,k}$, as in (2.284), the simple approximations (2.288, 289), etc. (with subscripts k appended) can be used in (2.354) to provide a more explicit solution.

Reintroducing all of the factors necessary to convert (2.348) to (2.343) gives the result

$$k_{nr}(s0 \to l) = |C_{sl}^k|^2 \omega_k \exp(-G) \left[\omega_M \Delta E_k f_k \left(1 + \sum_{\alpha \neq M} \Omega_{\alpha,k} \right. \right.$$

$$\left. \left. \times P_\alpha f_k^{p_\alpha - 1} \right) / 2\pi \right]^{-1/2} \exp(-\gamma_k' \Delta E_k / \hbar \omega_M)$$

(2.356)

as the general energy-gap law for the displaced potential model. Eq. (2.356) is only valid in the limit of large energy gaps

$$\Delta E_k > \hbar \omega_M$$

(2.357)

a condition for the existence of a high enough density of l-vibronic levels for the statistical limit to ensue. When (2.357) is violated, alternative forms of the energy gap law must be employed. As noted in Section 2.7, this alternative form has been derived for the single frequency approximation where the remaining modes have negligible accepting power. For this case, (2.348) reduces to [2.90]

$$I_k = (2\pi/\omega_M) X_M^{\Delta E_k / \hbar \omega_M} / \Gamma[1 + (\Delta E_k / \hbar \omega_M)],$$

(2.358)

a result which can often prove useful in evaluation of relative decay rates from single vibronic levels.

The power of the generating function method lies in the fact that it can also incorporate the effects of non-zero vibrational frequency changes in the

$s \rightarrow l$ transition as well as treating the general case of non-zero temperature. Needless to say, the algebra in the general case gets somewhat messy; however, the basic content remains the same as in the displaced potential model. Rather than getting enmeshed in all these details, we merely note some of the basic results. For zero temperature the generating function with frequency changes becomes

$$f_j(t) = [2(\omega_j^l \omega_j^s)^{1/2}/(\omega_j^s + \omega_j^l)] \exp[-i(\omega_j^s - \omega_j^l)t/2][1 - \xi_j^2 \exp(2i\omega^l t)]^{-1/2}$$

$$\times \exp\{-(1-\xi_j)X_j[1 - \exp(i\omega_j^l t)]/[1 + \xi_j \exp(i\omega_j^l t)]\}. \qquad (2.359)$$

Earlier work assumed that for small ξ_j series expansions in ξ_j may be employed. Later numerical work showed that this approximation is hazardous if applied directly to (2.359), so it should be reserved for the final rate expressions where applicable.

When (2.279) is substituted into (2.333) and that result into (2.317), the factors of $\exp[-i(\omega_j^s - \omega_j^l)t/2]$ in (2.359) can be combined with the energy gap ΔE to produce the new effective energy gap

$$\Delta E_k' = \Delta E_k + \tfrac{1}{2} \sum_{j=1}^{N} \hbar(\omega_j^s - \omega_j^l) \qquad (2.360)$$

which simply states that the available energy must now be corrected for the zero-point energy difference in the $s \rightarrow l$ transition.

The time integrals can again be evaluated by saddle point integration. For the simple model case of two modes M and α, where $X_M \neq 0$, $\xi_M = 0$ and $X_\alpha = 0$, $\xi_\alpha \neq 0$, the saddle-point equations just become those quoted in (2.306–308) after use of the substitution (2.351). (Of course, ΔE in (2.306) should be replaced by $\Delta E_k'$.) The case of many non-totally symmetric modes follows upon introduction of a summation sign $\left(\sum_\alpha\right)$ before $\hbar\omega_\alpha^l n_\alpha^0$ in (2.306), and the more general case of $X_\alpha \neq 0$, $\xi_\alpha \neq 0$ can likewise be easily treated. When the dominant mode M is taken to have $X_M \neq 0$ but $\xi_M = 0$, for simplicity, the rate is again expressible as an energy-gap law as in (2.291) and (2.356) with a different γ_k' and pre-exponential factor. Rather than displaying the cumbersome formulas, we note that for small frequency changes, $|\xi_\alpha| \ll 1$, the change in γ' due to non-totally symmetric modes (i.e., $X_\alpha \equiv 0$) is of order ξ_α^2 and can safely be ignored. Thus, these modes only affect the pre-exponential factor. When M is the dominant mode, so the simple approximation (2.285) can be invoked, this extra prefactor is primarily

$$\prod_\alpha \{1 - \Omega_{\alpha,k}' + \mathcal{O}[(\Omega_{\alpha,k}')^2]\}^{-1/2}, \qquad (2.361)$$

with $\Omega'_{\alpha,k}$ defined in (2.310) when $\Delta E'_k$ is substituted for ΔE. The extra factor (2.361) can have a considerable effect on absolute rates. For instance, assigning average values of ≈ 1.2 to $(1 - \Omega'_{\alpha,k})^{-1/2}$ leads to a value of (2.361) of 165 for 28 non-totally symmetric vibrations in benzene. No wonder that neglect of these modes, anharmonicities, and nuclear coordinate dependences of electronic matrix elements leads to predicted rates which are too low by several orders of magnitude! Likewise, a change in $(1 - \Omega'_{\alpha,k})^{-1/2}$ of an average of 10% on deuteration for, say, half of the modes in benzene leads to an isotope effect in (2.361) of a factor of 0.21. Consequently, the neglect of these poor accepting modes leads to absolute isotope effects which have considerable errors.

The case relative decay rates, however, is greatly simplified. The partitioning technique of (2.302) can be employed for each of the individual decay rates. Poor accepting non-totally symmetric modes α, which are not optical modes, enter into (2.302) only through the pre-exponential factor (2.361) with the effective energy gap

$$\Delta E'_k(m_a, n_a) = \Delta E(m_a, n_a) - \hbar \omega_k \tag{2.362}$$

replacing ΔE. Since $\Omega'_{\alpha,k}$ of (2.311) is slowly varying with moderate changes of $\Delta E_k(m_a, n_a)$, the prefactor (2.361) can be removed from the summation in (2.302), evaluating it at, say, $\Delta E'_k$. Consequently, when ratios of (2.302) are taken to obtain relative rates, the effects of these non-totally symmetric poor acceptors cancels to a good approximation. Hence, even though these modes are important for the establishment of the statistical limit, they can be neglected in evaluating the relative rates. The relative rates are then only a function of the dominant accepting modes and the optical modes in question. This enormous simplification is, in part, what makes analysis of the relative rates possible.

In this section we have considered the general generating function method for evaluating the non-radiative decay rate in order to provide the required pre-exponential factors A in (2.298) and to introduce the frequency changes in the modes. The latter is only briefly sketched herein as the interested reader can consult the original references for more of the details and for generalizations of the small energy gap forms (2.358). We now have considered sufficient material to return to the problem of the relative rates, the energy dependence of electronic relaxation phenomena.

2.9 Single Vibronic Level Decay Rates

The preceding two sections consider the general theory of the evaluation of the density-of-states weighted Franck-Condon factor by both the Boltzmann and generating function methods. These approaches lead to generalized energy-gap law expressions for the non-radiative decay rate of vibrationless molecules. Although our principal interest involves molecules with vibrational excitation, the general partitioning method (2.299–302) enables the separation of the excited modes (and any other special vibrations) from the remainder of the vibrations that are initially in their respective ground states [2.4, 25–29]. The generalized energy gap law is then employed for these remaining modes as if they were a hypothetical molecule which initially is in its vibrationless state. Thus, for any given molecule, the generalized energy gap law need only be determined once and for all. Very often the saddle point equations (for t_0 or b) can be solved analytically, leading to a rather simplified analysis. In other cases, the saddle point equations resulting from W of (2.264), or more generally I_k of (2.348) and its modifications for frequency changes, must be solved numerically to yield a single function $I_k[\varDelta E_k(m_a, n_a)]$ of a single variable $\varDelta E_k(m_a, n_a)$.

Since only relative rates are considered, it is permissible to drop all constant factors in the general rate expression which are independent of vibronic levels. Thus, the general non-radiative decay rate is written as[19]

$$k_{nr}(sm_a \to l) \propto \sum_{\{n_a\}} \left[\prod_a |\langle m_a | n_a \rangle|^2 \right] I_k[\varDelta E_k(m_a, n_a)]. \tag{2.363}$$

The modes $\{a\}$ encompass all vibrations which may be optically excited. They may also include any others which have large anharmonicities, whereupon the anharmonic Franck-Condon factors $|\langle m_a | n_a \rangle|^2$ are to be employed [2.134]. Likewise, when the promoting mode has frequency or geometry changes, we are free to either include it among the optical modes $\{a\}$ or to obtain appropriate generalized energy-gap law expressions which incorporate these features. Lastly, the modes $\{a\}$ must contain any vibrations for which the harmonic description breaks down. Examples of this involve the torsional vibration in systems that can undergo cis-trans isomerization [2.90] and the inversion vibrations in molecules like aniline [2.98] (the analog of the ammonia inversion).

2.9.1 Assumptions of the Model

The general assumptions inherent in (2.363) are associated with the neglect of any m_a-dependence of the electronic matrix elements, the assumption of parallel modes, and the use of the single promoting mode hypothesis. The

---

[19] For notational convenience, the prime on E_k is dropped, so it refers to (2.360) or (2.349) as the particular situation dictates.

latter is easily rectified by taking $\sum_k w_k$ of (2.363) to weight the contributions from various promoting modes. If, say, there are two promoting modes k and k' for which $\omega_k \approx \omega_{k'}$, (and even $\xi_k \approx \xi_{k'}$, $X_k \approx X_{k'}$) then we automatically have

$$I_k[\Delta E_k(m_a, n_a)] \approx I_{k'}[\Delta E_{k'}(m_a, n_a)], \quad k \approx k', \tag{2.364}$$

when k and k' are not amongst the optical modes $\{a\}$. Eq. (2.364) results from the fact that I_k depends on mode k only through ΔE_k (and ξ_k, X_k) which is taken in this example to be the same for the two modes. When (2.404) is satisfied for the promoting modes, the general result of taking the weighted sum over the promoting modes is to multiply (2.363) by an overall factor of $\sum_k w_k$ which is independent of vibronic level and can be dropped. Hence, the final rate expression can be written as if there were a single promoting mode. If, on the other hand, one of the promoting modes were optically excited or if there were promoting modes of sufficiently disparate frequencies that (2.364) were violated, it would be necessary to retain the weighted sum over all promoting modes. This feature must constantly be remembered in the analysis of experiments as simple theoretical arguments alone are not sufficient to unambiguously assign promoting modes. Indeed, even with considerable experimental data, no promoting mode has yet been determined. The multiplicity of possible promoting modes is not necessarily a serious problem as one might dominate over the others. For instance, if $\omega_k \not\approx \omega_{k'}$ (and/or $\xi_k \not\approx \xi_{k'}$, $X_k \not\approx X_{k'}$), the generalized energy gap laws can usually be expressed in the form

$$I_k[\Delta E_k(m_a, n_a)] \approx \Gamma_{kk'} I_{k'}[\Delta E_{k'}(m_a, n_a)] \tag{2.365}$$

with $\Gamma_{kk'}$ independent of, or slowly varying with, m_a and n_a. Thus, if

$$w_k \Gamma_{kk'} \gg w_{k'}, \tag{2.366}$$

the effects of promoting mode k' can be ignored with respect to those from k.

The other assumptions inherent in (2.363) all center about the validity of the energy-gap law itself. This law has been experimentally verified for $T_1 \rightarrow S_0$ and $S_1 \rightarrow S_0$ transitions in aromatic hydrocarbons [2.82, 132], for $S_2 \rightarrow S_1$ relaxation in substituted azulenes [2.83], and for electronic relaxation of various rare earth ions in $LaCl_3$ [2.84]. For a particular system, an empirical energy gap law, with a few adjustable parameters, can be employed for I_k in (2.363) if there is sufficient experimental data to overdetermine them. Alternatively, and more generally, theoretical analyses of increasing degree of sophistication can be employed for I_k. A useful avenue of pursuit would be the development of forms of I_k that include anharmonicities in some of the modes.

A proper theory requires the use of time-dependent anharmonic oscillator Green's functions (2.325), but this has yet to be accomplished[20].

As discussed in Section 2.8, an important aspect of (2.363) is the fact that, when relative rates are considered, the effects of the poor accepting modes (not in $\{a\}$) essentially cancel out. Symbolically, we can express this by writing the full I_k as

$$I_k[\Delta E_k(m_a, n_a)] \simeq \lambda_k \tilde{I}_k[\Delta E_k(m_a, n_a)] \tag{2.367}$$

where I_k is the generalized energy gap law for only on the dominant accepting modes. The factor λ_k is sufficiently weakly dependent on $\Delta E_k(m_a, n_a)$ that it can be removed outside the summation over $\{n_a\}$ and, hence, eliminated as a constant overall factor when forming relative rates. The determination of the dominant modes employs the criteria presented in Sections 2.7 and 2.8. These criteria involve knowledge of ξ_α and X_α which are often rather difficult to obtain. This is one reason why a good deal of experimental and theoretical study has been devoted to benzene and substituted benzenes as in this case there is a good deal of data available [2.26, 60, 100–105] in comparison with other possible candidates. It is highly desirable to generate more spectroscopic data for molecules amenable to single vibronic level studies (even benzene!) in order to facilitiate comparisons between theory and experiment.

For completeness, it is perhaps useful to quote the vibrational overlap for the general oscillator with ξ and X both non-zero. This can be shown to be

$$\langle m | n \rangle = N_m N_n \exp[-X f/(1+f)]$$
$$\times \sum_{l=0} \frac{(m!)(n!)}{(l!)(m-l)!(n-l)!} \left(\frac{4 f^{1/2}}{1+f} \right)^{l+1/2} \left(\frac{|1-f|}{1+f} \right)^{(m+n)/2-l} \tag{2.367a}$$
$$\times F_{m-l} \left[f \left(\frac{2X}{|1-f^2|} \right)^{1/2} \right] H_{n-l} \left[\left(\frac{2Xf}{|1-f^2|} \right)^{1/2} \right], \quad f < 1,$$

---

[20] In this regard, it may be noted that Fischer and co-workers [2.94, 95] have employed an anharmonic theory; however, it suffers from serious deficiencies as only cubic anharmonicities are retained [2.27]. Fischer's rate expression contains the anharmonic coupling constants in an exponential, so an expansion of this exponential would produce an infinite power series expansion in these anharmonic couplings, implying that the rate is an approximation to the exact cubic anharmonic decay rate. Unfortunately, the cubic anharmonic oscillator has no discrete energy levels, only a continuum of energy eigenstates that range between $-\infty \leq E \leq \infty$, so it is not clear what the meaning is of Fischer's approximation to this non-physical problem. Perhaps, this unphysical nature of the exact cubic anharmonic oscillation is responsible for Fischer's assignment of a large fraction of the available energy ΔE_k to anharmonic couplings [2.94, 95] Although the exact cubic anharmonic oscillator cannot represent normal vibrations, the customary procedure of retaining only the lowest orders in the cubic perturbation does lead to physically reasonable results [2.99, 133]. Mikami et al. [2.133] have also noted that Fisher's rate expression incorrectly is independent of the sign of the displacement of the equilibrium position of the oscillator between the two electronic states.

where

$$f = \omega^l/\omega^s$$
$$N_m = (2^{m+1/2}/m!)^{-1/2}$$
$$F_n(y) \equiv i^n H_n(y) = -2y F_{n-1}(y) + 2(n-1) F_{n-2}(y),$$

with $H_n(y)$ the n-th Hermite polynomial. When $f > 1$, Eq. (2.367a) is modified by the interchange $F_{m-l} \to H_{m-l}$ and $H_{n-l} \to F_{n-l}$. For $f = 1$, (2.367a) can be reduced to the well known limiting form

$$|\langle m|n\rangle|^2 = \exp(-X)(m!/n!) X^{n-m} [L_m^{n-m}(X)]^2, \qquad n > m, \qquad (2.368)$$

where L_m^{n-m} is the generalized Laguerre polynomial.

2.9.2 Experimental Problems

Before considering the detailed comparison of theory and experiment, it is useful to discuss some of the experimental problems associated with the measurement of single vibronic level decay rates. In order to be able to measure single vibronic level decay rates, it must first be necessary to be able to excite individual vibronic levels [2.101, 102]. This condition already limits experiments to the regions of lower vibrational excitation where the density of levels χ_{si} is low enough to permit spectral assignments. (In many instances the vibrational analysis of the absorption spectrum is not available, and it must therefore precede any treatment of the decay rates.) Because vibronic transitions occur as bands involving changes in rotational quantum numbers, the rotational contours of the vibronic bands are often up to 10–$30\,\mathrm{cm}^{-1}$ in width. Consequently, just the problem of overlapping bands places an upper limit to the vibrational excitation that can be probed in single vibronic level experiments. In practice, this implies that the bandpass of the exciting light must be $\lesssim 1\,\text{Å}$. Were this the only source of spectral congestion, matters would be reasonably favorable in many polyatomic molecules; however, it is also necessary to contend with the phenomena of sequence congestion [2.106, 107]. Sequence congestion arises because of the thermal distribution of vibronic levels in the ground electronic state ϕ_0 and because of low frequency vibrations which have very small ξ_α in the $\phi_0 \to \phi_s$ transition. If $\delta\omega_\alpha$ is the frequency shift, then associated with each fundamental $\phi_0 \to \phi_s$ vibronic band is a set of sequence bands shifted by $n\delta\omega_\alpha$ which have an intensity proportional to $\exp(-\beta\hbar\omega_\alpha)$ and arise from transitions from levels in ϕ_0 having n quanta in mode α. For instance, in naphthalene there are sequences with $\delta\omega_\alpha/c$ of -10 and $-55\,\mathrm{cm}^{-1}$, that arise from vibrational modes with ground state energies of $181\,\mathrm{cm}^{-1}$ (b_{2u}) and $195\,\mathrm{cm}^{-1}$ (b_{3g}), as well as one with a shift of $-6\,\mathrm{cm}^{-1}$ that is yet unassigned [2.108]. Thus, single vibronic level studies require the excitation bandpass to be narrow enough that the $-6\,\mathrm{cm}^{-1}$ sequence and the strong first member of the $-10\,\mathrm{cm}^{-1}$ sequence not be excited to any appreciable degree. For naphthalene KNIGHT et al. found that a bandpass of $12\,\mathrm{cm}^{-1}$ is sufficient for most transitions [2.108]. It is clear that the development of tunable lasers for the ultraviolet region would be of considerable utility in enabling the use of narrow bandpasses without total loss of intensity. The work of BOESL et al. [2.135] on naphthalene with a doubled dye laser, having an excitation bandwidth of $0.5\,\mathrm{cm}^{-1}$, provides an indication of the possibilities now becoming available. Furthermore, the use of supersonic nozzles to rotationally and vibrationally cool the molecules would a) aid in spectral assignments, b) narrow the vibronic bands through rotational cooling, and c) eliminate much of the cumbersome sequence congestion through vibrational cooling[21].

It is interesting to note that the presence of rotational bands introduces the possibility of rotational dependence of non-radiative decay rates due to a rotational dependence of the coupling strength. Such rotationally dependent non-adiabatic couplings are often observed in spectra of diatomic molecules [2.99]. Perhaps a more likely source of a variation of decay rates with rotational level arises from centrifugal distortion effects. The latter are greater in the higher rotational levels and cause an increasing admixture of other "harmonic" vibrational states. To the extent that the non-radiative decay rates are strongly dependent on the vibrational states, these effects are, in principle, observable. They require that the mixing of the other vibrational states, also be appreciable enough. Since this centrifugal induced mixing is proportional to the rotational energy divided by the vibrational quantum $\hbar\omega_d$ of energy, this yields a typical mixing of $kT/\hbar\omega_d$. Thus, the effect is most likely for mixings that involve low frequency vibrations, but these are often ones with weaker energy dependence. One exception would arise if the promoting mode were of low frequency. To date no experimental observation of such a rotationally dependent rate has been observed in conformity with the above estimates [2.105][22].

In order that the measurements probe the properties of individual molecules, it is, of course, necessary that the pressure be low enough that molecules do not collide during the lifetime of the excited level $|s\,i\rangle$. For benzene as a typical example, gas kinetic cross-sections give collision rates which are of the order of $10^7\,\mathrm{s}^{-1}\,\mathrm{Torr}^{-1}$. Thus, with lifetimes of $\sim 10^{-7}\,\mathrm{s}$ for the lower levels

---

[21] Of course, the problem of sequence congestion can be readily handled theoretically by just considering the thermally averaged rate for the sequence with additional weighting factors for any intensity differences in the intensity of sequence bands [2.107].

[22] SCHLAG and coworkers [2.135] have observed a substantial variation in k_{nr} across a single vibronic band in benzene upon the use of high resolution laser excitation. By analogy with high resolution experiments on the giant resonances in nuclear physics, this structure may be due to corresponding structure in the density of states weighted Franck-Condon factors arising from the strong coupled levels ϕ_b. In addition, high intensity, high resolution lasers excite a superposition of the resonances χ of Sections 2.3 and 2.5, and the particular superposition may yield an interference effect which may alter quantum yields even when the observed decays are exponential.

of S_1, pressures of 10^{-2} Torr ($\approx 10^{-5}$ atm) are required to insure that $\lesssim 1\%$ of the molecules collide during the S_1 lifetime. This low pressure complicates intensity problems beyond those already incurred by spectral congestion. For long lived states, such as triplet levels in aromatic hydrocarbons, it is impossible to perform experiments at low enough pressures to eliminate the effects of collisions. Hence, it is necessary to disentangle the effects of vibrational relaxation from the radiative and non-radiative decay. When individual levels are observable, a full kinetic analysis is, in principle possible. Otherwise, simple stochastic models can be employed to determine the average energy dependence of the decay rates as well as the important vibrational relaxation rates [2.31–34]. This interesting subject is, however, beyond the scope of our treatment of individual molecules.

The most serious problem associated with single vibronic level studies, when a molecule has passed all the other requirements, involves the determination of absolute quantum yields. The decay lifetimes (2.2) are measurable to rather good accuracy, but gas phase quantum yield standards are unavailable for most of the ultraviolet region of interest. Without measurements of the quantum yields (2.1), the radiative and non-radiative decay rates cannot generally be disentangled to enable further analysis. The assumption of a constant radiative decay rate, as a function of vibrational excitation in ϕ_s, can mask interesting variations, especially for vibronically induced optical absorptions.

The above discussion of the experimental difficulties associated with single vibronic level studies explains, in part, the paucity of experimental data. Hopefully, advances in technology will present new experimental opportunities.

2.9.3 Benzene $S_1 \rightarrow T_1$ Decay

Benzene is a good candidate for single vibronic level studies as it represents a prototype for the aromatic hydrocarbons. It is also small enough to permit spectral analysis and it is free of spectral congestion because the lowest vibrational frequency (in S_0) is $405 \, \text{cm}^{-1}$ [23]. Furthermore, benzene is among the best studied of the polyatomic molecules, and the C—C and C—H bond length shifts in the $S_1 \rightarrow T_1$ transition have been determined from experimental data, as noted in Section 2.7 [2.60]. We, therefore, consider the case of benzene in great detail in this review.

The measurement of single vibronic level lifetimes and fluorescence quantum yields were carried out by SPEARS and RICE [2.101] for over 20 initial vibronic levels in S_1 of benzene. ABRAMSON et al. [2.102] subsequently performed similar experiments on perdeuterobenzene and fluorobenzene, while GUTTMAN and RICE [2.103] considered partially deuterated benzene and difluorobenzene

---

[23] This is the v_{16} mode which is $203 \, \text{cm}^{-1}$ in S_1, then comes the $606 \, \text{cm}^{-1}$ v_6 mode which shifts to $521 \, \text{cm}^{-1}$ in S_1. There is therefore no problem with sequence congestion.

[2.109]. Other systems studied to date by RICE's and other groups involve aniline [2.98, 110, 111], various disubstituted benzenes [2.112, 113], styrene [2.114], and naphthalene [2.108, 115, 116]. The cases of formaldehyde [2.76] and chloro- and bromoacetylene [2.73] involve examples of photodissociation wherein the rate limiting step appears to be the $S_1 \to S_0$ internal conversion. (Here the dissociation broadens the S_0 levels, so they form a true continuum that can lead to irreversible internal conversion.) The acetylenes have been analyzed [2.73] as corresponding to the intermediate case described in Section 2.3, so the statistical limit theory of Sections 2.6–8 is not valid in this case. Calculations using the above theory have been applied to the formaldehyde internal conversion [2.76], but only the simple energy gap law (2.354) involving only mode M was applied in this case (i.e., $f_k = 0$, $\Omega_{\alpha,k} \equiv 0$). Because of the large C—O bond length and frequency change in the $S_1 \to S_0$ transition, the C—O vibration is, in fact, the best accepting mode, and it is necessary to employ the appropriate energy gap law (2.358) that is valid for small energy gaps[24]. Despite this fact, the calculated slope of the variation of the $S_1 \to S_0$ decay rate with excess vibrational energy is only off by a factor of two. A recalculation using (2.358) would be of some interest. As the calculation for formaldehyde with (2.358) essentially parallels that for benzene, the description of the benzene case is sufficient.

As noted in earlier sections, it was already well established by the earlier theories that the dominant accepting mode in aromatic hydrocarbons are the totally symmetric (a_{1g}) C—H vibrations of maximum frequency, with the highest frequency a_{1g} C—C skeletal stretches of lesser importance. In benzene there is only one vibration of each category, the v_1 C—C vibration and the v_2 C—H vibration. The simplest possible theoretical model involves the retention of these two vibrations, as the dominant modes, and the use of the simple displaced potential model. Calculations of this nature were first performed using a generalization of the Boltzmann statistics method, and they were obtained prior to a knowledge of the experimental results [2.26]. The subsequent more detailed analysis employs the full generating function theory to include the correct pre-exponential factors and frequency changes in modes [2.25–29].

The $S_0 \to S_1$ absorption in benzene is vibronically induced by the e_{2g} v_6 C-bending vibration. Consequently, it is possible to optically excite S_1 with excitations in the v_6 mode. Because of the large C—C bond length change in the $S_0 \to S_1$ transition (largest amongst the aromatic hydrocarbons) and the

---

[24] The theory of HELLER et al. [2.25] has also been applied to single vibronic level non-radiative decay rates in propynal [2.136], but, unfortunately, some results presented in Table VI of Ref. 2.136 are in conflict with the theory. For instance, the v_{12} vibration is taken as the promoting mode, but the calculated decay rate decreases with excitation in v_{12}. Figs. 11 and 12 show that this behavior cannot be a consequence of the theory of HELLER et al. [2.25]. Furthermore, unless the triplet levels are sufficiently broadened by predissociation, intramolecular intersystem crossing cannot occur, and a full theory of the collision induced crossing must be used [2.32a].

large ξ for the v_{16} mode, vibronic components of S_1 with v_1 and v_{16} excitations can also be optically excited. In a few cases v_{10} excitations were considered by SPEARS and RICE, and one instance involved v_5 excitation [2.101].

Because v_1 is by far the best acceptor of the available optical modes and because of the knowledge of the C—C bond length change in the $S_1 \rightarrow T_1$ transition [2.60], it is natural to first consider the relative non-radiative decay rates for excitations in the v_1 vibration in S_1. The simplest calculation involves the use of the displaced potential model wherein v_1 is the optical mode and the energy-gap law is the primitive one generated by v_2. Specifically, the non-radiative decay rates in this model are given by[25]

$$k_{nr}(S_1 m_{v_1} \rightarrow T_1) \propto \sum_{n_{v_1}} |\langle m_{v_1} | n_{v_1} \rangle|^2 [\Delta E_k(m_{v_1}, n_{v_1})]^{-1/2}$$

$$\times \exp[-\gamma_k(m_{v_1}, n_{v_1}) \Delta E_k(m_{v_1}, n_{v_1})/\hbar \omega_M] \qquad (2.369)$$

where

$$\gamma_k(m_{v_1}, n_{v_1}) = \log[\Delta E_k(m_{v_1}, n_{v_1})/X_M \hbar \omega_M] - 1 . \qquad (2.370)$$

All parameters that enter into (2.369) and (2.370) are known from spectroscopic data apart from the frequency of the ellusive promoting mode—more about that in a moment. Table 2.1 shows a comparison of experimental and theoretical

Table 2.1. Relative non-radiative decay rates for the totally symmetric carbon-carbon (v_1) optical mode

m_{v_1} (Quanta of v_1 optically excited)	$k_{nr}(S_1 m_{v_1} \rightarrow T_1)/k_{nr}(S_1 m_{v_1}=0 \rightarrow T_1)$			
	C_6H_6 Experimental[a] [2.101]	Calculated[b]	C_6D_6 Experimental[c] [2.102]	Calculated[d]
1	1.22	1.14	1.55	1.28
2	1.73	1.30	2.69	1.59
3	2.42	1.46	3.64	1.94

[a] (Error $\sim 20\%$)
[b] Eq. (2.369) with the following parameters:
$\Delta E = 8200$ cm^{-1} $\omega_k = 1500$ cm^{-1} (estimated)
$\omega_a = 923$ cm^{-1} $\omega_M = 3130$ cm^{-1}
$X_a = 0.025$ $X_M = 0.002$
[c] (Error $\sim 20\%$)
[d] Same as b) but with the following parameters:
$\Delta E = 8200$ cm^{-1} $\omega_k = 1440$ cm^{-1} (estimated)
$\omega_a = 879$ cm^{-1} $\omega_M = 2340$ cm^{-1}
$X_a = 0.025$ $X_M = 0.0028$

---

[25] Note that HELLER et al. [2.25] inadvertently dropped the pre-exponential factor of $[\Delta E_k(m_{v_1}, n_{v_1})]^{-1/2}$ in (2.369), so their results for the more highly excited levels have rates which are smaller than (2.369) would yield.

relative decay rates for the v_1 progression in benzene and perdeuterobenzene. The calculated results are in qualitative agreement with experiment. They predict the increase in decay rate for increasing excitation in v_1, and they correctly reproduce the more rapid increase in the perdeutero than the perhydro case.

This increase in rate with vibrational excitation and this deuterium isotope effect is a typical example of a general phenomenon for cases in which I_k of (2.363) displays energy-gap behavior. *Excitation in totally symmetric modes with $\xi_\alpha = 0$ increases the accepting ability of that vibration relative to all the other initially unexcited vibrations.* This feature can be seen by examining the contributions made to the total rate by each final state $|n_{v_1}\rangle$ in the summation in (2.369). Table 2.2 displays this breakdown for $m_{v_1} = 0$ and 3. The non-rad-

Table 2.2. Contributions of final state distributions to the total non-radiative decay rate

m_a	n_a	$|\langle m_a|n_a\rangle|^2$ [b]	C_6H_6 [a]		C_6D_6 [a]	
			$\Delta E_k(m_a, n_a)$	$k_{nr}(S_1\, m_a \rightarrow T_1\, n_a)$ [c]	$\Delta E_k(m_a, n_a)$	$k_{nr}(S_1\, m_a \rightarrow T_1\, n_a)$ [c]
0	0	9.75×10^{-1}	6700	2.72×10^{-6}	6760	3.38×10^{-8}
	1	2.44×10^{-2}	5777	5.21×10^{-7}	5881	1.12×10^{-8}
	2	3.05×10^{-4}	4854	4.76×10^{-8}	5002	1.74×10^{-9}
	3	2.54×10^{-6}	3931	2.74×10^{-9}	4123	1.70×10^{-10}
	4	1.59×10^{-8}	3008	1.10×10^{-10}	3244	1.14×10^{-11}
	5	7.94×10^{-11}	2085	3.23×10^{-12}	2365	5.56×10^{-13}
		$k_{nr}(m_a=0)$		3.29×10^{-6}	$k_{nr}(m_a=0)$	4.69×10^{-8}
3	0	2.54×10^{-6}	9469	1.26×10^{-14}	9397	2.91×10^{-17}
	1	8.99×10^{-4}	8546	3.80×10^{-11}	8518	1.55×10^{-13}
	2	6.96×10^{-2}	7623	2.43×10^{-8}	7639	1.74×10^{-10}
	3	8.36×10^{-1}	6700	2.33×10^{-6}	6760	2.90×10^{-8}
	4	9.04×10^{-2}	5777	1.93×10^{-6}	5881	4.14×10^{-8}
	5	2.90×10^{-3}	4854	4.52×10^{-7}	5002	1.66×10^{-8}
	6	4.89×10^{-5}	3931	5.27×10^{-8}	4123	3.27×10^{-9}
	7	5.39×10^{-7}	3008	3.73×10^{-9}	3244	3.88×10^{-10}
	8	4.33×10^{-9}	2085	1.76×10^{-10}	2365	3.03×10^{-11}
	9	2.72×10^{-11}	1162	5.68×10^{-12}	1486	1.60×10^{-12}
		$k_{nr}(m_a=3)$		4.80×10^{-6}	$k_{nr}(m_a=3)$	9.08×10^{-8}
			$k(3)/k(0) = 1.46$		$k(3)/k(0) = 1.94$	

[a] Values for totally symmetric carbon-carbon optical mode $(a = v_1)$ parameters as given in Table 2.1.
[b] Eq. (2.368).
[c] Term of $k_{nr}(S_1\, m_a \rightarrow T_1\, n_a)$ of (2.363) with I_k determined only by v_2.

iative rate is primarily determined by contributions for which the change in the number of quanta in the optical mode, $|m_{v_1} - n_{v_1}|$ is small, indicating rapid convergence of the summation in (2.369). Notice how the $n = 4$ and 5 levels make considerably greater contributions to the rate for $m = 3$ than $m = 0$. This trend is accentuated in the perdeutero case.

The predicted deuterium isotope effect also follows from the above general feature of the energy dependence of non-radiative decay rates. The accepting power of v_1 is determined by the value of Ω_{v_1} of (2.282) (with ΔE_k instead of ΔE). For C_6H_6, $\Omega_{v_1} = 0.03$, while for C_6D_6, $\Omega_{v_1} = 0.05$. Consequently, the lowering of the dominant v_2 mode frequency in going from the perhydro to the perdeutero case makes v_1 a better acceptor in C_6D_6. Excitation in v_1 further enhances its accepting ability, and this enhancement is greater in C_6D_6 because v_1 begins as a better acceptor in C_6D_6.

As noted above, the one parameter that enters into the above calculations is the unknown frequency of the promoting mode ω_k. The effects of variation in the promoting mode frequency over the whole possible range are displayed in Table 2.3. Table 2.3 shows how the predicted rates are fairly insensitive to

Table 2.3. Dependence of relative rates on effective energy gap

C_6H_6 [a]

ω_k	ΔE_k [b]	$k(1)/k(0)$	$k(2)/k(0)$	$k(3)/k(0)$
500	7700	1.15	1.31	1.49
1000	7200	1.15	1.30	1.47
1500	6700	1.14	1.30	1.46
2000	6200	1.14	1.29	1.44
2500	5700	1.13	1.27	1.42
3130	5070	1.13	1.26	1.40

C_6D_6 [a]

ω_k	ΔE_k [b]	$k(1)/k(0)$	$k(2)/k(0)$	$k(3)/k(0)$
500	7700	1.29	1.63	2.01
1000	7200	1.28	1.61	1.97
1500	6700	1.27	1.58	1.93
2000	6200	1.26	1.56	1.89
2340	5860	1.26	1.54	1.86

[a] Values for totally symmetric carbon-carbon optical mode $(a = v_1)$ as calculated by (2.369) and (2.368)—parameters not specified as given in Table 2.1.
[b] Effective energy gap $\Delta E_k = \Delta E - \hbar \omega_k$ in cm^{-1}.

variations in ω_k—the absolute rates vary considerably (over a factor of $\exp(5)$) with this range of ΔE! This result indicates that the unknown value of ω_k is not sufficient to explain the discrepancy between theory and experiment for the displaced potential model. It also displays another hazard in the attempt to calculate absolute non-radiative decay rates, and it emphasizes some of the insensitivities of the relative rates.

Another potential source of errors in the theoretical calculation involves any errors in the experimental values of X_{v_1} and X_{v_2}. A variation of both of these by $\pm 10\%$, as given in Table 2.4, shows that small errors of this sort are not the source of the experimental-theoretical discrepancy in Table 2.1.

Table 2.4. Variation of relative rates with C—C and C—H geometry changes

C_6H_6 [a]

X_a [b]	X_M [b]	$k(1)/k(0)$	$k(2)/k(0)$	$k(3)/k(0)$
0.0225	0.0018	1.13	1.28	1.43
0.0225	0.0020	1.13	1.26	1.41
0.0225	0.0022	1.12	1.26	1.39
0.0250	0.0018	1.15	1.31	1.48
0.0250	0.0020	1.14	1.30	1.46
0.0250	0.0022	1.14	1.28	1.44
0.0275	0.0018	1.16	1.34	1.53
0.0275	0.0020	1.16	1.32	1.50
0.0275	0.0022	1.15	1.31	1.49

C_6D_6 [a]

X_a [b]	X_M [b]	$k(1)/k(0)$	$k(2)/k(0)$	$k(3)/k(0)$
0.0225	0.00252	1.26	1.55	1.88
0.0225	0.00280	1.25	1.53	1.84
0.0225	0.00308	1.24	1.50	1.80
0.0250	0.00252	1.29	1.62	1.98
0.0250	0.00280	1.28	1.59	1.94
0.0250	0.00308	1.26	1.56	1.89
0.0275	0.00252	1.32	1.68	2.09
0.0275	0.00280	1.30	1.65	2.04
0.0275	0.00308	1.29	1.62	1.99

[a] Values for totally symmetric carbon-carbon optical mode $(a = v_1)$ as calculated by (2.369) and (2.368)—parameters not specified as given in Table 2.1.
[b] X defined in (2.257).

Furthermore, the addition of the effects of moderate frequency changes in non-totally symmetric vibrations, including the promoting mode, as modeled in Table 2.5, do not resolve the discrepancy. These effects of non-totally symmetric modes are included exactly in Table 2.5 by treating these modes in the category of "optical" modes. The final results confirm the basic theory in (2.401) as to the overall effect of these poor acceptors.

Model calculations can also be performed under the assumption that there are much larger frequency shifts than those used in Table 2.5, but these are not readily supported by theory or experiment. Since S_1 and T_1 are naively iso-configurational, in zeroth-order they should have very similar potential surfaces. Actually, S_1 has a larger contribution from ionic valence bond structure than does S_0 and T_1 [2.117] so the T_1 vibrational frequencies cannot be

Table 2.5. Dependence of relative rates on bending and promoting mode frequency changes

C_6H_6 [a, b]

$\Delta\omega$	$k(1)/k(0)$	$k(2)/k(0)$	$k(3)/k(0)$
$\Delta\omega_{bend} = +100$	1.13	1.27	1.41
$\Delta\omega_{bend} = -100$	1.14	1.30	1.46
$\Delta\omega_k \quad = + 25$	1.14	1.29	1.45
$\Delta\omega_k \quad = - 25$	1.14	1.29	1.45

C_6D_6 [a, c]

$\Delta\omega$	$k(1)/k(0)$	$k(2)/k(0)$	$k(3)/k(0)$
$\Delta\omega_{bend} = +100$	1.24	1.51	1.80
$\Delta\omega_{bend} = -100$	1.28	1.59	1.94
$\Delta\omega_k \quad = + 20$	1.25	1.53	1.84
$\Delta\omega_k \quad = - 20$	1.26	1.54	1.86

[a] Values for the totally symmetric carbon-carbon optical mode $(a = v_1)$ as calculated by (2.369) with the prefactor (2.361) and (2.368). $(\Delta\omega \equiv \omega^{(T_1)} - \omega^{(S_1)})$
[b] $\omega_{bend}^{(S_1)} = 600$ cm^{-1}, $d_{bend} = 9$, $\omega_k^{(T_1)} = 1500$ cm^{-1}.
[c] $\omega_{bend}^{(S_1)} = 400$ cm^{-1}, $d_{bend} = 9$, $\omega_k^{(T_1)} = 1440$ cm^{-1}.

expected to grossly differ from those in S_0. Hence, large ξ_α in the $S_1 \to T_1$ transition can be ruled out (see analyses of v_6 and v_{16} below)[26].

Within the harmonic approximation, the last approximation, which can be lifted, is the neglect in frequency changes in v_1. Spectroscopic information regarding the v_1 frequency shift

$$\Delta\omega_1 \equiv \omega_{v_1}(T_1) - \omega_{v_1}(S_1) \qquad (2.371)$$

is rather limited. If the known C—C bond length is used in Badger's rule [2.118], an estimate of $\Delta\omega_1 \approx 13$ cm^{-1} is obtained. This value assumes the v_1 vibration to be pure C—C motion, and therefore errors are introduced beyond those inherent in the use of Badger's rule. BURLAND and ROBINSON [2.60] estimated $\Delta\omega_1$ from phosphorescence $(T_1 \to S_0)$ intensities. With a slight readjustment of their analysis, it yields $\Delta\omega_1 \approx 42 \pm 15$ cm^{-1}. Thus, a preliminary estimate of $\Delta\omega_1$ between about 5 and 50 cm^{-1} represents a reasonable guess [i.e. (2.371) is positive]. The results of calculations of relative rates in Table 2.6 are in line with these estimates as a value of $\Delta\omega_1 \approx 25$ cm^{-1} fits the experimentally

[26] If benzene were to be non-hexagonal in T_1, as it is in the crystalline phase, some modes may have non-negligible parameters ξ_α and/or X_α. The principal candidate in this case is the v_8 vibration. The decrease in molecular symmetry in T_1 in the crystal arises, in large measure, from the lack of hexagonal site symmetry in the crystal [2.137]. There is no compelling reason to advocate this reduction in symmetry in the gas phase molecule, so it is not presently reasonable to assume such large frequency shifts in v_8.

Table 2.6. Dependence of relative rates on C—C optical mode frequency changes

	C_6H_6 [a, b]			C_6D_6 [a, c]		
$\Delta\omega_a$	$k(1)/k(0)$	$k(2)/k(0)$	$k(3)/k(0)$	$k(1)/k(0)$	$k(2)/k(0)$	$k(3)/k(0)$
+100	2.13	4.50	9.1			
+ 50	1.46	2.17	3.21	2.19	4.62	9.26
+ 25	1.27	1.62	2.06	1.59	2.50	3.85
+ 20	1.24	1.54	1.90	1.51	2.24	3.29
+ 10	1.19	1.40	1.65	1.38	1.86	2.46
+ 5	1.16	1.35	1.55	1.32	1.71	2.17
+ 0	1.14	1.30	1.46	1.28	1.59	1.94
− 5	1.12	1.25	1.38	1.24	1.49	1.75
− 10	1.10	1.21	1.31	1.20	1.40	1.61
− 20	1.07	1.14	1.20	1.15	1.28	1.40
− 25	1.06	1.11	1.15	1.12	1.24	1.34
− 50	1.00	1.00	1.00	1.07	1.15	1.24
−100	0.946	0.924	0.930	1.12	1.34	1.67

[a] Values for totally symmetric carbon-carbon optical mode ($a = \nu_1$) as calculated by (2.369) and (2.367a). ($\Delta\omega = \omega^{(T_1)} - \omega^{(S_1)}$)

[b] $\omega_a^{(S_1)} = 923 \text{ cm}^{-1}$.

[c] $\omega_a^{(S_1)} = 879 \text{ cm}^{-1}$.

observed rates *for both* C_6H_6 *and* C_6D_6. Since ν_1 is primarily a C—C motion, its S_1 frequency changes from 923 cm^{-1} in C_6H_6 to 879 cm^{-1} in C_6D_6, a 5% shift. Thus, the isotope effect on $\Delta\omega_1$ should also be in the neighborhood of 5% which is negligible considering the experimental and theoretical uncertainties which induce errors in $\Delta\omega_1$ of $\pm 15 \text{ cm}^{-1}$.

Table 2.6 is rather instructive as a model of cases which might pertain to other molecules. Notice how a value of $\Delta\omega_1 = 100 \text{ cm}^{-1}$ for C_6H_6 leads to predicted relative rates which increase faster than exponentially. As this represents $\approx 11\%$ frequency shift, it is not an unreasonable shift for other molecules. Indeed, the observed exponential increasing rate in β-naphthalamine [2.119] likely arises from large geometry and/or frequency shifts for one or more vibrations between the two electronic states. Physically, it is clear that β-naphthalamine [2.119] should have some modes (involving amine vibrations) for which this large change occurs. Similarly, ROCKLEY and PHILLIPS [2.112] found that parafluorotoluene exhibits a more sharply (but not exponential) increasing rate with ν_1 excitation than does unsubstituted benzene. In this case methyl group motions are probably included in the ν_1 mode, and its shift parameters are larger.

Note how a shift $\Delta\omega_1 = -100 \text{ cm}^{-1}$ leads to relative rates which *decrease* with increasing optical excitations for C_6H_6. After this prediction appeared, the phenomenon was observed for the ν_{14} vibration in the S_1 state of benzene where $\Delta\omega_{14} = -257 \text{ cm}^{-1}$ [2.138].

When the optical mode is a non-totally symmetric vibration, the analysis proceeds just as in the case of the ν_1 mode above, except that there is only one

Table 2.7. Dependence of relative rates on frequency change in non-totally symmetric optical mode (v_6)

C_6H_6 [a, b]			C_6D_6 [a, c]		
$\Delta\omega_a$	$k(1)/k(0)$	$k(2)/k(0)$	$k(3)/k(0)$	$k(1)/k(0)$	$k(2)/k(0)$	$k(3)/k(0)$
+100	1.43	2.16	3.38	1.88	3.90	8.54
+ 50	1.15	1.34	1.58	1.24	1.57	2.05
+ 25	1.06	1.14	1.22	1.09	1.20	1.33
+ 20	1.05	1.11	1.17	1.07	1.15	1.24
+ 10	1.02	1.05	1.08	1.03	1.07	1.10
+ 5	1.01	1.02	1.04	1.02	1.03	1.05
0	1.00	1.00	1.00	1.00	1.00	1.00
− 5	0.989	0.978	0.967	0.985	0.971	0.957
− 10	0.978	0.957	0.937	0.972	0.945	0.920
− 20	0.958	0.920	0.884	0.948	0.901	0.860
− 25	0.949	0.903	0.861	0.937	0.882	0.835
− 50	0.909	0.833	0.770	0.894	0.814	0.753
−100	0.851	0.745	0.670	0.845	0.755	0.708

[a] Values for non-totally symmetric carbon-carbon optical mode ($a = v_6$) as calculated by (2.369) and (2.367a). ($\Delta\omega = \omega^{(T_1)} - \omega^{(S_1)}$)
[b] $\omega_a^{(S_1)} = 521 \text{ cm}^{-1}$.
[c] $\omega_a^{(S_1)} = 499 \text{ cm}^{-1}$.

parameter $\Delta\omega$ (or ξ) to be considered. An example is given in Table 2.7 where calculations for the v_6 vibration are presented. Agreement with experimental results for both C_6H_6 and C_6D_6 is obtained for $\Delta\omega_6 \approx 50-65 \text{ cm}^{-1}$, but the

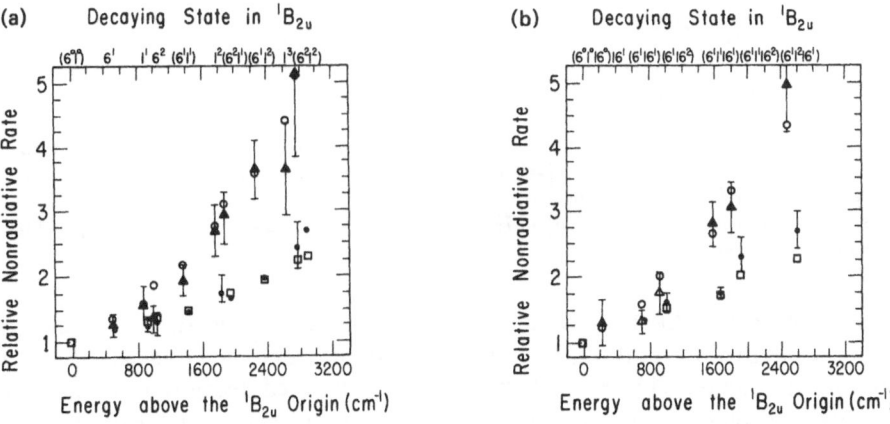

Fig. 2.10a and b. Vibronic level dependence of the relative $S_1 \to T_1$ decay rates in C_6H_6 and C_6D_6. The experimental values, ●—C_6H_6 and ▲—C_6D_6 were taken from [2.101, 102]. The calculated values [2.25, 29] □—C_6H_6 and ○—C_6D_6 assumed the following parameters: (a) C_6H_6: $\omega_1(S_1) = 923 \text{ cm}^{-1}$, $\omega_1(T_1) = 948 \text{ cm}^{-1}$, $\omega_6(S_1) = 521 \text{ cm}^{-1}$, $\omega_6(T_1) = 571 \text{ cm}^{-1}$; $\omega_2 = 3130 \text{ cm}^{-1}$, $X_1 = 0.025$, $X_2 = 0.002$, $\omega_k = 1500 \text{ cm}^{-1}$, $\Delta E(S_1, T_1) = 8200 \text{ cm}^{-1}$. C_6D_6: $\omega_1(T_1) = 904 \text{ cm}^{-1}$, $\omega_6(S_1) = 879 \text{ cm}^{-1}$, $\omega_2 = 2340 \text{ cm}^{-1}$, $\omega_6(S_1) = 499 \text{ cm}^{-1}$, $\omega_6(T_1) = 564 \text{ cm}^{-1}$, $X_1 = 0.025$, $X_2 = 0.003$, $\omega_k = 1440 \text{ cm}^{-1}$, $\Delta E(S_1, T_1) = 8200 \text{ cm}^{-1}$. (b) C_6H_6: $\omega_{16}(S_1) = 243 \text{ cm}^{-1}$, $\omega_{16}(T_1) = 303 \text{ cm}^{-1}$. C_6D_6: $\omega_{16}(S_1) = 208 \text{ cm}^{-1}$, $\omega_{16}(T_1) = 253 \text{ cm}^{-1}$

experimental uncertainties are considerably larger in this case. (Notice again how Table 2.7 shows that $\Delta\omega_a \simeq 100\,\text{cm}^{-1}$ leads to a faster than exponential rate increase.) The estimates given by BURLAND and ROBINSON [2.60] for $\Delta\omega_6$ are $\approx 100\,\text{cm}^{-1}$ for C_6H_6 and $\approx 75\,\text{cm}^{-1}$ for C_6D_6 which are in general in line with the analysis of non-radiative decay rates given the experimental errors involved.

Given the fitted values of $\Delta\omega_1$ and $\Delta\omega_6$ from the decay rates of 1^n and 6^n states of S_1, respectively, consistency requires that the relative decay rates of the mixed progression $1^n 6^m$ also be correctly reproduced by these same parameters. Table 2.8 and Fig. 2.10a displays this agreement between theory

Table 2.8. Relative non-radiative rates for the $6^m 1^n$ optical progression

C_6H_6 $\qquad\qquad\qquad\qquad\qquad\qquad\qquad\qquad\qquad\qquad$ C_6D_6

Optically selected stated	E_{ex} [a]	Experiment Average[b]	Range[c]	Calculated[d]	E_{ex} [a]	Experiment[e]	Calculated [f]	[g]
$6^0 1^0$	0	1.00		1.00	0	1.00	1.00	1.00
$6^1 1^0$	521	1.19	(1.1–1.35)	1.15	498	1.24 ± 0.19	1.23	1.35
$6^0 1^1$	923	1.22	(1.2–1.4)	1.30	879	1.55 ± 0.31	1.66	1.66
$6^2 1^0$	1040	1.27	(1.1–1.45)	1.34	996	1.36 ± 0.20	1.56	1.91
$6^1 1^1$	1444	1.42		1.50	1377	1.92 ± 0.23	2.04	2.23
$6^0 1^2$	1846	1.73	(1.6–2.0)	1.70	1758	2.69 ± 0.40	2.73	2.72
$6^2 1^1$	1965	1.64		1.74	1875	2.89 ± 0.38	2.56	3.12
$6^1 1^2$	2367	1.94		1.95	2256	3.64 ± 0.47	3.33	3.62
$6^0 1^3$	2769	2.42	(2.1–2.8)	2.21	2637	3.64 ± 0.73	4.37	4.35
$6^2 1^2$	2882	2.67		2.27	2754	5.15 ± 1.30	4.16	5.02

[a] Energy above S_1 electronic origin ($m_a = 0$).
[b] [Ref. 2.101, Table 3], $\tau_{nr}^0 = 128$ ns.
[c] [Ref. 2.101, Table 5].
[d] Eqs. (2.363) (I_k from (2.369)) and (2.367a) with $\Delta E = 8200\,\text{cm}^{-1}$, $\omega_k = 1500$, $\omega_1 = 923\,\text{cm}^{-1}$, $\omega_6 = 521\,\text{cm}^{-1}$, $\omega_M = 3130\,\text{cm}^{-1}$; $X_1 = 0.025$, $X_6 = 0$, $X_M = 0.0020$; $\Delta\omega_1 = 25\,\text{cm}^{-1}$, $\Delta\omega_6 = 50\,\text{cm}^{-1}$.
[e] [Ref. 2.102, Table 7], $\tau_{nr}^0 = 309$ ns.
[f] Same as [d] but $\omega_k = 1440\,\text{cm}^{-1}$, $\omega_1 = 879\,\text{cm}^{-1}$, $\omega_6 = 499\,\text{cm}^{-1}$, $\omega_M = 2340\,\text{cm}^{-1}$; $X_M = 0.0028$.
[g] Same as [f] but $\Delta\omega_6 = 65\,\text{cm}^{-1}$.

and experiment. Analysis of the $1^m 6^n 16^p$ progressions leads to a value of $\Delta\omega_{16} \approx 60\,\text{cm}^{-1}$ for both C_6H_6 and C_6D_6, and the calculated and experimental results are compared in Fig. 2.10b. There are observed decay rates of S_1 levels involving excitation in ν_5 and ν_{10}, but there are not yet enough experimental observations to sufficiently overdetermine the frequency shifts.

2.9.4 Search for Promoting Modes

The above treatment of the experimental data is made under the assumption that none of the optical modes is the promoting mode. If one of them were a promoting mode, it would be rather easy to predict the relative rates for the simple case of $\xi_k = 1$—note that in benzene a promoting mode must be non-totally symmetric so $X_k \equiv 0$. Introduction of the general matrix elements

$$|\langle m_k | \partial/\partial Q_k | n_k \rangle|^2 = (\omega^s/2\hbar) | (m_k)^{1/2} \langle m_k - 1 | n_k \rangle$$
$$- (m_k + 1)^{1/2} \langle m_k + 1 | n_k \rangle|^2 \qquad (2.372)$$

into the rate expression (2.363), use of $\xi_k = 0$ and of the energy gap law like (2.356) for I_k, and neglect in the minor variation of the pre-exponential factor leads to the result that in benzene

$$1.1\, m_k + 1 \lesssim k_{\mathrm{nr}}(S_1\, m_k \to T_1)/k_{\mathrm{nr}}(S_1\, 0_k \to T_1) \lesssim 1.2 m_k + 1 , \qquad (2.373)$$

which is somewhat less than the $(2m_k + 1)$-dependence predicted by workers who inadvertently omitted the disfavor which the energy-gap law places on the $n_k = m_k - 1$ contribution to (2.363) in benzene. This feature is also exhibited by non-promoting modes as seen by examining Table 2.2. The predicted rate increase in (2.373) is considerably sharper than those observed in benzene. However, by analogy with the case of non-promoting modes, it is possible that the allowance of frequency shifts, $\xi_k \neq 0$, might considerably reduce the rate increase with m_k to bring predicted rates in line with experimental observations.

Model calculations, depicted in Figs. 2.11a and 11b, compare the predicted relative rates for excitation in v_6 and v_{16}, respectively, under the assumption that they are promoting modes and have $\xi_k \neq 0$. The Figs. 2.11a,b also include the predictions for the case that they are not promoting modes. Figs. 2.11a,b show that the estimate (2.373) is, in fact, a lower bound to the rate increase when frequency shifts in a possible promoting mode are included. Thus, it is clear that v_6 and v_{16} (as well as v_5 and v_{10}) cannot be the (dominant) promoting for the $S_1 \to T_1$ decay. Although the v_1 mode is prevented from being a promoting mode in benzene, model calculations comparing its behavior as a promoting mode, as opposed to a simple accepting mode, are given in Fig. 2.11c as a guide for other cases where symmetry restrictions are not operative.

2.9.5 Generalizations for Moderate Energy Gaps and Other Molecules

The case of benzene is a very favorable one for the theoretical calculations as values for some of the relevant parameters are known from independent spectroscopic data [2.60]. This available data, however, does possess considerable experimental uncertainty. Thus, if new experiments alter the accepted values of X_1 and X_2, the fitted frequency shifts will also suffer changes, although small changes in the former imply small ones in the latter. This lack of precise spectroscopic data becomes accentuated with most other polyatomic mole-

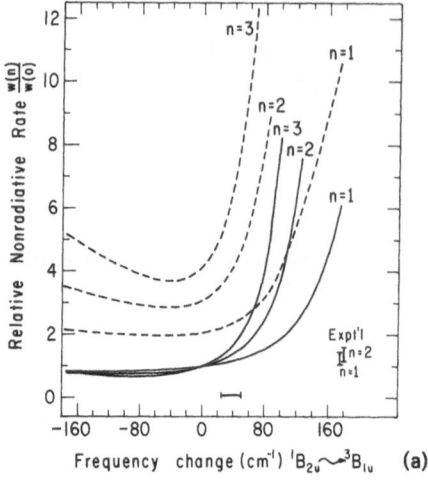

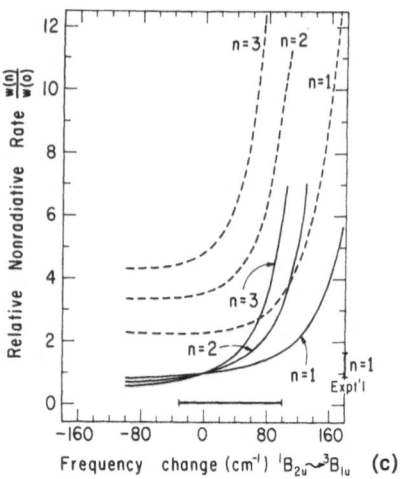

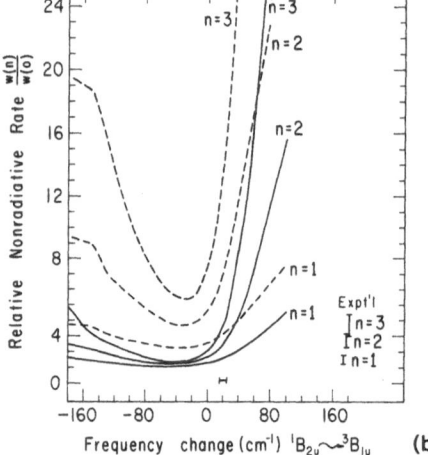

Fig. 2.11a–c. Dependence of the calculated relative $S_1 \rightarrow T_1$ decay rates [2.29] in C_6D_6 on the triplet frequency of the optically excited mode. The plotted frequency change is $\Delta\omega \equiv \omega(T_1) - \omega(S_1)$. The broken curves are for the case that the excited mode promotes the decay process, while the solid curve is for an ordinary optical mode. All parameters listed in Fig. 2.10. (a) Non-totally symmetric optical mode ν_6. (b) Non-totally symmetric optical mode ν_{16}. (c) Totally symmetric optical mode ν_1

cules, but in these cases, the calculations for benzene can serve as useful guides to theoretically expected results for molecules with moderate energy gaps as the calculations present a range of different possible frequency shifts. Table 2.9 exploits this modelistic aspect of benzene to provide predicted relative rates for non-totally symmetric modes (promoting or not) with various initial state frequencies and frequency shifts. It is clear from inspection of Table 2.9 that a much richer energy dependence of the non-radiative decay rates is possible for the case of a predominately weak coupling energy gap law I_k and moderate energy gaps. Typical calculations would likewise be useful for cases of somewhat larger X_M and a concomitant weaker (i.e., smaller γ'_k) energy-gap law. In a similar manner, the case of benzene can serve to generate model theoretical behavior for large energy-gap situations. Before

Table 2.9. C_6D_6 relative $S_1 \to T_1$ decay rates for non-totally symmetric optical modes[a, b]

$\Delta\omega$ [c] / $\omega_{(S_1)}$	Non-promoting optical modes											
	200			500			1000			1500		
-200				0.82,	0.77,	0.79	2.0,	3.4,	9.3	2.7,	4.4,	5.9
-150				0.82,	0.75,	0.73	1.8,	3.0,	4.6	2.9,	5.5,	8.3
-100	0.82,	0.70,	0.60	0.84,	0.75,	0.71	1.4,	2.0,	3.0	2.9,	6.0,	10.2
$-\ 50$	0.89,	0.80,	0.72	0.89,	0.81,	0.75	1.0,	1.2,	1.4	2.6,	5.0,	8.5
$-\ 25$	0.94,	0.88,	0.82	0.94,	0.88,	0.84	0.98,	0.98,	1.0	1.2,	1.6,	2.0
0	1.0,	1.0,	1.0	1.0,	1.0,	1.0	1.0,	1.0,	1.0	1.0,	1.0,	1.0
$+\ 25$	1.1,	1.2,	1.3	1.1,	1.2,	1.3	1.2,	1.4,	1.7	1.5,	2.3,	3.5
$+\ 50$	1.2,	1.5,	1.9	1.2,	1.6,	2.0	1.6,	2.6,	4.4	2.4,	4.9,	8.9
$+100$	1.7,	3.2,	6.2	1.9,	3.8,	8.2	3.5,	10.6,	28.2	3.6,	9.2,	20.0

$\Delta\omega$ [c] / $\omega_{(S_1)}$	Promoting optical modes											
	200			500			1000			1500		
-200				2.2,	3.7,	5.7	4.2,	11.2,	23	3.4,	7.0,	11
-150				2.1,	3.3,	4.8	3.6,	8.5,	28	3.9,	8.9,	16
-100	2.3,	3.4,	4.8	2.0,	3.0,	4.1	3.1,	7.2,	14	4.2,	10.7,	21
$-\ 50$	2.3,	3.4,	4.4	2.0,	2.8,	3.7	2.3,	4.2,	6.9	6.3,	16.2,	34
$-\ 20$	2.3,	3.5,	4.6	2.0,	2.9,	3.8	2.0,	3.2,	4.5	2.6,	4.9,	8.2
0	2.3,	3.6,	4.9	2.0,	3.1,	4.1	2.0,	3.0,	4.0	2.0,	3.0,	4.0
$+\ 25$	2.4,	3.9,	5.6	2.2,	3.6,	5.2	2.4,	4.3,	7.1	3.1,	6.9,	13
$+\ 50$	2.5,	4.5,	7.3	2.5,	4.7,	8.1	3.3,	8.5,	19	4.0,	10.9,	24
$+100$	3.5,	9.6,	24.7	4.1,	13.2,	40.0	6.1,	24.0,	76	5.0,	16.2,	42

[a] Values given are for $k(v^n)/k(v^0)$ $n = 1, 2, 3$.
[b] Parameters (see [2.25, 60] $\Delta E_{(S_1 T_1)} = 8200\ \text{cm}^{-1}$, $\omega_k = 1440\ \text{cm}^{-1}$, $\omega_1^{(S_1)} = 879\ \text{cm}^{-1}$, $\omega_2 = 2340\ \text{cm}^{-1}$, $X_1 = 0.025$, $X_2 = 0.003$, $\Delta\omega_1 = \Delta\omega_2 = 0$.
[c] $\Delta\omega = \omega_{(T_1)} - \omega_{(S_1)}$.

considering this interesting situation, it is of interest to complete our study of the moderate energy-gap case.

The lack of importance of non-optical poor accepting modes in the evaluation of relative non-radiative decay rates has already been thoroughly discussed. A corollary of this involves the energy dependence of k_{nr} when a poor acceptor is an optical mode. For simplicity, consider the case of two optical modes 1 and 2 as the generalization to more is quite straightforward. So long as 1 and 2 are both not promoting modes and one of them is a poor acceptor, the rates for mixed optical excitation obey the relation [2.29]

$$k_{nr}(m_1, m_2)/k_{nr}(m_2) \approx k_{nr}(m_1)/k_{nr}(0),\qquad(2.374)$$

Eq. (2.374) is valid when the summation in (2.363) converges for some n_α ($\alpha = 1$ or 2) such that

$$n_\alpha < N_\alpha \quad \text{and} \quad N_\alpha \hbar \omega_\alpha \ll \Delta E(n_\alpha = m_\alpha),\qquad(2.375)$$

which implies that mode α is a poor acceptor. Rearrangement of (2.375) leads to [2.29]

$$\tau_{nr}(m_1, m_2) \approx \tau_{nr}(m_1)\,\tau_{nr}(m_2)/\tau_{nr}(0)\,. \tag{2.376}$$

Indeed, the empirical observation that the incremental change in the non-radiative decay rate per quantum of excitation is approximately constant [2.101, 102] derives from the special case of (2.376) for modes 1 and 2 the same and for $m_2 = 1$,

$$\tau(m_1)/\tau(m_1 - 1) \approx \tau(1_1)/\tau(0)\,. \tag{2.377}$$

These rules (2.374–377) are, of course, valid for the case of moderate energy gaps, and they begin to break down rather rapidly with increasing m_α, since in that case mode α becomes a much better acceptor. When the non-radiative decay rate greatly exceeds the radiative decay rates of the excited levels, (e.g., in benzene [2.101, 102]), Eqs. (2.374–377) may be used to relate the experimentally observed lifetimes directly.

In lieu of detailed relative rate calculations, the simple relations (2.374–377) can be very helpful in making vibronic assignments. Scheps et al. [2.98] used (2.377) as an empirical rule to aid in the assignment of vibronic levels in various deuterated and undeuterated anilines. Prais et al. [2.29] showed that non-radiative decay rate of the level assigned as 10^2 violates (2.374) and in more likely $6^1 1^1 10^1$ both on the basis of (2.374) and on the agreement with the experimental excitation energy.

The introduction of substituents, such as fluorine, $-NH_2$, $-CH_3$, etc., leads to a lowering of the molecular symmetry from that in benzene and the introduction of some additional vibrations. The symmetry lowering increases the number of totally symmetric vibrations that can act as good acceptors. The introduction of the substituents also alters the nature of the vibrations that correlate with v_1 and v_2 in benzene, so the use of the same parameters X_{v_1} and X_{v_2} as in benzene may often be invalid. The simplest possible cases involve those in which the substituents are deuteriums, the partially deuterated benzenes. For these molecules, the C—C and C—H (or C—D) bond length changes and force constant changes are identical to those in benzene (apart from non-adiabatic corrections), so no new "molecular parameters" are introduced. However, the change in molecular symmetry and the mass effect upon deuteration both conspire to alter the nature of the normal modes v_1, v_2, v_6, v_{16}, etc., from what they were in benzene. A complete analysis would therefore require a new normal coordinate analysis for each of the partially substituted benzenes using the known bond lengths and force constants from benzene. This is clearly a tedious and uninteresting task which was circumvented by Guttman and Rice [2.103] who employ a simple local mode model to analyze their experiments on C_6H_5D, p-$C_6H_4D_2$, m-$C_6H_4D_2$, and C_6HD_5.

The local mode model is based on the observation of HENRY and SIEBRAND [2.120] that a local C—H model adequately enables the description of the observed overtone bands ground state of benzene. Consequently, GUTTMAN and RICE [2.103] employ a model involving six C—H local modes. The shift parameters associated with each is taken as $X_{v_2}^{C_6H_6}/6$ if the mode is a C—H one and $X_{v_2}^{C_6D_6}/6$ if it is a C—D. The parameters associated with v_1 are approximated as being unchanged. As v_1 is one of the optical modes, the dominant acceptors are the local high frequency C—H and C—D modes. Thus, the derivation of the effective energy-gap law to be inserted in (2.363) requires the solution of the two-mode energy conservation equations

$$\Delta E_k = \hbar \omega_H d_H X_H \exp(b\hbar \omega_H) + \hbar \omega_D d_D X_D \exp(b\hbar \omega_D), \tag{2.378}$$

where ω_H, X_H, d_H denote, respectively, the local mode C—H frequency and shift parameter and the number of hydrogen atoms in the molecule (etc. for D). Invoking the approximation (2.297) and solving (2.378) to second-order in the parameter $\Omega_{k,D}$ gives

$$\gamma' = \ln[\Delta E_k/d_H X_H \hbar \omega_H] - 1 + p_D^{-1} \Omega_{k,D} + \Omega_{k,D}^2/2 + \mathcal{O}(\Omega_{k,D}^3), \tag{2.379}$$

where

$$p_D \equiv \omega_D/\omega_H \tag{2.380}$$

and $\Omega_{k,D}$ can be written in terms of d_D/d_H via

$$\Omega_{k,D} = (d_D/\Delta E d_H)(\Delta E/d_H X_H \hbar \omega_H)^{p_D} < 1. \tag{2.381}$$

For the $S_1 \to T_1$ transition in C_6D_5H, $\Omega_{k,D} = 0.52$, so the third-order terms in (2.379) may be needed for numerical accuracy. Since

$$\gamma(C_6H_6) = \ln[\Delta E_k/6 X_H \hbar \omega_H] - 1 = 6 \tag{2.382}$$

for this transition, (2.379) shows that deuteration leads to a slight variation of γ' over γ, but the overall change is small and gradual through C_6D_5H. Eq. (2.379) is, of course, not valid for C_6D_6, but we already know that γ' is the same for C_6D_6 and C_6H_6 when (2.297) is used. The large change in going between C_6D_5H and C_6D_6 then comes from the change to $\gamma' \Delta E_k/\hbar \omega_D$ from $\gamma'(C_6H_5H)\Delta E_k/\hbar \omega_H$. The observed relative decay rates are presented in Table 2.10 where the close similarity of the C_6H_5D and p-$C_6H_4D_2$ relative rates to those in C_6H_6 are evident. The difference between the m-$C_6H_4D_2$ and p-$C_6H_6D_2$ relative rates may be primarily due to errors in the measurement of the vibrationless 1^06^0 decay rate, as noted by GUTTMAN and RICE. The C_6D_5H relative rates are intermediate between those for C_6H_6 and C_6D_6. The overall trend follows from the above analysis, and detailed computations using

Table 2.10. Observed relative non-radiative rates of benzenes[a]

Optically selected state	H_6	D_1	p-D_2	m-D_2	D_5	D_6
$6^0 1^0$	1.00	1.00	1.00	1.00	1.00	1.00
$6^0 1^1$	1.22		1.27			1.55
$6^0 1^2$	1.73		2.02			2.69
$6^0 1^3$	2.42		2.52			3.64
$6^1 1^0$	1.19	0.89	0.98	1.11	0.90	1.24
$6^1 1^1$	1.42	1.43	1.43	1.83	1.46	1.92
$6^1 1^2$	1.94	2.71	2.82	3.38	3.17	3.64
$6^2 1^0$	1.27	1.07	1.16		1.39	1.36
$6^2 1^1$	1.64	1.74	1.93		2.26	2.89
$6^2 1^2$	2.67	3.47	3.83			5.15

[a] Taken from [2.103].

the parameters determined for C_6H_6 and C_6D_6 gives the results in Table 2.11 which show a somewhat smoother variation than the experimental data.

The trends in the variation in the relative decay rates can be qualitatively predicted by rather simple arguments. In going from C_6D_6 to C_6D_5H the one C—H stretch that is added to the molecule competes well with the five other C—D stretches for the available energy because of its higher frequency, despite a lower shift parameter X_H. Consequently, in C_6D_5H most of the energy winds up in the C—H stretch. This then allows the partially deuterated rates to be represented, cf. (2.379), as a perturbation on the pure C—H case. Upon excitation of ν_1 or ν_6, these low frequency motions are competing

Table 2.11. Predicted relative rates of non-radiative decay of deuterated benzenes[a]

State	H_6 [b]	D_1 [c]	D_2 [c]	D_5 [c]	D_6 [b]	D_6 [b]
$6^0 1^0$	1.00	1.00	1.00	1.00	1.00	1.00
$6^1 1^0$	1.15	1.15	1.16	1.20	1.23	1.35
$6^0 1^1$	1.30		1.34		1.66	1.66
$6^2 1^0$	1.34	1.35	1.37	1.50	1.56	1.91
$6^1 1^1$	1.50	1.52	1.55	1.76	2.04	2.23
$6^0 1^2$	1.70		1.79		2.73	2.72
$6^2 1^1$	1.74	1.77	1.82	2.14	2.56	3.12
$6^1 1^2$	1.95	2.00	2.07	2.58	3.33	3.62
$6^0 1^3$	2.21		2.39		4.37	4.35
$6^2 1^2$	2.27	2.33	2.43		4.16	5.02

[a] $\Delta\omega_6 = +50\ \text{cm}^{-1}$.
[b] Taken from Table 2.8. The second column for D_6 corresponds to $\Delta\omega_6 = +65\ \text{cm}^{-1}$.
[c] Taken from [2.103].

primarily with C—D(C—H) vibrations in $C_6D_6(C_6D_5H)$, making the relative rate increase much slower in C_6D_5H than in C_6D_6. Upon going from C_6D_5H to $C_6D_4H_2$, the introduction of yet another local C—H mode does not alter much as the other C—H mode was already dominant.

As noted above, the local mode approach ignores changes in the normal modes upon deuteration, but it also neglects the "creation" of additional a_{1g} vibrations with possible shift parameters $X_\alpha \neq 0$ that would affect γ'. Thus, it is not unreasonable that the agreement between theory and experiment for the partially deuterated benzenes is not as good as for the limiting cases of C_6H_6 and C_6D_6.

Some of the other molecules that have been studied by single vibronic level fluorescence experiments are C_6H_5F, $C_6H_5NH_2$ (and deuterated analogs), $C_6H_4F_2$, naphthalene, and a series of disubstituted benzenes. As the molecules become larger or the added substituents introduce low frequency vibrations, the problems with spectral congestion make it exceedingly difficult to excite individual vibronic levels. For instance, C_6F_6 would be a very interesting case for study as the C—C skeletal vibrations become the modes of highest frequency, and they, consequently, should dominate the non-radiative decay; however, the $S_0 \rightarrow S_1$ absorption in C_6F_6 is a broad featureless spectrum under fairly high resolution [2.121]. The basic trends of the relative rates in most of the observed cases are rather similar to benzene. Some differences can be rationalized as due to the possible increase in the number of good acceptors, when the symmetry is lowered by substituent addition, leading to relative rates which increase faster with excitation than in benzene. In some cases, changes in the energy gap and/or the nature of the optical and accepting modes probably account for departures from the benzene behavior. Unfortunately, until more spectroscopic data is obtained for these systems, detailed calculations of their relative rates are not really possible with any degree of certainty.

All of the calculations described above explicitly assume that the S_1 nonradiative decay involves T_1 as the final state (of the S_1 relaxation). There has been some speculation in the literature that the S_1 relaxation proceeds through some higher triplet states T_n, contrary to the above assumptions. Consider what would transpire *if we could prepare a molecule initially in* T_n. The internal conversion rate to lower triplets and ultimately to T_1 must be in the range of $10^{10} - 10^{12} s^{-1}$, so T_1 is virtually "non-existent" on the timescales of the nanosecond flash experiments used to date. On this timescale, it is more realistic to consider the mixed state arising from the coupling together of the various interacting triplet manifolds. The mixed triplet levels that are isoenergetic with S_1 vibronic components have mixed electronic character, but by far the dominant nature of the vibrational levels in this energy region is primarily governed by the T_1 potential surface. In effect, this means that the electronic matrix elements may be affected by the T_n, but the above use of $S_1 - T_1$ Franck-Condon factors is quite reasonable. Actually, a proper analysis probably requires the use of a dynamical Jahn-Teller calculation. The mixing between $n-\pi^*$ and $\pi-\pi^*$ triplets in nitrogen containing heterocyclics represents a similar situation[27].

Possible T_n mixing is an important consideration in the evaluation of absolute decay rates, a hazard that is eliminated by considering relative rates. It should be noted that the "guesstimates" of T_n lifetimes indicate that *if they participate in the S_1 decay*, picosecond experiments could, in principle, catch benzene in these states on their way to T_1.

Table 2.12. C_6D_6 relative $T_1 \to S_0$ decay rates for non-totally symmetric optical modes[a, b]

$\Delta\omega$ [c]	$\omega_{(S_0)}$ Non-promoting optical modes											
	500			1000			1500			2000		
-100	0.86,	0.76,	0.69	0.99,	1.1,	1.2	2.0,	4.4,	9.3	6.2,	25,	76
-60	0.90,	0.82,	0.75	0.94,	0.92,	0.92	1.2,	1.6,	2.3	3.2,	9.0,	22
-40	0.93,	0.86,	0.81	0.94,	0.91,	0.89	1.05,	1.2,	1.4	1.8,	3.4,	6.5
-20	0.96,	0.92,	0.89	0.96,	0.93,	0.91	1.0,	1.0,	1.0	1.1,	1.4,	1.7
0	1.0,	1.0,	1.0	1.0,	1.0,	1.0	1.0,	1.0,	1.0	1.0,	1.0,	1.0
$+20$	1.05,	1.1,	1.2	1.1,	1.1,	1.2	1.1,	1.2,	1.3	1.2,	1.6,	2.3
$+40$	1.1,	1.2,	1.4	1.1,	1.3,	1.5	1.3,	1.7,	2.4	2.2,	5.0,	11
$+60$	1.2,	1.4,	1.7	1.2,	1.6,	2.1	1.6,	2.9,	5.4	4.2,	15.5,	49
$+100$	1.4,	2.0,	3.1	1.6,	2.9,	5.5	3.5,	12.5,	42	9.3,	53,	233

$\Delta\omega$ [c]	$\omega_{(S_0)}$ Promoting optical modes											
	500			1000			1500			2000		
-100	2.0,	2.9,	3.8	2.2,	4.0,	6.4	4.4,	14,	38	7.7,	35,	122
-60	2.0,	2.9,	3.8	2.0,	3.1,	4.4	2.7,	5.8,	11.3	5.5,	2.0,	58
-40	2.0,	3.0,	3.8	2.0,	2.9,	3.9	2.2,	4.0,	6.4	3.8,	10.5,	25
-20	2.05,	3.1,	4.2	2.0,	2.9,	3.8	2.0,	3.1,	4.3	2.4,	4.5,	7.8
0	2.1,	3.2,	4.3	2.0,	3.0,	4.0	2.0,	3.0,	4.0	2.0,	3.0,	4.0
$+20$	2.2,	3.4,	4.8	2.1,	3.3,	4.7	2.2,	3.6,	5.5	2.6,	5.4,	10
$+40$	2.3,	3.8,	5.5	2.3,	3.9,	6.2	2.7,	5.6,	10.6	4.5,	15,	42
$+60$	2.4,	4.2,	6.7	2.5,	5.0,	8.9	3.5,	9.9,	25	7.0,	32,	119
$+100$	2.8,	6.0,	12	3.5,	9.7,	25	7.1,	35,	146	11.1,	72,	345

[a] Values given are for $k(v^n)/k(v^0)$, $n = 1, 2, 3$.
[b] Parameters: $\Delta E_{(T_1 S_0)} = 29{,}847\ \mathrm{cm}^{-1}$, $\omega_k = 1440\ \mathrm{cm}^{-1}$, $\omega_1 = 945\ \mathrm{cm}^{-1}$, $\omega_2 = 2303\ \mathrm{cm}^{-1}$, $X_1 = 1.1$, $X_2 = 0.052$, $\Delta\omega_1 = \Delta\omega_2 = 0$.
[c] $\Delta\omega = \omega^{(S_0)} - \omega^{(T_1)}$.

---

[27] Since the relative positions of the $n-\pi^*$ and $\pi-\pi^*$ states can be altered by the introduction of substituents and by the use of various solvents, it would be interesting to attempt to obtain situations where these triplet levels are degenerate. Because of the differing geometries in these two states, there might still be a barrier to interconversion between them. Hence, low temperature EPR experiments may be able to follow the non-radiative decay between these states (as a function of temperature) to provide a model system for isomerization reactions.

2.9.6 Large Energy Gaps: $T_1 \to S_0$ and $S_1 \to S_0$ Decays in Benzene

Benzene $S_1 \to T_1$ decay has been used as a model of general weak coupling molecules with moderate energy gaps. Although single vibronic level studies of $T_1 \to S_0$ decay are not possible in benzene, the study of this system provides a model for the expected large energy gap behavior. Table 2.12 displays this dependence of relative rates in C_6D_6 for progressions in non-totally symmetric vibrations. Notice how the larger energy gap ($\approx 30,000\,\text{cm}^{-1}$) produces rates that increase faster upon vibrational excitation than do corresponding rates for smaller energy gaps, cf. Table 2.9. This behavior is displayed graphically in Fig. 2.12a which represents the case of initial excitation in the ν_6 mode.

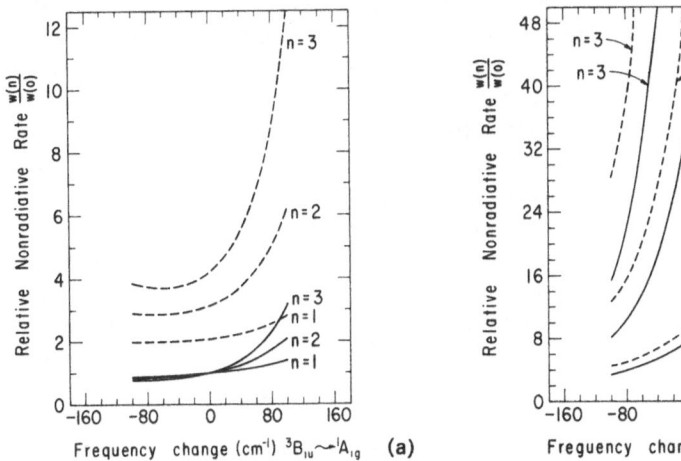

Fig. 2.12a and b. Dependence of the calculated $T_1 \to S_0$ decay rates [2.29] in C_6D_6 on the triplet frequency of the optically excited mode. The plotted frequency change is $\Delta\omega \equiv \omega(S_0) - \omega(T_1)$. The broken lines are for the case that the excited mode promotes the decay process, while the solid curve is for an ordinary optical mode. (a) Non-totally symmetric mode ν_6. Parameters are $\omega_1(S_0) = 945\,\text{cm}^{-1}$, $\omega_2 = 2303\,\text{cm}^{-1}$, $X_2 = 0.05$, $\omega_6(S_0) = 579\,\text{cm}^{-1}$, $\omega_k = 1440\,\text{cm}^{-1}$, $\Delta E(T_1, S_0) = 29,847\,\text{cm}^{-1}$, $X_1 = 1.1$. (b) Totally symmetric optical mode ν_1

Promoting modes have rates which rise much more sharply with excitation than do ordinary accepting modes. Fig. 2.12b shows how the relative rates vary with ν_1 as the "optical mode" in T_1. Notice how the rates rise exponentially with increasing excitation in ν_1 because of the combination of a large value of X for this transition ($X^{T_1 \to S_0}/X^{S_1 \to T_1} \sim 10$) and of the increased energy gap over the $S_1 \to T_1$ decay.

Substantially more electronic energy must be converted into vibrational energy on the lower surface for large energy gaps. This makes the two totally symmetric modes (ν_1 and ν_2) more favorable as accepting modes, as compared to low frequency modes like ν_6. Comparison of Figs. 2.12a and 11a shows that the $T_1 \to S_0$ ν_6 "optical mode" rates would increase more slowly than their

$S_1 \rightarrow T_1$ counterparts, even given identical frequency shifts. This feature is a reflection of the greater emphasis on the v_1 and v_2 modes as dominant acceptors.

While single vibronic level experiments have not yet been possible in the large energy gap case, the above results are useful for the analysis of lower resolution experiments involving $S_1 \rightarrow S_0$ internal conversion in aromatic hydrocarbons. Since the S_1 decays from the low lying vibronic levels proceed through intersystem crossing, the S_1 non-radiative decay rates should be expected to increase linearly (or slightly faster) with increasing excess vibrational energy. The $S_1 \rightarrow S_0$ internal conversion rates are negligible for the lowest vibronic levels in S_1, but because of their exponential increase with excitation, they should ultimately provide the primary decay mechanism for S_1 [2.29, 122].

Experimentally, the high lying vibronic levels of S_1 are populated by optical excitation of higher singlets S_n which are known to undergo very rapid internal conversion to S_1. (Hence, the assumption of similarly rapid T_n decays.) The resulting radiative and non-radiative decays are characteristic of the high lying S_1 levels that are reached in the internal conversion process. LIM and coworkers have indeed demonstrated this to be the case in their experiments on naphthalene, β-naphthol, β-naphthalamine, anthracene, 9,10-dimethylanthracene, phenanthrene, and fluorene [2.123, 124]. Typical experimental data for naphthalene are reproduced in Figs. 2.13 and 14. Fig. 2.13 shows (bottom) the initial linear and subsequent exponential increase in the non-radiative decay rate. The upper part of Fig. 2.13 gives the yield of the sensitized bracetyl phosphorescene, a direct measure of the triplet yield in naphthalene. Note that when the exponential increase in k_{nr} begins, the triplet yield drops off significantly, verifying a transition from predominately $S_1 \rightarrow T_1$ decay at lower energies to $S_1 \rightarrow S_0$ decay at higher energies.

LIM and coworkers also show how the energy dependence of the deuterium isotope effect is accentuated by large energy gap situations that make it harder for low frequency vibrations to compete with the dominant high frequency

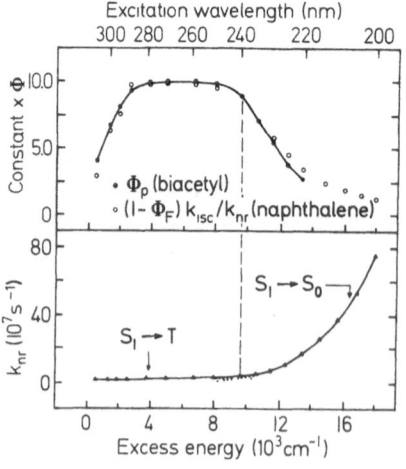

Fig. 2.13. Relative quantum yield of sensitized phosphorescence in biacetyl (top) and non-radiative decay rate of naphthalene (bottom), as a function of excess energy. The top figure also displays the energy dependence of $(1 - \Phi_F) k_{isc}/k_{nr}$ in naphthalene vapor assuming that the total radiationless decay rate (k_{nr}) can be decomposed into $S_1 \rightarrow T$ intersystem crossing (k_{isc}) and $S_1 \rightarrow S_0$ internal conversion in the manner shown in the bottom figure. (Φ_F is the fluorescence quantum yield.) Figure taken from [2.123c]

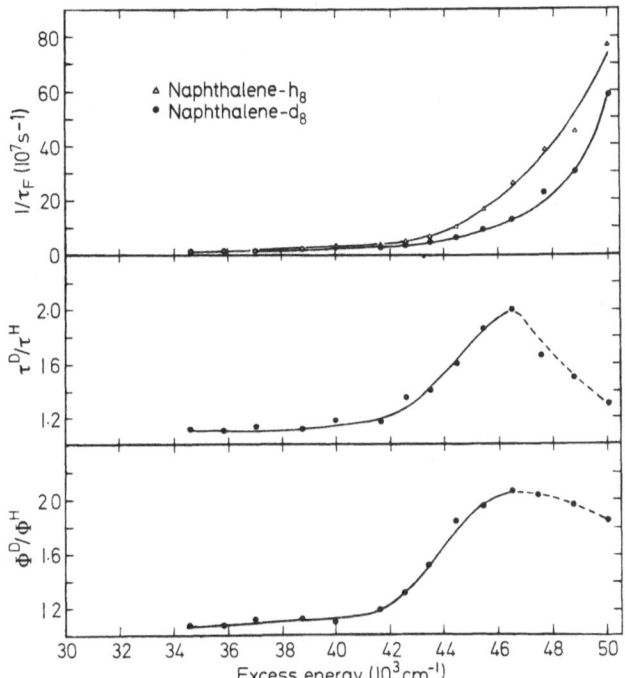

Fig. 2.14. Isotope dependence of fluorescence lifetimes τ and quantum yields Φ in naphthalene vapor. Figure taken from [2.123b]

totally symmetric stretches than in the lower energy-gap cases. Fig. 2.14 exhibits the experimental data for naphthalene. It is also relevant to note here that HEMMINGER and LEE [2.125] have shown that the photochemical decomposition of cyclobutanone has two decay channels one with a linearly increasing rate as the excess vibrational energy is increased and the other with an exponentially increasing rate. These observations are consistent with decomposition following $S_1 \rightarrow T_1$ and $S_1 \rightarrow S_0$ transitions, respectively.

Optical excitation in the S_1 state of naphthalene produces excitation of primarily totally symmetric skeletal vibrations, so the observed $S_1 \rightarrow T_1$ energy dependence of the decay rates is characteristic of excitation in these modes. At lower wavelengths optical excitation produces naphthalene in the S_2 state, whereupon internal conversion to S_1 produces molecules with considerable excitation in high frequency C—H stretches. Thus, *if* this vibrational energy resides in these C—H modes for times on the order of the reciprocal decay rate of the states, the energy dependence of the $S_1 \rightarrow T_1$ decay rates, that proceed through S_2, reflects these C—H vibrations and *may* differ from the lower energy rates which involve optical excitation of S_1. The experimental results of Fig. 2.15 indeed show a change in the energy dependence of k_{nr} when absorption is to S_2. This provides a striking example of the lack of vibrational relaxation in S_1 on the nanosecond timescale, thereby heightening fundamental

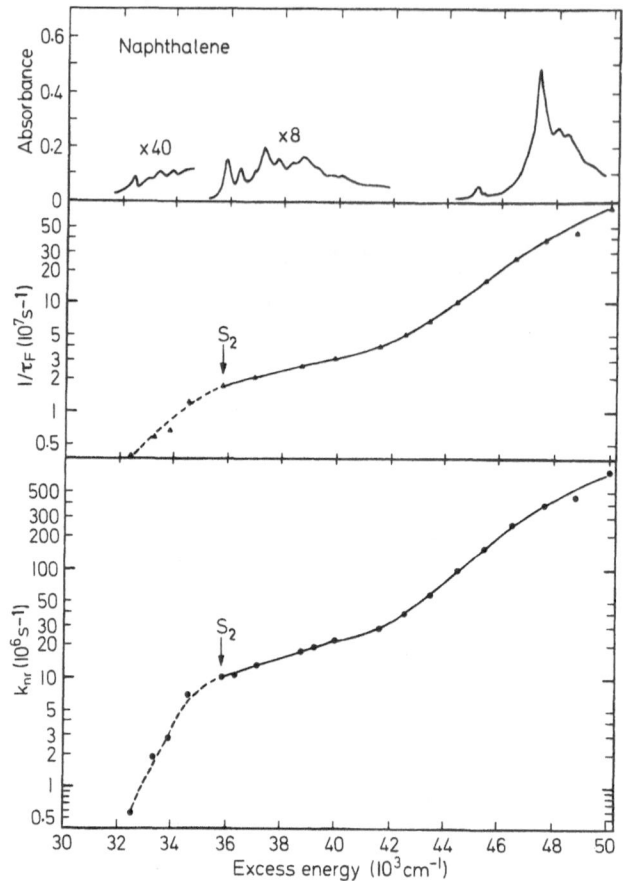

Fig. 2.15. Low-resolution vapor phase absorption spectra of naphthalene-h_8, and excitionation energy (E_{exc} is the excess vibrational energy) dependence of total ($1/\tau$) and non-radiative decay rates. Figure taken from [2.123b]

questions concerning basic theories of unimolecular chemical reactions. This is an important subject which should attract a good deal of attention in the near future, but it is beyond the scope of the present review.

The above discussion is predicated upon the use of a harmonic model for describing the vibronic states of interest. However, with increasing vibrational excitation and increasing size of the molecule, the density of vibronic states grows rapidly. Anharmonic couplings between nearly degenerate vibronic levels of a single electronic state then lead to strong Fermi resonance. Consequently, the higher lying vibronic levels become strongly mixed [2.139]. The effects of this state mixing on the energy dependence of the non-radiative decay rate can be dramatic [2.140]. Optical excitation above the S_1 threshold in aromatic hydrocarbons results in the excitation primarily of totally symmetric C—C stretching vibrations because of the considerable C—C bond length shift in the $S_0 \to S_1$ transition. This excitation in the C—C stretches enhances the non-radiative decay rates for both $S_1 \to T_1$ and $S_1 \to S_0$ [2.25, 29]. But if the actual vibronic states are strongly mixed, the individual vibrational eigenstates each contain only a small portion of the decay enhancing C—C stretches. These mixed states presumably contain a large component of excitations in low frequency bending vibrations since the density of this type of states is greatest at any given excitation energy. Even though the mixed states contain only small pieces of the excited C—C stretches, the absorption spectrum still displays a strong progression in this vibration provided the anharmonic couplings are weak enough that the purely harmonic excited C—C basis states are only spread amongst vibrational eigenstates within a narrow energy range. (If this condition is violated, these zeroth-order vibronic peaks are quite spread out in the absorption spectrum. Thus, this additional source of spectral broadening can even lead to the loss of structure.) Hence, for weak anharmonic coupling, the absorption spectrum does not manifest the mixed nature of the eigenstates. On the other hand, the non-radiative decay rates are very sensitive to the exact composition of each of the individual vibrational eigenstates. Because the excitation of low frequency bending vibrations is expected to result in very small changes in non-radiative decay rates ($S_1 \to T_1$ or S_0), a mixed vibrational state with mostly excited low frequency bends has decay rates which are much closer to those characteristic of the vibrationless level than of purely harmonic levels with excited C—C stretches. In moderate sized molecules the strong vibrational mixing enters only above a threshold excess vibrational energy, E_t. Below E_t, optical excitation leads to the population of fairly uncontaminated excited C—C stretching levels with a concomitant increase in non-radiative decay rates. Near E_t, the proportion of excited C—C stretches in a given vibrational eigenstate begins to decrease, possibly resulting even in a *decrease* in the non-radiative decay rates. High above E_t, the increase in non-radiative decay rate with excess vibrational energy corresponds to the incremental growth associated with the low frequency vibrations which comprise the dominant portion of the eigenstates. In tetracene the total non-radiative decay rate is found to first increase rapidly with excess vibrational energy and then to level off to a much slower incremental increase [2.140]. This is conformity with the expectations of having a large percentage of excited C—C stretches at lower energies and a small percentage at higher ones. In pentacene there are possibly complications due to sequence congestion, so, even at low levels of excitation, the vibrational eigenstates are fairly mixed. The non-radiative decay rate is observed [2.140] to initially decrease and then increase with higher energy excitation. This result is consistent with a decrease in C—C character of individual vibrational eigenstates with increasing energy.

In a pure harmonic picture, internal conversion $S_n \to S_1$ is expected to yield excitation primarily in the totally symmetric C—H stretches (as well as some in the C—C stretches). But in large molecules with a lot of vibrational energy, the eigenstates produced after internal conversion are again strongly mixed.

Consequently, the excited C—H (or C—C) stretching component of a given vibrational eigenstate is small, and non-radiative decay rates of these mixed levels are again characteristic of ones containing excitation in poor accepting, low frequency vibrations. Thus, in the highly mixed limit the non-radiative decay rates should not display a discontinuity at the S_2, S_3, etc., absorption thresholds. Likewise, the high lying S_0^* levels, populated by internal conversion from S_1 in pentacene, are strongly mixed. Because these mixed eigenstates are ones with the parentage of the good accepting C—H and C—C stretches with the more numerous poor accepting vibrations, the absorption spectrum from these S_0^* states is again characteristic of that to be expected as arising from pure harmonic C—H and C—C stretches [2.141].

The theoretical treatment of these mixed states follows just as in the discussion of electronic couplings in the small molecule limit. The zeroth-order harmonic levels correspond to the ABO basis, and the vibrational eigenstates correspond to the molecular eigenstates. However, now coherent excitation of an apparent C—C vibrationally excited spectral peak in the electronic spectrum could lead to something resembling the initial excitation of the non-stationary pure harmonic C—C level whose subsequent time evolution displays an apparent "intramolecular vibrational relaxation" which is just a consequence of the reversible time evolution, or dephasing, of this initial non-stationary state. True intramolecular vibrational relaxation, an irreversible process, necessitates a sufficiently large final density of states [see (2.394) below] to ensure effective irreversibility on the relevant timescales just as in the case of electronic relaxation. The literature, unfortunately, fails to distinguish between the purely quantum mechanical dephasing of the initial non-stationary state and the irreversible relaxation phenomenon which is associated with the irretrievable loss of phase memory into an infinite sink of final states.

2.9.7 Channel 3 Decay in Benzene

Since LIM and coworkers have shown how $S_1 \rightarrow S_0$ decay ultimately surpasses the $S_1 \rightarrow T_1$ decay channel at high excess vibrational energies, it should be expected that benzene also manifest this decay pattern. Benzene, however, has provided a rather perplexing decay scheme at higher vibrational energies within S_1, i.e., even below the S_2 threshold. Below about $3000 \, \text{cm}^{-1}$ of vibrational excitation, the region probed by fluorescence studies, k_{nr} increases roughly linearly with increasing energy. Above $3000 \, \text{cm}^{-1}$, the decay rates, as deduced by CALLOMON et al. [2.104] from broadening of the rotational bands, jump precipitously by a few orders of magnitude. For instance, after three quanta of v_1 are excited in S_1, the further excitation of one quantum of v_6 or v_{16} makes the decay rate increase by a factor of nearly 10^3. This extreme behavior could not even arise if v_6 or v_{16} were promoting modes. Interestingly, after the additional excitation of a single quantum of either v_6 or v_{16}, further excitation in v_1, v_6, or v_{16} does not appear to affect the decay rate anomalously; the additional increase is consistent with those expected for $S_1 \rightarrow S_0$ decays. (The approximate incremental decay rate rises are a factor of 5 for v_1, 2 for v_6, and 2 for v_{16}.) Periodically, there is a level above $3000\ cm^{-1}$ of excess energy which has little v_1 excitation and appears sharp, corresponding to a level with $S_1 \rightarrow T_1$ type nanosecond timescale decay rates.

Table 2.13 shows calculated relative $S_1 \rightarrow S_0$ decay rates for excitation in v_1 and v_6 vibrations. (Excitation of v_{16} is qualitatively similar to that in v_6

Table 2.13. $S_1 \rightarrow S_0$ relative decay rates[a, b] for $6^m 1^n$ single vibronic levels

C_6H_6	1^0	1^1	1^2	1^3	1^4	1^5	C_6D_6	1^0	1^1	1^2	1^3	1^4
6^0	1	9.1	51.4	228	866	2937	6^0	1	15.7	140	926	4876
6^1	1.2	10.8	61.2	271	1025	3472	6^1	1.25	19.5	174	1144	6130
6^2	1.45	13.1	73.8	326	1231	—	6^2	1.60	24.8	220	1440	—

[a] Values given are $k(6^m 1^n)/k(6^0 1^0)$.

[b] Parameters as given in earlier tables and [2.60]:

C_6H_6: $\Delta E_{(S_1 S_0)} = 37{,}853\ cm^{-1}$,
$\omega_k = 1500\ cm^{-1}$, $\omega_1^{(S_0)} = 990\ cm^{-1}$,
$\omega_1^{(S_1)} = 923\ cm^{-1}$,
$\omega_6^{(S_0)} = 606\ cm^{-1}$, $\omega_6^{(S_1)} = 521\ cm^{-1}$,
$\omega_2 = 3130\ cm^{-1}$, $X_1 = 1.4$, $X_2 = 0.05$.

C_6D_6: $\Delta E = 38{,}086\ cm^{-1}$, $\omega_k = 1440\ cm^{-1}$,
$\omega_1^{(S_0)} = 945\ cm^{-1}$, $\omega_1^{(S_1)} = 879\ cm^{-1}$,
$\omega_6^{(S_0)} = 579\ cm^{-1}$, $\omega_6^{(S_1)} = 499\ cm^{-1}$,
$\omega_2 = 2340\ cm^{-1}$, $X_1 = 1.3$, $X_2 = 0.08$.

as can be inferred from Table 2.9.) In order for the $S_1 \rightarrow S_0$ rates to overtake the $S_1 \rightarrow T_1$ rates, the initial disparity of the decay rates cannot be too large (unless the $S_1 \rightarrow S_0$ electronic matrix elements increase with S_1 vibrational excitation) because for high excitation in S_1, the $S_1 \rightarrow T_1$ rates also increase rapidly. Table 2.13 indicates that the $S_1 \rightarrow S_0$ rates may begin to taper-off at high energies. Unfortunately, the harmonic approximation begins to break down for high vibrational excitation in S_1—it is, of course very poor for the S_0 levels iso-energetic with S_1. Table 2.13 shows the expected exponential increase in the rate upon v_1 excitation.

Although the calculated rates do not exhibit the same sudden jump as the experimental ones, they at least are consistent with the observed deuterium isotope effect. However, before considering this agreement, an explanation must be given for the dramatic rate increase in benzene. This rate increase has been mysterious enough that the "new" decay has been ascribed to some unknown "channel 3". It is perhaps useful to begin the analysis by asking why benzene differs substantially from all of the other aromatic hydrocarbons which do not have channel 3 decays. On the other hand, benzene does not show the simple transition from $S_1 \rightarrow T_1$ to $T_1 \rightarrow S_0$ decay as does a number of its higher homologs. What then makes benzene so different from its "brothers?" One feature which singles benzene out is the fact that it has an anomalously large $S_0 \rightarrow S_1$ shift parameter X_{v_1}. Since the channel 3 decay is enhanced by v_1 excitation, the difference in benzene's behavior is very likely tied to this large

X_{v_1}. We already know that excitation of v_1 in S_1 makes the v_1 vibration a better accepting mode for both $S_1 \rightarrow T_1$ and $S_1 \rightarrow S_0$ decays. The increased rate in the later case, cf. Table 2.13, is enormous, so by 1^4, v_1 is definitely a dominant acceptor for the $S_1 \rightarrow S_0$ decay. Hence, because of the large value of X_{v_1}, the v_1 mode jumps from being a moderate acceptor to being a dominant one upon excitation of just a few quanta in this mode.

Up until this point, we have assumed that the electronic matrix elements are slowly varying with vibrational excitation. All cases considered heretofore involve vibrations which remain either poor or good acceptors. But when one mode undergoes a transition in acceptor power, is the assumption of a weakly varying electronic factor tenable? The answer, as discussed below, is no. All of the observed channel 3 behavior in benzene is qualitatively consistent with the hypothesis that the $S_1 \rightarrow S_0$ decay is enhanced by a sharp increase in the electronic factor with excitation in certain modes due to real or incipient $S_1 \rightarrow S_0$ surface crossings. In the case of surface crossings, Franck-Condon factors, as calculated from approximate diabatic surfaces may no longer be appropriate, so quantitative calculations are not possible.

The electronic matrix elements can depend on vibrational excitation through the normal coordinate dependence of the multi-dimensional adiabatic potential surface crossings, cf. (2.292–293) and (2.330–332). FREED and LIN have developed the Q-centroid method to include the effects of this nuclear coordinate variation of the electronic matrix elements [2.23]. Although the Q-centroid calculations of FREED and LIN are only for the vibrationless level, the dependence of the v_1 Q-centroid $Q_{v_1}^0(1^m)$ on 1^m can be inferred from their results [2.29]. As a totally symmetric vibration becomes a better acceptor, $Q^0(0)$ for that mode begins to deviate considerably from the equilibrium position for that mode and tends towards the surface crossing along that normal coordinate as it becomes the dominant acceptor. As noted above, the v_1 mode begins as a moderately good accepting mode for the $S_1 \rightarrow S_0$ transition from 1^0, but by 1^4, v_1 is definitely a dominant acceptor. Thus, in this case $Q_{v_1}^0(1^m)$ should be in the neighborhood of the surface crossing along the v_1 coordinate for $m \gtrsim 4$, with the $m=3$ case questionable. (Actually, because v_2 is also a dominant acceptor, it is more likely that $Q_{v_1}^0$ and $Q_{v_2}^0$ lie on the surface crossing between S_0 and S_1 along the coordinates of both v_1 and v_2.) Thus, the electronic factor should be substantially enhanced for $m \gtrsim 4$, while it is unaffected by isoenergetic excitations which place the energy in poor accepting modes rather than in v_1. This is in accord with experimental observations [2.104]. The $A_3^0(6_0^1 1_0^3)$ line in C_6H_6 shows signs of the onset of diffuseness, while $Q_0^0(7_0^1)$, which is nearby in energy, has sharp rotational structure. $Q_1^0(7_0^1 1_0^1)$ has sharp structure and is close in energy to the already diffuse $A_4^0(6_0^1 1_0^4)$ band. The next member $Q_2^0(7_0^1 1_0^2)$ has a little diffuseness, but the nearby $A_5^0(6_0^1 1_0^5)$ band is very diffuse. As v_7 is an e_{2g} C—H stretch with only a slight ($\approx 1\%$) frequency shift in the $S_1 \rightarrow S_0$ transition, [2.60] it is expected to be a poorer acceptor than three quanta of v_1, and it should not lead to an enhanced electronic factor.

If the proposed mechanism for "channel three" is correct, the deuterium isotope effect should be marked because the v_1 mode is a substantially better acceptor in C_6D_6 than in C_6H_6. As the accepting strength of v_1 in C_6D_6 increases faster with excitation (see Table 2.13) than in C_6H_6, $Q_{v_1}(1^m)$ should reach the surface crossing region at lower excitation in the d_6 than in the h_6 case. Again this is consistent with experiment. Because of isotope shifts in C_6D_6, Q_0^0 is close in energy to A_2^0, and both bands are observed to be sharp. A_3^0 is already diffuse because of the enhanced accepting power of v_1 in C_6D_6, but the nearly isoenergetic Q_1^0 band still has structure. A_4^0 is very diffuse and the nearby Q_2^0 band is fairly diffuse.

Further verification of the above mechanism for "channel three" is probably best obtained from experimental studies of the energy dependence of triplet yields in benzene in order to extract the energy dependence of the $S_1 \rightarrow S_0$ rates. Because triplet yields must generally be measured at pressures for which vibrational relaxation is not negligible, the stochastic model of FUNG and FREED [2.32] should be useful for this analysis. It should also be noted that after crossing from S_1 to S_0, the benzene molecule would find itself far from the equilibrium geometry in S_0. Thus, it is not surprising to find benzevalene formation in solutions due to the rapid quenching of the initially formed S_0 molecules into a "metastable" geometry (just as a metal can be quenched into a metastable phase on rapid cooling). In the gas phase, however, the initially formed hot S_0 relax slowly by vibrational collisions, so it is not surprising to find that they all reach the equilibrium S_0 geometry.

2.9.8 Other Theories

Some other closely related theories of the energy dependence of non-radiative decay rates have been developed, although FISCHER's is the only one for which extensive numerical calculations have been presented [2.93–95, 124]. It is, perhaps, worthwhile to briefly mention certain aspects of these theories.

All of the theories of the energy dependence of non-radiative decay rates begin with either the general Golden-Rule rate expression

$$k_{nr}(si \rightarrow l) = (2\pi/\hbar) \sum_j |v_{si,lj}|^2 \delta(E_{si} - E_{lj}) \tag{2.383}$$

or the equivalent expression that is obtained from the equivalent generating function form

$$k_{nr}(si \rightarrow l) = \hbar^{-2} \int_{-\infty}^{\infty} dt \sum_j |v_{si,lj}|^2 \exp[i(E_{si} - E_{lj})t/\hbar]$$

$$\equiv \hbar^{-2} \int_{-\infty}^{\infty} dt \exp(-i\Delta E_k t/\hbar) f_{si}(t), \tag{2.384}$$

where $v_{si,lj}$ is some effective coupling constant which may include higher-order contributions as in (2.205), (2.79) or (2.80). All then employ some Condon approximation, Q-centroid approximation, or some explicit representation of a parametric dependence of the electronic matrix element on the nuclear coordinates as in (2.253). Using the relationships

$$H_N(s)\chi_{si} = E_{si}\chi_{si} \tag{2.385a}$$

$$H_N(l)\chi_{lj} = E_{lj}\chi_{lj}, \tag{2.385b}$$

where

$$H_N(s) = T_N + U_s(Q) \tag{2.386a}$$

$$H_N(l) = T_N + U_l(Q) \tag{2.386b}$$

and T_N is the nuclear kinetic energy operator, Eq. (2.384) can be rearranged in the form

$$k_{nr}(si \to l) = \hbar^{-2}|C_{s,l}|^2 \int_{-\infty}^{\infty} d(t-t')\langle M(t)M(t')\rangle_{si} \tag{2.387}$$

with the correlation function defined by

$$\langle M(t)M(t')\rangle_{si} = \langle \chi_{si} | \exp[i H_N(s)t/\hbar] M(Q) \exp[-i H_N(l)(t-t')/\hbar]$$
$$\times M(Q) \exp[-i H_N(s)t'/\hbar] | \chi_{si}\rangle . \tag{2.388}$$

In (2.388), $M(Q)$ is the remaining Q-dependent coupling operator, as in (2.253). By expressing (2.388) in the second quantized representation, by introducing a thermal average, and by changing the notation somewhat, it is readily found that the result is identical to FISCHER's correlation function expression which represents the starting point in his theory. Thus, in summary, all theories of the vibronic state dependence of k_{nr} begin from the same basic equation (2.383), (2.384), or one of their many equivalent forms.

The only differences between theories then arise from the choice of different models to represent $U_s(Q)$, $U_l(Q)$, $\chi_{si}(Q)$, $\chi_{lj}(Q)$, and $M(Q)$ as well as any additional approximations of a purely mathematical (or empirical nature).

BRAILSFORD and CHANG [2.126] employed the adiabatic Born-Oppenheimer representation for the $U(Q)$, a harmonic model for the $\chi(Q)$, and the electronic factor as in the work of ENGLMAN and JORTNER [2.88]. The normal modes are also taken to be parallel, the same in s and l, and involve only bond length changes between s and l. The crucial step in the BRAILSFORD-CHANG [2.126] treatment is the expansion of each vibrational overlap integral in a power series in the displacement of the equilibrium position. They retain terms only through those linear in X_α of (2.257) to yield the approximation

$$
\begin{aligned}
|\langle n_\alpha | m_\alpha \rangle|^2 \doteq \; & [1 - (2n_\alpha + 1) X_\alpha] \delta(n_\alpha, m_\alpha) \\
& + (n_\alpha + 1) X_\alpha \delta(n_\alpha + 1, m_\alpha) \\
& + n_\alpha X_\alpha \delta(n_\alpha - 1, m_\alpha),
\end{aligned}
\tag{2.389}
$$

where δ is the Kronecker delta. Given (2.389), the integral (2.384) can be evaluated by use of the saddle point methods, discussed in Section 2.8, to give an analytic representation of the non-radiative decay rate. Rather than reproducing the final formula, we note that the Brailsford-Chang theory is not even applicable to perdeuterobenzene since a glance at Table 2.2 shows that $|n_\alpha - m_\alpha| > 1$ terms contribute substantially in contrast with the assumption (2.389). Their theory is also lacking in its inability to include frequency shifts in the modes.

The theory of SIEBRAND [2.127] differs fundamentally from all the rest as it is based upon the crude factorization method (2.7-8) with a single effective oscillator, the single optical mode, and an "effective" density of states for the remaining vibrations. In order to utilize these undefined "effective" densities of states, SIEBRAND introduced a purely empirical parameterization in an exponential form to mimic the energy-gap law. The resulting expressions for successive rates $k_{nr}(m+1)/k_{nr}(m)$ are "very complicated and therefore not easy to interpret" [2.127]. As the parameters ρ_{eff} and $\langle F \rangle$, see (2.7) and (2.8), of Siebrand's theory are not readily relatable, even in principle, to molecular properties, this crude treatment is grossly inferior to the generating function theories, or their equivalents, where all quantities which enter into the theories are molecular properties that are subject to independent verification. The lack of a precise definition of ρ_{eff} and $\langle F \rangle$ led SIEBRAND [2.127] to omit the requirement of energy conservation that leads to (2.373) for promoting modes in benzene rather than his incorrect results of $(2 m_k + 1)$.

NITZAN and JORTNER [2.128] employed basically the same model, as discussed in Section 2.8, except they attempt to evaluate a closed form expression for the generating function $f_{si}(t)$ of (2.384) which depends on each initial vibronic level i. Using the elegant Feynman operator algebra they succeed in this analysis only for the displaced potential model. The generating functions $f_{si}(t)$ in this case are rather involved expressions (many summations), as compared to the simple partitioning approach (2.299–302) generated from the "Golden-Rule" form of the rate. However, the method should be amenable to actual computations, and, apart from any differences in the numerical approximations, it should yield comparable results to the partitioning method described above. Indeed, model calculations by NITZAN and JORTNER [2.128] for small and large energy gaps are in general qualitative agreement with the results we have already presented.

When frequency shifts are included, Nitzan and Jortner [2.128] are unable to exactly obtain the $f_{si}(t)$ by use of the elegant Feynman operator method[28]. On the other hand, the partitioning approach (2.299–302) trivially leads to the exact rate expression in the harmonic expression. Lin [2.129] has very cleverly derived $f_{si}(t)$ by the following method: The microcanonical rate $k_{nr}(s\rightarrow l; E)$ is defined as the sum of all the individual single level rates $k_{nr}(si\rightarrow l)$ for which $E_{si} = E$, and can be evaluated as the inverse Laplace transform (on β) of the thermal rate $k_{nr}(s\rightarrow l; \beta)$. By choosing the coefficient of $\delta(E_{si} - E)$ in $k_{nr}(s\rightarrow l; E)$, Lin [2.129] obtained a closed form expression for the required generating function, but he only derived its form for the case of small frequency shifts. When $\xi_\alpha = 0$, his results reduce to those of Nitzan and Jortner [2.128] and are

$$f_{si}(t) = \exp\left[-\sum_\alpha X_\alpha + \sum_\alpha X_\alpha \exp(i\omega_\alpha t)\right] \hat{f}_{si}(t),$$

$$\hat{f}_{si}(t) = \left[\exp(i\omega_k t) \sum_{n_k=0}^{m_k} \sum_{n_k'=0}^{m_k-n_k} \frac{(m_k-n_k)!}{(m_k-n_k-n_k')!(n_k'!)^2} \Delta_k^{2n_k'}\right.$$

$$\times (\cos\omega_k t - 1)^{n_k} + \exp(-i\omega_k t) \sum_{n_k=0}^{m_k} \sum_{n_k'=0}^{m_k-n_k} \frac{(m_k-1-n_k)!\Delta_k^{2n_k'}}{(m_k-1-n_k-n_k')!(n_k'!)^2}$$

$$\left.\times (\cos\omega_k t - 1)^{n_k}\right] \sum_{n_1=0}^{m_1} \cdots \sum_{n_N=0}^{m_N} \prod_{\alpha\neq k} \frac{m_\alpha!\Delta_\alpha^{2m_\alpha}}{(m_\alpha-n_\alpha)!(n_\alpha!)^2} \tag{2.390}$$

$$\times (\cos\omega_\alpha t - 1)^{n_\alpha}. \tag{2.391}$$

The integral (2.384) is then to be evaluated by saddle point integration using (2.390) or the generalization thereof involving non-zero frequency shifts. Lin [2.129] circumvented the necessity for evaluating this complicated saddle point equations anew for each different vibronic level χ_{si} by introducing a further numerical approximation. This approximation involves the determination of the saddle point t_0 for the vibrationless level only as in (2.390) for the displaced potential surface model. The non-radiative decay rate is then written approximately as

$$k_{nr}(si\rightarrow l) \approx k_{nr}(s0\rightarrow l)\hat{f}_{si}(t_0) \tag{2.392}$$

where $k_{nr}(s0\rightarrow l)$ is the value of the vibrationless decay rate obtained by using t_0 as in Section 2.8. Eq. (2.392) neglects any variation in the full saddle point due to the time-dependence of $\hat{f}_{si}(t)$ as well as any correction to the pre-ex-

---

[28] Metz [2.142] has successfully completed this difficult derivation. However, by employing other methods, the same results can be obtained in a simple few line derivation [2.143]. This simple derivation also yields a closed form analytic expression for (2.390) which can be used to generate exact recursion relations for the vibronic state dependence of the non-radiative decay rates [2.143].

ponential factor. For the case of the $6^n 1^m$ levels in benzene, the $S_1 \to T_1$ relative non-radiative decay rates obtained by LIN [2.129] are in excellent agreement with those of HELLER et al. [2.25] that are described earlier in this section. Consequently, in this case the approximation (2.392) is valid. More study of the range of validity of (2.392) is clearly desirable. Thus, the approach of NITZAN and JORTNER [2.128] and of LIN [2.129] is essentially identical to that described in previous sections. Because of its apparently different form, however, it results, perforce, in the use of different numerical approximations than the partitioning method. Hence, the two methods complement each other in serving as checks on these numerical approximations.

FISCHER and co-workers have departed from the above approaches by introducing additional auxiliary physical approximations [2.94, 95]. They invoked the usual assumption of unimolecular reaction theory to regard the vibrational energy

$$E_{ex} = E_{si} - E_{s0} \tag{2.393}$$

as being randomly distributed over all the internal degrees of freedom on a timescale that is very short compared to any electronic relaxation times. They took the anharmonic interactions to be responsible for this intramolecular vibrational relaxation in isolated molecules, and he has even applied his theory (with the additional approximations described below) to the case of the $S_1 \to T_1$ decay in naphthalene with $E_{ex} \leq 1600 \, \mathrm{cm}^{-1}$. The naphthalene S_1 vibronic levels in this region are sufficiently sparsely distributed to permit spectroscopic assignment. Consequently, an initially excited vibrational eigenstates level χ_{si} (with sufficiently narrow band light) cannot undergo intramolecular vibrational relaxation to other vibronic levels of S_1 in isolated molecules. The necessary, but definitely not sufficient, requirement for the occurrence of any *irreversible* relaxation, be it vibrational or electronic, is that the decay rate $\hbar/\tau_d$ be much greater than the Poincaré recurrence rate, or

$$\tau_d < \hbar \rho. \tag{2.394}$$

In the region of spectroscopically well separated levels which are individually excited, the effective recurrence times are zero as the levels are described in the vibrational molecular eigenstates basis. This fact has been verified in numerous instances which show that the luminescence originating from the excitation of low lying, single, well isolated levels in isolated molecules is characteristic only of the excited levels, not of any other levels [2.130]. This provides an experimental proof of the absence of intramolecular vibrational relaxation in isolated molecules with sparse level distributions ($\rho \lesssim 1 \, \mathrm{cm}$) when optical excitation selects a single level. The experiments of LIM and coworkers [2.123] have further shown that this intramolecular vibrational relaxation does not even occur at high energies (2.393) where (2.394) would presumably be valid.

These results have led Fischer et al. [2.124] to drop his communicating states model for low excess vibrational energies E_{ex} of (2.393).

Fischer et al. [2.94, 95] also introduced some additional approximations which are of sufficient interest to merit comment[29]. Given the communicating states model, the decay rate should be taken as the microcanonical rate expression considered by Lin [2.129]. However, Fischer employed the additional assumption that the non-radiative decay be represented instead by a thermal rate expression $k_{nr}(s \rightarrow l; \beta^*)$ with the "temperature parameter" β^* being determined as that temperature which results in the average energy E_{ex},

$$E_{ex} = \sum_j \hbar \omega_j^s [\exp(\beta^* \hbar \omega_j^s) - 1]^{-1}. \tag{2.395}$$

The additional assumption (2.395) and the use of a thermal rate implies a distribution of initially decaying levels with the correct average value of E_{ex}, but where there is a dispersion

$$\sigma = [\langle E_{ex}^2 \rangle - \langle E_{ex} \rangle^2]. \tag{2.396}$$

Using the parameters in Fischer's calculations, it is readily shown that [2.27]

$$\sigma = \langle E_{ex} \rangle [(\hbar \omega_{eff}/\langle E_{ex} \rangle) + L^{-1}]^{1/2} \tag{2.397}$$

where ω_{eff} and L are the "effective" frequency and number of oscillators, respectively. For the S_1 state of naphthalene Fischer et al. [2.95] employ the values $\omega_{eff} = 760 \, cm^{-1}$ and $L = 35$, in which case $\sigma \approx 1200 \, cm^{-1}$ when $\langle E_{ex} \rangle$ has the maximum value $1600 \, cm^{-1}$; they employ molecules with a distribution of energies ranging from $400 \, cm^{-1}$ to $2800 \, cm^{-1}$ [2.27]. Even at high energies, the use of an "effective" β^* is poor, e.g. for $E_{ex} = 3 \times 10^4 \, cm^{-1}$ and the same ω_{eff} and L, the dispersion is $\sigma \approx 7000 \, cm^{-1}$ [2.27]. Lin [2.29] has shown how the thermal rate with β^*, $k_{nr}(s \rightarrow l, \beta^*)$ given by (2.395), can be derived as a mathematical approximation to the microcanonical rate $k_{nr}(s \rightarrow l; E)$. Further study is necessary to determine the numerical accuracy of this approximation as it may yield useful results despite its apparently unphysical nature.

For the low lying levels where the communicating states model is invalid, Fischer et al. [2.124] and coworkers have developed a retention model for the evaluation of single vibronic level decay rates. The basic assumptions of this model are that the modes are parallel in both electronic states, the electronic coupling does not vary with vibronic level, the initial excess vibrational energy is retained in the same vibrational modes after the electronic relaxation, and the effective density of states varies slowly enough that the most probable

---

[29] See footnote 20 on p. 128 discussing Fischer's treatment of anharmonicity.

final values $n_j^0(m_j)$ can be evaluated solely in terms of the $n_j \equiv n_j^0(0)$, that are calculated from the decay of the vibrationless level, and the m_j via [2.124]

$$n_j^0(m_j) = n_j + m_j. \tag{2.398}$$

Note that (2.398) simply states that mode j contains the m_j quanta of the initial state plus the n_j that would wind up in vibration j upon the decay of the vibrationless level. For the simple case of the displaced potential surface model, the theory leads to the remarkably simple formula for the relative rates [2.124]

$$k_{nr}(\{m_j\})/k_{nr}(0) = \prod_j \frac{\Gamma(m_j + n_j + 1)}{\Gamma(m_j + 1)\Gamma(n_j + 1)}. \tag{2.399}$$

We can employ the two mode (v_1 and v_2) model of benzene as an example. In this case, Table 2.1 can be taken to give the exact results for this model. Eq. (2.399) is simply evaluated for the v_1 optical mode to give

$$k(1)/k(0) = 1 + n_1 \tag{2.400a}$$

$$k(2)/k(0) = 1 + \tfrac{3}{2}n_1 + n_1^2 \tag{2.400b}$$

$$k(3)/k(0) = 1 + \tfrac{11}{6}n_1 + \tfrac{3}{2}n_1^2 + \tfrac{1}{3}n_1^3 \tag{2.400c}$$

$$k(4)/k(0) = 1 + \tfrac{25}{12}n_1 + \tfrac{47}{24}n_1^2 + \tfrac{17}{34}n_1^3 + \tfrac{1}{12}n_1^4. \tag{2.400d}$$

The two-mode saddle-point equations are solved in Section 2.7 for this model, so, after using (2.326) for v_1, it is found that $n_1 \approx 0.25$. Hence, the relative rates are in the ratio of

$$k(0):k(1):k(2):k(4) = 1:1,25:1,44:1,55:1,65, \tag{2.401}$$

which does not agree with the almost linear dependence of the exact results in Table 2.1. In order to rectify these errors of the retention model, (if it is indeed possible, within the model) it may be necessary to include extra terms in the expansion of ratios of Franck-Condon factors and/or to allow for a variation in the "effective" density of states. The enormous simplicity of the retention model, permitting the calculation of (2.400, 401) by hand, makes the study of these corrections a worthwhile goal.

Most recently METZ [2.142] has employed the single vibronic level generating functions along with the single saddle-point approximation of LIN [2.129] —from the vibrationless level—to consider non-radiative decay rates in benzene. Unfortunately, METZ treats the saddle point, t_0, as an ad hoc empirical parameter in order to force the theory to reproduce absolute rates. t_0 enters into the exponential part of energy gap-type expressions, but, as explained in (2.361) and in the discussion of corrections to the Condon approximation, these

large corrections to absolute rates enter as pre-exponential factors which
have very little to do with the saddle point, t_0, and which, therefore, yield
rather small dependences on the initial vibronic state. The large anharmonic
corrections may also affect the absolute rates in a pre-exponential form rather
than by an alteration of t_0, but more study of this question is necessary.
Thus, METZ incurs a qualitative error in absorbing all these pre-exponential
terms into an empirical saddle point as this procedure affects the vibronic
state dependence in an unwarranted fashion. A simple analogy may be helpful
to elucidate this point. Consider a thermal rate with an Arrhenius form. Let
us say that a given theory is defective in yielding a pre-exponential factor that
is too small by orders of magnitude because of the neglect of important effects.
The analog of METZ's approach is the unwarranted ad hoc adjustment of the
activation energy to reproduce experiments in a limited temperature range.
Despite the above deficiency, METZ does demonstrate that a large change in
X_{v_1} would alter the predicted $\Delta\omega_1$, but we must await further experimental
progress with difficult measurements of these bond lenght shifts.

Acknowledgements. Parts of this review were originally prepared for the 1971
Winter School in Theoretical Chemistry of the Danish Chemical Society, held
in Spätind, Norway as lecture notes for the lectures of Professor STUART RICE.
As these lecture notes were also used by Professor RICE in preparing his review
on photochemistry [2.9], there is a slight amount of overlap which should
emphasize the strong overlap between experimental and theoretical work
on the photophysical and photochemical processes.

Some of the work described in the review results from collaborative efforts
with Professors WILLIAM GELBART, JOSHUA JORTNER, SHENG LIN, and STUART
RICE and with DR. DONALD HELLER.

The research is supported, in part, by Grant # CHE 75-01549 from the
National Science Foundation and a Teacher-Scholar Grant from the Camille
and Henry Dreyfus Foundation.

References

2.1   B. R. HENRY, M. KASHA: Ann. Rev. Phys. Chem. 19, 161 (1968)
2.2   J. JORTNER, S. A. RICE, R. W. HOCHSTRASSER: Adv. Photochem. 7, 149 (1969)
2.3   E. W. SCHLAG, S. SCHNEIDER, S. F. FISCHER: Ann. Rev. Phys. Chem. 22, 465 (1971)
2.4   K. F. FREED: Topics Curr. Chem. 31, 105 (1972)
2.5   B. R. HENRY, W. SIEBRAND: Org. Mol. Photophys. 1, 153 (1973)
2.6   G. W. ROBINSON: Excited States 1, 1 (1974)
2.7   R. VOLZ: Publication F-67037, Centre de Recherches Nucléaires de Strasbourg, 1974
2.8   J. JORTNER, S. MUKAMEL: In *The World of Quantum Chemistry*, ed. by R. DAUDEL and
       B. PULLMAN (Reidel, Boston 1974); "Radiationless Transitions" (to be published)
2.9   S. A. RICE: In *Excited States*, ed. by EDWARD C. LIM (Academic Press, New York)
2.10   M. KASHA: Discuss. Faraday Soc. 9, 14 (1950)

2.11   G. W. ROBINSON, R. P. FROSCH: J. Chem. Phys. **37**, 1962 (1962); **38**, 1187 (1963)
2.12   G. R. HUNT, E. F. McKOY, I. G. ROSS: Aust. J. Chem. **15**, 591 (1962)
2.13   J. P. BYRNE, E. F. McKOY, I. G. ROSS: Aust. J. Chem. **18**, 1589 (1965)
2.14   K. F. FREED, W. M. GELBART: Chem. Phys. Lett. **10**, 187 (1971)
2.15   A. NITZAN, J. JORTNER: Mol. Phys. **24**, 109 (1972); Chem. Phys. Lett. **11**, 458 (1971)
2.16   O. ATABEK, A. HARDISSON, R. LEFEBVRE: Chem. Phys. Lett. **20**, 40 (1973)
2.17   W. G. BREILAND, C. B. HARRIS: Chem. Phys. Lett. **18**, 309 (1973)
2.18   K. K. DOCKEN, J. HINZE: J. Chem. Phys. **57**, 4928 (1972)
2.19   K. F. FREED: J. Chem. Phys. **45**, 4214 and 1714 (1966)
2.20   B. SHARF, R. SILBEY: Chem. Phys. Lett. **4**, 423 (1969); **4**, 561 (1970); **9**, 125 (1971)
2.21   A. NITZAN, J. JORTNER: J. Chem. Phys. **56**, 3360 (1972)
2.22   V. A. KOVARSKII: Fiz. Tverd. Tela **4**, 1636 (1962) [Sov. Phys.-Solid State **4**, 1200 (1963)];
        V. A. KOVARSKII, E. P. SMYAVSKII: *ibid.* **4**, 3202 (1962) [*ibid.* **4**, 2345 (1963)]
2.23   K. F. FREED, S. H. LIN: Chem. Phys. **11**, 409 (1975)
2.24   D. J. DIESTLER: Chapter 3
2.24a H. METIU, D. W. OXTOBY, K. F. FREED: Phys. Rev. A. (to be published)
2.25   D. F. HELLER, K. F. FREED, W. M. GELBART: J. Chem. Phys. **56**, 2309 (1972)
2.26   W. M. GELBART, K. SPEARS, K. F. FREED, J. JORTNER, S. A. RICE: Chem. Phys. Lett. **6**, 345
        (1970)
2.27   D. F. HELLER, K. F. FREED: Intern. J. Quant. Chem. **S6**, 267 (1972)
2.28   D. F. HELLER, W. M. GELBART, K. F. FREED: Chem. Phys. Lett **23**, 56 (1973)
2.29   M. G. PRAIS, D. F. HELLER, K. F. FREED: Chem. Phys. **6**, 331 (1974)
2.30   J. R. CHRISTIE, D. P. CRAIG: Mol. Phys. **23**, 345 and 352 (1972)
2.31   S. H. LIN: J. Chem. Phys. **56**, 4155 (1972)
2.32   K. F. FREED, D. F. HELLER: J. Chem. Phys. **61**, 3942 (1974);
        K. H. FUNG, K. F. FREED: Chem. Phys. **14**, 13 (1976)
2.32a K. F. FREED: J. Chem. Phys. **64**, 1604 (1976); Chem. Phys. Lett. **37**, 47 (1976)
2.33   G. S. BEDDARD, G. R. FLEMING, O. L. J. GIJZEMAN, G. PORTER: Proc. Roy. Soc. (London)
        A**340**, 519 (1974)
2.34   R. G. BROWN, M. G. ROCKLEY, D. PHILLIPS: Chem. Phys. **7**, 41 (1975)
2.35   Y. B. BAND, K. F. FREED: Chem. Phys. Lett. **28**, 328 (1974); J. Chem. Phys. **63**, 3382, 4479
        (1975)
2.36   G. B. KISTIAKOWSKY, C. S. PARMENTER: J. Chem. Phys. **42**, 2942 (1965)
2.37   E. M. ANDERSON, G. B. KISTIAKOWSKY: J. Chem. Phys. **48**, 4787 (1968);
        C. S. PARMENTER, A. H. WHITE: *ibid.* **50**, 1631 (1969)
2.38   A. E. DOUGLAS: J. Chem. Phys. **45**, 1007 (1967)
2.39   L. E. BRUS, J. R. McDONALD: J. Chem. Phys. **61**, 97 (1974); Chem. Phys. Lett. **21**, 283 (1973)
2.40   M. BIXON, J. JORTNER: J. Chem. Phys. **48**, 715 (1968); **50**, 3284, 4061 (1969)
2.41   D. CHOCK, S. A. RICE, J. JORTNER: J. Chem. Phys. **49**, 610 (1968)
2.42   R. S. BERRY, J. JORTNER: J. Chem. Phys. **48**, 2757 (1968)
2.43   W. RHODES, B. R. HENRY, M. KASHA: Proc. Nat. Acad. Sci. (U.S.), **63**, 31 (1969)
2.44   W. RHODES: J. Chem. Phys. **50**, 2885 (1969); Chem. Phys. Lett. **11**, 179 (1971)
2.45   K. F. FREED, J. JORTNER: J. Chem. Phys. **50**, 2916 (1969)
2.46   K. F. FREED: J. Chem. Phys. **52**, 1345 (1970)
2.47   M. BIXON, Y. DOTHAN, J. JORTNER: Mol. Phys. **17**, 109 (1969)
2.48   L. MOWER: Phys. Rev. **142**, 799 (1966); **165**, 145 (1968)
2.49   B. FRIEDLAND, O. WING, R. ASH: *Principles of Linear Networks* (McGraw-Hill, New York
        1961)
2.50   M. L. GOLDBERGER, K. M. WATSON: *Collision Theory* (Wiley, New York 1963)
2.51   A. MESSIAH: *Quantum Mechanics* (Wiley, New York 1961)
2.52   R. A. HARRIS: J. Chem. Phys. **39**, 978 (1963)
2.53   C. P. SLICHTER: *Principles of Magnetic Resonance* (Harper & Row, New York 1963)
2.54   C. TRIC: Chem. Phys. Lett. **21**, 83 (1973)
2.55   R. LEFEBVRE, J. SAVOLAINEN: J. Chem. Phys. **60**, 2509 (1974)

2.56   C. TRIC: J. Chem. Phys. **55**, 4303 (1971)

2.57   A. NITZAN, J. JORTNER: Chem. Phys. Lett. **14**, 177 (1972)

2.58   W. H. MILLER: Phys. Rev. **152**, 70 (1966); thesis, Harvard (1967)

2.59   B. R. HENRY, W. SIEBRAND: J. Chem. Phys. **54**, 1072 (1971)

2.60   D. M. BURLAND, G. W. ROBINSON: J. Chem. Phys. **51**, 4548 (1969); Proc. Natl. Acad. Sci. (U.S.) **66**, 257 (1970)

2.61   D. H. LEVY: J. Chem. Phys. **56**, 5493 (1972)

2.62   W. M. GELBART, K. F. FREED: Chem. Phys. Lett. **18**, 470 (1973)

2.63   M. H. HUI, S. A. RICE: Chem. Phys. Lett. **17**, 474 (1972)

2.64   R. E. SMALLEY, B. L. RAMAKRISHNA, D. H. LEVY, L. WHARTON: J. Chem. Phys. **61**, 4363 (1974)

2.65   R. E. SMALLEY, B. L. RAMAKRISHNA, D. H. LEVY, L. WHARTON (private communication);
       R. E. SMALLEY, L. WHARTON, D. H. LEVY: J. Chem. Phys. **63**, 4977 (1975)

2.66   S. BUTLER, C. KAHLER, D. H. LEVY: J. Chem. Phys. (in press)

2.67   K. SAKURAI, H. P. BROIDA: J. Chem. Phys. **50**, 2404 (1969)

2.68   R. SOLARZ, S. BUTLER, D. H. LEVY: J. Chem. Phys. **58**, 5172 (1973)

2.69   A. NITZAN, J. JORTNER, P. M. RENTZEPIS: Proc. Roy. Soc.(London) A**327**, 367 (1972)

2.70   A. NITZAN, J. JORTNER: J. Chem. Phys. **57**, 2870 (1972)

2.71   A. NITZAN, J. JORTNER, P. M. RENTZEPIS: Chem. Phys. Lett. **8**, 445 (1971)

2.72   P. WANNIER, P. M. RENTZEPIS, J. JORTNER: Chem. Phys. Lett. **10**, 102 and 193 (1971);
       G. E. BUSCH, P. M. RENTZEPIS, J. JORTNER: *ibid.* **11**, 437 (1971); J. Chem. Phys. **56**, 361 (1972)

2.73   K. EVANS, D. HELLER, S. A. RICE, R. SCHEPS: J. C. S. Faraday II **69**, 856 (1973)

2.74   C. A. LANGHOFF, G. W. ROBINSON: Mol. Phys. **26**, 249 (1973); Chem. Phys. **5**, 1 (1974); **6**, 34 (1974);
       J. O. BERG, C. A. LANGHOFF, G. W. ROBINSON: Chem. Phys. Lett. **29**, 305 (1974);
       J. FRIEDMAN, R. M. HOCHSTRASSER: Chem. Phys. **6**, 155 (1975); Chem. Phys. Lett. **32**, 414 (1975);
       H. K. HONG: Chem. Phys. **9**, 1 (1975)

2.75   J. M. DELORY, C. TRIC: Chem. Phys. **3**, 54 (1974)

2.76   C. B. MOORE, E. S. YEUNG: J. Chem. Phys. **58**, 3988 (1973); **60**, 2139 (1974);
       R. G. MILLER, E. K. C. LEE: Chem. Phys. Lett. **33**, 104 (1975)

2.77   O. ATABEK, R. LEFEBVRE: J. Chem. Phys. **59**, 4145 (1973)

2.78   K. MOROKUMA, K. F. FREED: J. Chem. Phys. **61**, 4342 (1974)

2.79   C. S. PARMENTER, H. M. POLAND: J. Chem. Phys. **51**, 1551 (1969)

2.80   A. VILLAEYS (unpublished work)

2.81   K. F. FREED, J. JORTNER: J. Chem. Phys. **52**, 6272 (1970)

2.82   W. SIEBRAND: In *The Triplet State*, ed. by A. B. ZAHLAN (Cambridge University Press, London 1967) p. 31

2.83   S. MURATA, C. IWANAGA, T. TODA, H. KOKUBUN: Chem. Phys. Lett. **13**, 101; **15**, 152 (1972)

2.84   F. K. FONG, S. L. NABERHUIS, M. M. MILLER: J. Chem. Phys. **56**, 4020 (1972);
       F. K. FONG: *Theory of Molecular Relaxation* (Wiley Interscience, New York 1975) Ch. 6

2.85   S. H. LIN: J. Chem. Phys. **44**, 3759 (1966);
       S. H. LIN, R. BERSOHN: J. Chem. Phys. **48**, 2732 (1968)

2.86   R. KUBO, Y. TOYOZAWA: Progr. Theoret. Phys. **13**, 160 (1955);
       R. KUBO: Phys. Rev. **86**, 929 (1952)

2.87   K. HUANG, A. RHYS: Proc. Roy. Soc. (London) A**204**, 406 (1950);
       J. J. MARKHAM: Rev. Mod. Phys. **31**, 956 (1959)

2.88   R. ENGLMAN, J. JORTNER: Mol. Phys. **18**, 145 (1970)

2.89   T. L. HILL: *Introduction to Statistical Thermodynamics* (Addison-Wesley, Reading, Mass. 1960)

2.90   W. M. GELBART, K. F. FREED, S. A. RICE: J. Chem. Phys. **52**, 2460 (1970)

2.91   W. M. VISSCHER: Ann. Phys. (N.Y.) **9**, 194 (1960);
       K. S. SINGWI, A. SJÖLANDER: Phys. Rev. **120**, 1093 (1960)

2.92   M. LAX: J. Chem. Phys. **20**, 1752 (1952);
       R. C. O'ROURKE: Phys. Rev. **91**, 265 (1953);

D. E. McCumber: Phys. Rev. **133**, A163 (1964);
Y. E. Perlin: Usp. Fiz. Nauk **80**, 553 (1963) [Sov. Phys. Usp. **6**, 542 (1964)]

2.93 S. Fischer: J. Chem. Phys. **53**, 3195 (1970)

2.94 S. Fischer: Chem. Phys. Lett. **4**, 333 (1969);
S. Fischer, E. W. Schlag: Chem. Phys. Lett. **4**, 393 (1969)

2.95 S. Fischer, E. W. Schlag, S. Schneider: Chem. Phys. Lett. **11**, 383 (1971);
S. Fischer: Chem. Phys. Lett. **11**, 577 (1971); **17**, 25 (1972);
S. Fischer, S. Schneider: Chem. Phys. Lett. **10**, 392 (1971)

2.96 G. R. Fleming, O. J. Gijzeman, S. H. Lin: Chem. Phys. Lett. **21**, 527 (1973)

2.97 A. Nitzan, J. Jortner: Theoret. Chim. Acta **30**, 217 (1973)

2.98 R. Scheps, D. Florida, S. A. Rice: J. Chem. Phys. **61**, 1730 (1974)

2.99 G. Herzberg: *Molecular Spectra and Molecular Structure I, Spectra of Diatomic Molecules*, 2nd ed. (Van Nostrand, Princeton 1950)

2.100 B. K. Selinger, W. R. Ware: J. Chem. Phys. **53**, 3160 (1970);
C. S. Parmenter, M. W. Schuyler: Chem. Phys. Lett. **6**, 339 (1970)

2.101 K. G. Spears, S. A. Rice: J. Chem. Phys. **55**, 5561 (1971)

2.102 A. S. Abramson, K. G. Spears, S. A. Rice: J. Chem. Phys. **56**, 2291 (1972)

2.103 C. Guttman, S. A. Rice: J. Chem. Phys. **61**, 651 (1974)

2.104 J. H. Callomon, J. E. Parkin, R. Lopez-Delgado: Chem. Phys. Lett. **13**, 125 (1972);
J. H. Callomon (private communication)

2.105 C. S. Parmenter, M. D. Schuh: Chem. Phys. Lett. **13**, 120 (1972)

2.106 J. P. Byrne, I. G. Ross: Aust. J. Chem. **24**, 1107 (1971)

2.107 A. Nitzan, J. Jortner: Chem. Phys. Lett. **13**, 466 (1972)

2.108 A. E. W. Knight, B. K. Selinger, I. G. Ross: Aust. J. Chem. **26**, 1159 (1973)

2.109 G. Loper, E. K. C. Lee: Chem. Phys. Lett. **13**, 140 (1972)

2.110 H. von Weyssenhoff, F. Kraus: J. Chem. Phys. **54**, 2387 (1971)

2.111 W. R. Ware, A. M. Garcia: J. Chem. Phys. **61**, 187 (1974)

2.112 M. G. Rockley, D. Phillips: Chem. Phys. Lett. **21**, 181 (1973)

2.113 C. Guttman, S. A. Rice: J. Chem. Phys. **61**, 661 (1974)

2.114 M. H. Hui, S. A. Rice: J. Chem. Phys. **61**, 833 (1974)

2.115 U. Laor, P. K. Ludwig: J. Chem. Phys. **54**, 1054 (1971);
J. C. Hsieh, U. Laor, P. K. Ludwig: Chem. Phys. Lett. **10**, 412 (1971)

2.116 E. W. Schlag, S. Schneider, D. W. Chandler: Chem. Phys. Lett. **11**, 474 (1971)

2.117 S. Iwata, K. F. Freed: J. Chem. Phys. **61**, 1500 (1974)

2.118 R. M. Badger: J. Chem. Phys. **2**, 128 (1934); **3**, 710 (1935)

2.119 E. W. Schlag, H. von Weyssenhoff: J. Chem. Phys. **53**, 3108 (1970)

2.120 B. R. Henry, W. Siebrand: J. Chem. Phys. **49**, 5369 (1968);
R. J. Hayward, B. R. Henry: Chem. Phys. **12**, 387 (1976)

2.121 R. Scheps, S. A. Rice (private communication)

2.122 G. S. Beddard, G. R. Fleming, O. L. J. Gijzeman, G. Porter: Chem. Phys. Lett. **18**, 481 (1973)

2.123 a) C. S. Huang, J. C. Hsieh, E. C. Lim: Chem. Phys. Lett. **28**, 130 (1974);
b) J. C. Hsieh, C. S. Huang, E. C. Lim: J. Chem. Phys. **60**, 4345 (1974);
c) J. C. Hsieh, E. C. Lim: J. Chem. Phys. **61**, 737 (1974);
d) C. S. Huang, J. C. Hsieh, E. C. Lim: Chem. Phys. Lett. **37**, 349 (1976)

2.124 S. F. Fischer, E. C. Lim: Chem. Phys. Lett. **26**, 312 (1974);
S. F. Fischer, A. L. Stanford, E. C. Lim: J. Chem. Phys. **61**, 582 (1974)

2.125 J. C. Hemminger, E. K. C. Lee: J. Chem. Phys. **56**, 5284 (1972)

2.126 A. D. Brailsford, T. Y. Chang: J. Chem. Phys. **53**, 3108 (1970)

2.127 W. Siebrand: J. Chem. Phys. **54**, 363 (1971)

2.128 A. Nitzan, J. Jortner: J. Chem. Phys. **55**, 1355 (1971); **56**, 2079 (1972)

2.129 S. H. Lin: J. Chem. Phys. **58**, 5760 (1973)

2.130 C. S. Parmenter: Adv. Chem. Phys. **22**, 365 (1972)

2.131 A. Villaeys, K. F. Freed: Chem. Phys. **13**, 271 (1976)

2.132 C. S. HUANG, E. C. LIM: J. Chem. Phys. **62**, 3826 (1975)
2.133 Y. MIKAMI, K. MIZUNOYA, T. NAKAJIMA: Chem. Phys. Lett. **30**, 373 (1975)
2.134 N. SHIMAKURA, Y. FUJIMURA, T. NAKAJIMA: Theoret. Chim. Acta **37**, 77 (1975)
2.135 U. BOESL, H. J. NEUSSER, E. W. SCHLAG: Chem. Phys. Lett. **31**, 1, 7 (1975)
2.136 C. A. THAYER, A. V. POCIUS, J. T. YARDLEY: J. Chem. Phys. **62**, 3712 (1975)
2.137 P. J. VERGRAT, J. VAN DER WAALS: Chem. Phys. Lett. **36**, 283 (1975)
2.138 L. WUNSCH, H. J. NEUSSER, E. W. SCHLAG: Chem. Phys. Lett. **32**, 210 (1975)
2.139 A. E. W. KNIGHT, C. M. LAWBURGH, C. S. PARMENTER: J. Chem. Phys. **63**, 4336 (1975)
2.140 S. OKAJIMI, E. C. LIM: Chem. Phys. Lett. **37**, 403 (1976)
2.141 B. SOEP: Chem. Phys. Lett. **33**, 108 (1975);
       R. K. SANDER, B. SOEP, R. N. ZARE: J. Chem. Phys. **64**, 1242 (1976)
2.142 F. METZ: Chem. Phys. **9**, 121 (1975)
2.143 M. PAGITSAS, K. F. FREED (to be published)

3. Vibrational Relaxation of Molecules in Condensed Media

D. J. DIESTLER

With 9 Figures

3.1 Introductory Remarks

Even a casual reading of the other chapters of this book should impart an appreciation of the important role played by vibrational relaxation in a wide variety of physical and chemical processes. As used just now, the term *vibrational relaxation* (which will henceforth frequently be abbreviated as VR) should be taken quite generally to mean the process by which energy associated with the *vibrational* degree(s) of freedom of a system is transported to *other* (perhaps also vibrational) degrees of freedom. However, throughout the remainder of this chapter our considerations will be essentially restricted, at least from a theoretical standpoint, to VR of diatomic molecules (or of the lowest-frequency modes of polyatomics) in their ground electronic states in dilute solutions (either solid or liquid) of monatomic solvents. We shall refer to such systems as *prototypic*.

The motivation for the apparently severe restriction to such ideal systems is, of course, the desire to simplify not only the theoretical analysis but also the interpretation of experimental data. From a theoretical viewpoint our prototypic system may be idealized as a harmonic oscillator coupled to the radiative continuum and to a heat bath (solvent). This model may be handled relatively easily. Unfortunately, it seems that the difficulty of executing a definitive measurement of VR is in proportion to the ideality of the system under investigation, and so there is a paucity of data on systems conforming to our prototype. Nevertheless, some interesting correlations between theory and experiment can be made.

This chapter is not intended to be a comprehensive review of the entire field of vibrational relaxation, but rather an extensive introduction to the subject for the novice who is moderately well versed in quantum and statistical mechanics. Since the author attempts to make a living doing theory, the representation is heavily biased in favor of the theoretical interpretation of experiments, especially spectroscopic experiments in which the population of a preferentially excited vibrational mode is monitored directly as a function of time. Of course, the experimental methods and the theoretical descriptions and models emphasized here are those with which the author is most conversant. Although he has made every effort to discuss alternative approaches at the appropriate junctures, undoubtedly some relevant works have escaped his attention and to the authors of these he apologizes in advance.

The remainder of the chapter is organized as follows. In Section 3.2 experimental techniques of investigating VR are reviewed, accentuation being given the direct spectroscopic methods, as mentioned above. Each of these techniques yields a characteristic time (τ) related more or less directly to the fundamental microscopic physical processes responsible for the relaxation. Since the measurement is performed on a macroscopic system, this characteristic time must involve an average of some microscopic dynamical quantity over the statistical-mechanical ensemble characterizing the system. By adopting suitable models for the system of interest, we may effect this ensemble average to obtain τ explicitly in terms of microscopic parameters and thermodynamic variables (e.g. temperature and density).

In Section 3.3 we employ the ZWANZIG [3.1] projection-operator formalism to describe the dynamics of relaxation in our prototypic model: a harmonic oscillator coupled to a heat bath. First, an exact equation of motion in generalized Langevin form is obtained for $G(t)$, the expectation value of the occupation number of the oscillator, known to be in a definite initial nonequilibrium state. This equation is then solved in the van Hove [3.2] weak-coupling limit for the case in which the oscillator-bath coupling is linear in the oscillator coordinate. In this case $G(t)$ is found to decay exponentially with a time constant (vibrational lifetime) $\tau' = \tau_v/2$, where τ_v is the vibrational correlation time, or dephasing time, i.e., the time constant associated with the exponentially decaying time-correlation function of the oscillator coordinate, $G_v(t) \equiv \langle Q(0) Q(t) \rangle$. The approach taken in Section 3.3 is easily generalizable and leads in a rather natural and straightforward way not only to relations between different characteristic relaxation times but also to useful explicit expressions for these times.

In Section 3.4 we derive a general expression for the rate $W_{\alpha \to \beta}$ of nonradiative vibrational transitions in our prototypic model. The zero-order states of the system are taken as direct products of the unperturbed internal states (rotational-vibrational states of the diatomic molecule labelled by α, β, etc.) and external states (translation-like states of the solvent), the external states depending, in general, upon the internal states. Transitions $(\alpha \to \beta)$ between the internal states are induced by off-diagonal elements of the Hamiltonian matrix in the zero-order representation. We express $W_{\alpha \to \beta}$ as the expectation value of an appropriately defined projection operator $\hat{P}_\beta$, i.e., as

$$W_{\alpha \to \beta} \propto \mathrm{Tr}\{\hat{\rho}(t)\hat{P}_\beta\},$$

where $\hat{\rho}(t)$ is the full density operator of the system. Evaluating $\hat{\rho}$ to first order in the off-diagonal perturbation, we obtain $W_{\alpha \to \beta}$ approximately as the Fourier transform of an autocorrelation function of the coupling potential. This form is then shown to be precisely equivalent to the usual Golden Rule expression for $W_{\alpha \to \beta}$.

In Section 3.5 we demonstrate that, for the two-state model, the Golden Rule rate derived in Section 3.4 is simply related to the inverse of the vibrational lifetime τ'.

We devote Sections 3.6 and 3.7 to the problem of obtaining approximate closed-form expressions for the vibrational transition rate $W_{\alpha \to \beta}$ and related characteristic times, in particular the vibrational lifetime τ', in terms of molecular properties and thermodynamic state variables. Section 3.6 deals with diatomics in rare-gas matrices and Section 3.7 with diatomics in rare-gas liquids. We compare theory with experiment for carbon monoxide in argon matrices. We also predict transition rates for selected diatomics (H_2, D_2, CO, N_2, and O_2) in liquid rare gases (Ne and Ar). Alternative methods of calculating $W_{\alpha \to \beta}$ are reviewed and discussed.

Finally, we conclude in Section 3.8 with a summary of our principal points and a discussion of the limitations and shortcomings of some of the methods and models considered.

3.2 Experimental Methods and Results

We are concerned here with outlining the principles upon which the experimental methods rest rather than the detailed procedures by which they are executed. The various techniques of observing VR are conveniently separable into two categories—those which employ the scattering and/or absorption of electromagnetic radiation, i. e., spectroscopic methods, and those which do not. The most commonly used techniques fall into the former class. Of those in the latter, ultrasonic absorption seems to be the method of choice.

3.2.1 Ultrasonic Absorption [3.3, 4]

Consider a polyatomic fluid at rest, in which the translation-like (external) degrees of freedom are in equilibrium with the internal degrees of freedom, each set of degrees of freedom being characterized by a single temperature. A sound wave propagating through the fluid creates alternately compressed and rarefied regions, in which the external and internal degrees of freedom are no longer in equilibrium with each other. As the system attempts to return to equilibrium, energy flows between the external and internal degrees of freedom and the amplitude of the sound wave is diminished, i. e., sound is absorbed. This mechanism of sound absorption is often referred to as thermal relaxation and it is characterized by the time τ required for the system to reduce its deviation from the equilibrium state to e^{-1} its original value.

Letting $\varepsilon_i(T_i)$ be the instantaneous value of the internal energy and $\varepsilon_i(T_{tr})$ the value it would have were the system to remain at equilibrium, we can write

$$\frac{d\varepsilon_i}{dt} = -\tau^{-1} \left[\varepsilon_i(T_i) - \varepsilon_i(T_{tr}) \right] . \tag{3.1}$$

For periodic sound waves of frequency $v=\omega/2\pi$ the effective specific heat at constant volume may be expressed as

$$C_{v\omega} = \tilde{C}_v + C_i/(1+i\omega\tau)$$
$$= C_v - i\omega\tau C_i/(1+i\omega\tau), \tag{3.2}$$

where $\tilde{C}_v$ is the contribution arising from external degrees of freedom and C_i that from internal degrees of freedom. Neglecting the "classical" mechanisms of absorption due to viscosity and heat conduction, one may show that the speed of sound c and the absorption per unit wavelength μ are given by

$$c^{-2} \simeq c_0^{-2}[1 - A\omega^2\tau'^2/(1+\omega^2\tau'^2)] \tag{3.3}$$

and

$$\mu = \pi A c^2 c_0^{-2}\omega\tau'/(1+\omega^2\tau'^2), \tag{3.4}$$

where

$$A \equiv \frac{C_i \Delta}{C_v(C_p - C_i)} \tag{3.5}$$

$$\tau' \equiv (C_p - C_i)\tau/C_p, \tag{3.6}$$

and

$$\Delta \equiv C_p - C_v. \tag{3.7}$$

In (3.3)–(3.7) c_0 and c are the speeds of sound at frequencies $v=0$ and $v=\omega/2\pi$, and C_v and C_p are specific heats at constant volume and pressure, respectively. From (3.4) we see that a plot of v vs ω has a maximum at

$$\omega_{max} = \tau'^{-1}, \tag{3.8}$$

from which we can deduce the relaxation time τ by (3.6).

In arriving at (3.4) we have implicitly assumed the existence of only a single relaxation mechanism. Such is certainly the case in diatomic liquids, which have only one internal vibrational mode. As an example, in Fig. 3.1 we show plots of μ vs v for liquid Cl_2 at its equilibrium vapor pressure at several different temperatures [3.5]. The experimental data are fitted well by (3.4) and the calculated τ'^{-1} may be taken as a measure of the rate at which vibrational-translational energy exchange occurs. Note that τ' decreases with increasing temperature, as we except.

There are two principal limitations of ultrasonic absorption. First, even for many systems in which only a single relaxation mechanism is operative $\omega\tau \ll 1$ for the entire range of frequencies accessible to presently available instrumentation. More important, in the majority of "interesting" cases, where

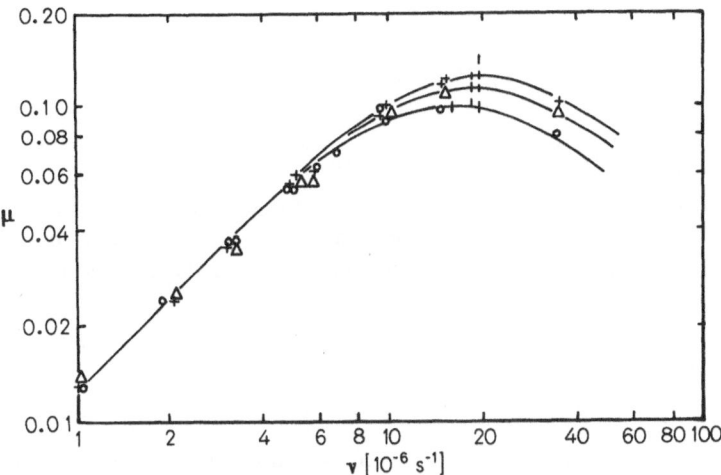

Fig. 3.1. Absorption per unit wavelength, μ, as a function of frequency, ν, for liquid Cl_2 at its vapor pressure. (O) 273 K; ($\triangle$) 293 K; (+) 303 K

several different modes of relaxation participate, theoretical analysis of the absorption curve is extremely difficult. However, there are a number of systems in which the τ's associated with various mechanisms are sufficiently different that the absorption curve may be described as a superposition of single-mechanism curves.

3.2.2 Rayleigh-Brillouin Scattering [3.6–8]

We turn now to the first of the methods utilizing the scattering of electromagnetic radiation.

The thermal motion of molecules in a fluid causes local fluctuations in the density and also in the orientation of molecules in regions of dimensions small compared to the wavelength of the incident light. These fluctuations in turn lead to fluctuations in the local dielectric constant, thus giving rise to light scattering. Orientation fluctuations result in depolarization of the scattered light, whereas density fluctuations do not. Hence the contributions to the scattered light can be separated experimentally. It can be shown [3.9] that the intensity of the polarized component of Brillouin-scattered light is

$$I(\mathbf{R},\omega) = N I_0 \alpha^2 k_i^4 \sin^2 \phi \mathscr{S}(k,\omega)/(2\pi R^2), \tag{3.9}$$

where $\mathscr{S}(k,\omega)$ is a generalized structure-factor, given by

$$\mathscr{S}(k,\omega) \equiv \int d\mathbf{r} e^{i\mathbf{k}\cdot\mathbf{r}} \int dt\, e^{-i\omega t} G(r,t) \tag{3.10}$$

and $G(r,t)$ is a density-density autocorrelation function

$$G(r,t) \equiv N^{-1} \int d\mathbf{r}' \langle \rho[\mathbf{r}' - r(0),0] \rho(\mathbf{r}',t) \rangle . \tag{3.11}$$

In (3.9) N is the number of molecules and α is their effective polarizability; k_i is the wavevector of the incident light, R is radius vector from the scattering center to the point of observation, and ϕ is the angle between R and the electric field vector associated with the incident light; ω is the *shift* in the frequency of the scattered light and k is the magnitude of the *change* in the wavevector of the scattered light, i. e.

$$k = |k_f - k_i| = 2k_i \sin(\theta/2), \tag{3.12}$$

where θ is the scattering angle.

For the case in which only one relaxation mechanism is present in the system of interest, Mountain [3.6–8] has solved the linearized hydrodynamic equations approximately to obtain an expression for $\mathscr{S}(k,\omega)$ in terms of a relaxation time τ' which characterizes a frequency-dependent contribution to the bulk viscosity

$$\eta_i = \eta_1 (1 + i\omega\tau')^{-1} . \tag{3.13}$$

Note the analogy between (3.2) and (3.13). The spectrum of the scattered light can be represented approximately as a sum of four Lorentzian components—two central (Rayleigh) components and two Brillouin components centered at $\omega = \pm vk$, where $v = v(k)$ is the phonon speed.

From material parameters determined independently, including τ' taken to be equal to τ measured by ultrasonic absorption, Mountain [3.6–8] has calculated the phonon speed and the ratio of the intensity of the central to the Brillouin components for CS_2 at $20\,°C$. The discrepancy between calculated and experimental values of $v(k)$ is within the uncertainty of the measured value, although the same is not true of the intensity ratio.

The drawbacks of the Rayleigh-Brillouin scattering technique are essentially the same as those inherent in ultrasonic absorption.

3.2.3 Optical Absorption and Emission Spectra

Figure 3.2 shows a schematic diagram of the potential energy curves corresponding to the electronic ground state and an excited state of a diatomic molecule. Note that the diagram can also refer to a polyatomic molecule in the harmonic approximation, where the abscissa is now a particular normal-mode coordinate[1]. Light of the appropriate wavelength propagating through

---

[1] See Ch. 2.

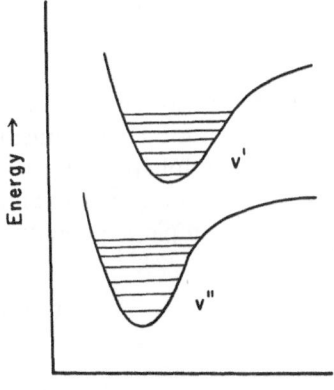

Fig. 3.2. Vibronic levels for a typical diatomic molecule

a solution of the diatomic molecules induces transitions $(v'' \rightarrow v')$ between specific electronic-vibrational (vibronic) levels of the diatomic, thereby giving rise to an absorption spectrum. It is also possible to produce molecules in the various excited states v' by exposing the solution to a beam of high-energy electrons or X-rays. The excited states may subsequently decay by a variety of modes (or channels), both radiative and non-radiative. The intensity of light emitted as a function of wavelength constitutes the emission spectrum, which is an important source of information on VR. In fact, the mere existence of emission from $v'>0$ implies that the non-radiative (vibrational) lifetime of the state v' is of the same order as or longer than the radiative (natural) lifetime, for if vibrational relaxation is very fast, i. e., the non-radiative lifetime is very short, emission only from $v'=0$ is observed. For example, SCHOEN and BROIDA [3.10] have studied the emission spectra of O_2 and N_2 in rare-gas matrices at ~ 4.2 K. For O_2 in Ar, Kr, and Xe they observed only $0 \rightarrow v''$ transitions in the Herzberg band $(C^3 \Sigma_u^+ \rightarrow X^3 \Sigma_g^-)$, whereas for N_2 in Xe they saw emissions from $v'=0, 1$, and 2 in the Vegard-Kaplan system $(A^3 \Sigma_u^+ \rightarrow X^1 \Sigma_g^+)$. As the radiative lifetime of N_2 is estimated to be about 10 s., the vibrational lifetimes of the states $v'=0, 1, 2$ of N_2 must be of the order of seconds in Xe. In contrast, the lifetime of the corresponding states of the $C^3 \Sigma_u^+$ manifold of O_2 must be considerably shorter.

A similar investigation by TINTI [3.11] of the $A^3 \Sigma^+ \rightarrow X^2 \pi_i$ bands of OH and OD in Ne matrices at 4.2 K has established that the vibrational lifetimes of the states $v'=0-4$ belonging to the $A^3 \Sigma^+$ manifold are longer than about 10^{-6} s.

Using an argon-ion laser as the excitation source, SHIRK and BASS [3.12] have obtained the emission spectra of CuO in rare-gas matrices. In Xe and Kr matrices they observed a $1 \rightarrow 0$ band[2], concluding that the vibrational lifetime of the upper state $v'=1$ is of the order of 10^{-8} s.

---

[2] This band was not assigned [3.12].

If the excitation source is shut off rapidly and the intensity of light emitted at a given wavelength is monitored as a function of time, a *decay curve* is obtained. It is possible in certain instances to calculate quantitatively the non-radiative lifetime of the vibronic state v' from the decay curve, if the radiative lifetime is known. For example, in case the excitation mechanism populates only a single well-defined state v' of the upper manifold (see Fig. 3.2) and this state decays by only two channels, namely emission of a photon and vibrational relaxation, then the decay curve assumes the simple exponential form

$$I(t) = I(0)\,e^{-t/\tau_F} \qquad\qquad (3.14)$$

where

$$\tau_F = 1/(\tau_{v'}^{-1} + \tau_r^{-1}) \qquad\qquad (3.15)$$

and τ_r and $\tau_{v'}$ are, respectively, the radiative and non-radiative lifetimes. A plot of $\log I(t)$ vs t is a straight line of slope $-\tau_F^{-1}$. Knowing τ_r, one can calculate $\tau_{v'}$ using (3.15). Even if τ_r is unknown, τ_F is a lower bound on τ_v, i.e. $\tau_v \geq \tau_F$. TINTI and ROBINSON [3.13] exploited this idea in a study of the Vegard-Kaplan $(A^3\Sigma_u^+ \rightarrow X^1\Sigma_g^+)$ and second positive $(C^3\pi_u \rightarrow B^3\pi_g)$ band systems of N_2 in rare-gas matrices over a range of temperatures. In agreement with SCHOEN and BROIDA [3.10] they found that the (lower bounds on the) lifetimes of the vibrational states $v' = 0-4$ belonging to the excited electronic manifolds of N_2 are indeed long, varying from 0.4 to 3.3 s. Interestingly, the lifetime seems to be relatively insensitive to temperature over the range 1.7–20 K.

3.2.4 Laser-Induced Vibrational Fluorescence (LIF)

The methods considered in Subsect. 3.2.3 are capable of yielding information only about VR of excited electronic states. Recently, however, DUBOST et al. [3.14] have developed a technique for measuring the lifetimes of vibrationally excited states of the ground electronic manifold. An intense infrared laser pulse is allowed to impinge upon a matrix containing the molecule of interest, thus populating the first excited level $v'' = 1$ (see Fig. 3.2). Fluorescence from this level is then monitored as a function of time to obtain the decay curve. DUBOST et al. [3.14] have applied this method to study VR of CO in Ne and Ar matrices. The CO was excited to $v'' = 1$ using a pulse from a Q-switched, frequency-doubled CO_2 laser operating on the R(8) line of the $00°1-02°0$ band of CO_2 at $v = 3.2091 \times 10^{13}\ \mathrm{s}^{-1}$, just one-half the fundamental frequency of CO in an Ne matrix. For these systems (i.e., CO in Ne or Ar) there are two modes of decay of $v'' = 1$—fluorescence and non-radiative VR. Hence the decay curve is described by (3.14). Using the gas-phase value of $\tau_r = 1$ ms for CO, DUBOST and LEGAY [3.15] calculated $\tau_{v''}$ as a function of temperature (see Table 3.1). Although the non-radiative lifetime is long, it decreases quite rapidly with increasing temperature.

Table 3.1. Non-radiative vibrational relaxation rates of CO in an Ar matrix as a function of temperature

Temperature [K]	Relaxation rate [s^{-1}]			ρ [c]
	Experiment [3.15]	Theory: weak-coupling limit[a]	Theory: general expression[b]	
6	72	77.2	78.0	0.138
8	76	78.5	79.1	0.138
10	91	83.0	82.9	0.138
13	101	100.4	98.9	0.138
18	179	179.3	179.6	0.136
20			247.7	0.135
50			6.47×10^4	0.110
100			1.11×10^7	0.070
150			1.08×10^8	0.046
200			3.60×10^8	0.032
300			1.20×10^9	0.018
350			1.67×10^9	0.015
400			2.13×10^9	0.012

[a] From least-squares fit of (3.215).
[b] From least-squares fit of (3.213).
[c] Ratio defined by (3.214): a measure of the adequacy of the saddle-point approximation.

3.2.5 Optical Double Resonance (ODR)

ALLAMANDOLA and NIBLER [3.16] have pointed out a number of disadvantages of the LIF technique:

a) the laser frequency (or twice the laser frequency) must be well matched with the fundamental absorption frequency of the molecule of interest, yet tunable lasers in the infrared spectral region are not readily available;

b) only molecules which absorb and emit in the infrared region of the spectrum and which have radiative lifetimes longer than their non-radiative lifetimes can be studied;

c) the sensitivity of fast infrared photon detectors is relatively poor;

d) vibrational fluorescence from low-frequency (below $v = 3 \times 10^7 \, s^{-1}$) modes is difficult to measure because of poor performance by infrared detectors at longer wavelengths.

ALLAMANDOLA and NIBLER [3.16] have devised an optical double resonance (ODR) technique which overcomes these problems. The general scheme may be understood by reference to Fig. 3.3, which displays the vibronic level diagram of a "typical" diatomic molecule. A pulse from a "pump" laser of frequency v_p is passed through a matrix containing the molecule of interest. The 1″ and 2″ levels are thus populated via the absorption-fluorescence sequence shown. If the molecule does not absorb or emit light in the infrared spectrum, then 1″ and 2″ decay only non-radiatively with rate constants k_{10} and k_{21}, respectively. After the variable delay time Δ a second pulse from a probe laser, tuned to excite selectively molecules in the 1″ level to the 0′ level, is passed through the sample, thereby giving rise to fluorescence at v_e. The fluorescence intensity is a measure of the population of the 1″ level. Hence, a plot of the fluorescence intensity I at v_e versus the delay time yields the decay curve for level 1″. Since fluorescence arising from the transitions $0'$, $1' \rightarrow v'' > 2$ is

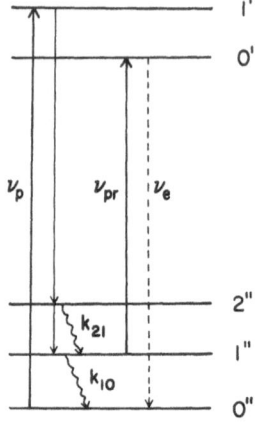

Fig. 3.3. Pump-fluorescence-probe sequence utilized in the optical double resonance (ODR) method

very weak, only the three levels 0″, 1″, and 2″ are essentially involved in the overall process. A straightforward analysis leads to the following integrated equation for the fluorescence intensity at v_e

$$\frac{I - I_\infty}{I_0 - I_\infty} = e^{-k_{10}t} + \frac{\alpha k_{21}}{k_{10} - k_{21}} (e^{-k_{21}t} - e^{-k_{10}t}), \qquad (3.16)$$

where I is assumed to be proportional to the population of the level 1″, I_0 is the fluorescence intensity at $t = 0$, I_∞ is the intensity at $t = \infty$ (i. e., the background), and α is the ratio of the population of level 2″ to that of level 1″ at $t = 0$.

The ODR method has been applied [3.16] to C_2^- generated photolytically from C_2H_2 in N_2 and Ar matrices. Measurements performed at two different temperatures (16 and 24 K) indicate that the lifetime $(k_{10}^{-1}$ or $k_{21}^{-1})$ of neither level depends significantly upon the temperature. It is interesting, however, that the lifetime of level 2″ is 3 to 5 times that of level 1″. This latter result seems to be discordant with existing models for VR of diatomics in rare-gas matrices discussed below in Section 3.6.

3.2.6 Infrared Absorption and Spontaneous Raman Scattering

Nafie and Peticolas [3.17], and Bratos and coworkers [3.18–20], have developed theories of ordinary infrared absorption and spontaneous Raman scattering that take into account the effects of vibrational relaxation of the absorbing (scattering) molecule. Neglecting any correlation between rotational and vibrational motions in the absorbing (scattering) molecule, NAFIE and PETICOLAS derived the following expressions

$$G_{r1}(t) = \langle \hat{\mu}(0) \cdot \hat{\mu}(t) \rangle_{tr} = \int_{-\infty}^{+\infty} e^{-i\omega t} \hat{I}_{ir}(\omega) d\omega / \int_{-\infty}^{+\infty} e^{-i\omega t} \hat{I}_{vib}(\omega) d\omega \qquad (3.17)$$

$$G_{r2}(t) = \langle \text{Tr}\,[\boldsymbol{\beta}^v(0) \cdot \boldsymbol{\beta}^v(t)] \rangle_{tr} = \int_{-\infty}^{+\infty} e^{-i\omega t} \hat{I}_{\perp}(\omega) d\omega / \int_{-\infty}^{+\infty} e^{-i\omega t} \hat{I}_{vib}(\omega) d\omega \quad (3.18)$$

$$G_v(t) = \langle Q^v(0) Q^v(t) \rangle = \int_{-\infty}^{+\infty} e^{-i\omega t} \hat{I}_{vib}(\omega) d\omega \qquad (3.19)$$

$$\hat{I}_{vib}(\omega) \equiv [I_{\parallel}(\omega) - \tfrac{4}{3} I_{\perp}(\omega)] / \int_{-\infty}^{+\infty} [I_{\parallel}(\omega) - \tfrac{4}{3} I_{\perp}(\omega)] d\omega \qquad (3.20)$$

$$\hat{I}_{\perp}(\omega) \equiv I_{\perp}(\omega) / \int_{-\infty}^{+\infty} I_{\perp}(\omega) d\omega \qquad (3.21)$$

$$\hat{I}_{ir}(\omega) \equiv I_{ir}(\omega) / \int_{-\infty}^{+\infty} I_{ir}(\omega) d\omega. \qquad (3.22)$$

In (3.17–22) $\hat{\mu}$ is a unit vector parallel to the transition moment for the normal coordinate Q^v and $\boldsymbol{\beta}^v$ is the traceless part of the electric polarizability tensor; $I_{\parallel}(\omega)$ and $I_{\perp}(\omega)$ are the intensities of Raman-scattered light having polarizations parallel and perpendicular, respectively, to that of the incident light; $I_{ir}(\omega)$ is the infrared absorption intensity. Note that implicit in the derivation of (3.17–22) is the assumption that the trace of the spherical part of the polarizability is non-zero. Thus, v must be a totally symmetric mode.

The theory of NAFIE and PETICOLAS [3.17] may be put into practice in the following fashion. Preferably one deals with dilute solutions of the molecule of interest. First, an isolated infrared absorption band arising from a totally symmetric vibrational mode is recorded. Next, the corresponding Raman bands (for both $\parallel$ and $\perp$ scattering geometries) are recorded. All of these bands are then normalized according to (3.20–22) and Fourier inverted by (3.17–19) to give autocorrelation functions of μ, $\boldsymbol{\beta}^v$, and Q^v. Plots of $\log G_x$ vs t yield straight lines (at least at long times) of slope $-\tau_x^{-1}$, where τ_x is a correlation time. In particular, τ_v, the vibrational correlation time, tells one something about the rate and mechanism of VR.

A number of studies of vibrational relaxation in liquids have been based on this approach, liquid methyl iodide having received an extraordinary amount of attention [3.21–23]. Other organic liquids that have been investigated by this method include TCO [3.24] and quinoline [3.25]. LE DUFF [3.26] has reported an interesting study of N_2 dissolved in various "inert" solvents.

3.2.7 Stimulated Raman Scattering (SRS) [3.91]

A very powerful and direct technique for studying VR experimentally utilizes the following scheme. An intense picosecond pump laser pulse propagating through the sample of interest undergoes Stokes scattering, thereby creating vibrationally excited states (in the ground electronic manifold) at a rate which is nonlinear with the intensity of the pulse. In essence, this is stimulated Raman scattering (SRS). It is important to note that the vibrations of the molecules in the sample are excited in phase. Now a second probe laser pulse of different frequency, suitably delayed with respect to the pump pulse, is scattered by anti-Stokes vibrational transitions of the excited molecules. The *incoherent* anti-Stokes Raman intensity, observed at 90° with respect to the incident (forward) direction, is a direct measure of the population of the excited vibrational state. On the other hand, the *coherent* anti-Stokes signal, observed under phase-matching conditions close to the forward direction, provides a measure of the rate of dephasing of the molecular vibrations.

MAIER et al. [3.27] have derived the following set of coupled differential equations describing the SRS process

$$\frac{\partial E_s}{\partial z} + v_s^{-1} \frac{\partial E_s}{\partial t} = \frac{i\pi\omega_s^2 N}{c^2 k_s} \frac{\partial \alpha}{\partial Q} Q^* E_{L_1} \tag{3.23}$$

$$\frac{\partial Q}{\partial t} + \frac{Q}{2\tau} = \frac{i}{4m\omega_0} \frac{\partial \alpha}{\partial Q} E_{L_1} E_s^* (1 - 2n) \tag{3.24}$$

$$\frac{\partial n}{\partial t} + \frac{n}{\tau'} = \frac{i}{8\hbar} \frac{\partial \alpha}{\partial Q} (E_{L_1} E_s^* Q^* - E_{L_1}^* E_s Q). \tag{3.25}$$

Here E_{L_1} and E_s are the electric fields associated with the pump pulse and Stokes pulse, respectively; Q is the expectation value of the relevant normal mode coordinate; v_s, k_s, and ω_s are the group "velocity", wavenumber, and frequency of the Stokes light; ω_0 and m are the fundamental frequency and effective mass of the molecular normal mode; n is the population of the excited vibrational state, N is the number density of the molecules and α is their polarizability. From (3.24) and (3.25) we infer that the characteristic time τ governs the rate of dephasing of the coherently excited vibrations whereas τ' determines the lifetime of the excited vibrational state. Note that τ' and τ are analogous to T_1 and T_2 in the Bloch equations [3.28]. Imposing specific initial and boundary conditions, one may solve (3.23–25) and compare the solution with observed intensities to deduce τ and τ'. In many cases, both the incoherent and coherent scattering signals decay exponentially, so that τ and τ' may be determined simply by plotting the logs of the respective signals versus the delay time of the probe pulse.

A wide range of very interesting investigations of VR in solids, pure liquids, and solutions have been carried out by the SRS technique; τ and τ'

for various systems are given in Table 3.2. We shall consider just a few representative examples. LAUBEREAU et al. [3.29] have determined the "lifetime" of phonons in highly populated ("hot") TO modes of diamond. These same workers have also measured τ and τ' for molecular normal modes in CCl_4 [3.88] and in C_2H_5OH and CH_3Cl_3 [3.30].

Table 3.2. Vibrational lifetimes τ' and dephasing times τ for various systems

System	Mode	Temperature [K]	τ' [ps]	τ [ps]	Reference
Diamond	TO	77	—	3.4 ± 0.3	[3.29]
		295	—	2.9 ± 0.3	[3.29]
CCl_4	$459\,cm^{-1}$	ambient room	—	4.0 ± 0.5	[3.88]
C_2H_5OH	CH stretch $(2923\,cm^{-1})$	ambient room	20 ± 5	~ 0.25	[3.30]
CH_3Cl_3	CH stretch $(2939\,cm^{-1})$	ambient room	5 ± 1	~ 1.3	[3.30]
N_2		73	10^9	—	[3.36]
		77	—	75 ± 8	[3.35]
		78	1.5×10^{12}	—	[3.37]
O_2		73	5.0×10^7	—	[3.36]

From SRS measurements ALFANO and SHAPIRO [3.31] have speculated that excitation of the pumped mode v_H ($v = 1$) [C—H stretching, $2928\,cm^{-1}$] in C_2H_5OH probably decays via either an intra- or intermolecular mechanism, in which two modes δ_H ($v = 1$) [C—H bending, $1460\,cm^{-1}$] of approximately half the frequency receive the excitation. However, a recent SRS study by LAUBEREAU et al. [3.32] provides support for the following VR mechanism. The pumped mode shares its energy rapidly by a nearly resonant V–V (vibration-vibration) transfer to the δ_H ($v = 2$) mode, both modes subsequently relaxing to δ_H ($v = 1$) by a slower V–T (vibration-translation) transfer.

LAUBEREAU et al. [3.33] have studied VR in binary mixtures of organic polyatomic liquids, finding evidence for the importance of near resonance for the efficient intermolecular transfer of vibrational energy.

By studying the dependence of τ for C—H stretching mode in a series of liquid hydrocarbons, MONSON et al. [3.34] have concluded that VR in these molecules takes place primarily through the methyl groups.

LAUBEREAU [3.35] has measured τ for liquid N_2 at 77 K, and RENNER and MAIER [3.36] have determined τ' for liquid N_2 and O_2 at 73 K. Very recently CALAWAY and EWING [3.37] have reported a value of τ' for $N_2(l)$ at 78 K considerably longer than that found by RENNER and MAIER. The discrepancy is apparently due to the presence of impurities in the samples of RENNER and MAIER [3.38]. It is interesting to note (see Table 3.2) the vast difference between τ and τ' for the low-temperature diatomic liquids in contrast to the room-temperature polyatomic liquids, for which τ and τ' are quite similar in magnitude. This difference, of course, is due to the very different mechanisms leading to dephasing (τ) and vibrational energy loss (τ').

3.3 Theoretical Description of the Dynamics of Vibrational Relaxation

We turn now to the development of a model by which the experiments discussed in Section 3.2 can be interpreted. Let us focus our attention first on the idealized, prototypic system—a dilute solution of diatomic solute molecules in a monatomic solvent (matrix). This system may be idealized as a single harmonic oscillator coupled to a heat bath and to the radiation field.

The nexus between our model and the "real-world" population-decay studies by LIF, ODR, and SRS can be established as follows. Suppose the real system of interest to be in thermal equilibrium at large negative times. An initial high-power laser pulse effectively inverts the population of vibrational states of the diatomic, leaving the matrix (or solvent) in thermal equilibrium so that at time $t=0$ an abnormally large fraction of the diatomic molecules are in a definite vibrational state n. We note in passing that under the conditions of most experiments selection rules permit the population of only the first vibrationally excited state. As time increases beyond $t=0$, the population (taken to be proportional to the expectation value of the occupation number of the oscillator, $\langle \hat{n}(t) \rangle$) of vibrationally excited diatomics decays by two channels: 1) coupling to the radiation field accompanied by creation of *photons*; 2) coupling to the matrix (solvent) accompanied by production of "*phonons*". The population decay is monitored by measuring the intensity either of fluorescence (in the case of LIF or ODR) or of the anti-Stokes Raman signal (in the case of SRS), either of which is proportional to the population of excited diatomics.

To make the connection between our model and the infrared absorption and spontaneous Raman scattering experiments, we simply note from (3.17–19) that it is necessary to calculate the time-correlation functions G_{r1}, G_{r2}, and G_v.

By additional, rather more involved, considerations we could establish a link between our model and the ultrasonic absorption and Rayleigh-Brillouin scattering experiments. However, in view of the facts that these techniques are of limited applicability and that few data on prototypic systems are available, we shall not pursue such considerations here.

At least two alternative approaches to the determination of $\langle \hat{n}(t) \rangle$ and $G_x(t)$ can be envisaged. On the one hand, we may use classical mechanics to compute the coordinates and conjugate momenta of every particle in the system as a function of time, i. e., we may determine a "representative" trajectory of the phase point of the system. Then the ensemble averages implied in $\langle \hat{n}(t) \rangle$ and $G_x(t)$ are expressed as time averages over this trajectory. For example

$$G_{r1}(t_i) = (J\,N)^{-1} \sum_{j=1}^{J} \sum_{k=1}^{N} \hat{\mu}_k(t_j) \cdot \hat{\mu}_k(t_j + t_i)\,, \tag{3.26}$$

where j labels time origins and k labels molecules. This approach is generally known as the molecular dynamics method [3.39]. On the other hand, we may attempt to solve the equations of motion for $\langle \hat{n}(t) \rangle$ and $G_x(t)$ approximately by adopting specific forms for the oscillator-bath coupling. It is the latter approach that we follow here, calculating in particular $\langle \hat{n}(t) \rangle$ and $G_v(t)$.

3.3.1 Description of the Model[3]

The Hamiltonian[4] for our model is

$$\mathscr{H} = \hat{H}^0 + \lambda \hat{V} = \hat{H}^0_{\text{osc}} + \hat{H}^0_{\text{B}} + \lambda \hat{V}\,, \tag{3.27}$$

where $\hat{H}^0_{\text{osc}}$ is the Hamiltonian of the free harmonic oscillator, given in the number-operator ($\hat{n}$) representation by

$$\hat{H}^0_{\text{osc}} = \hbar \omega_0 (\hat{n} + 1/2) = \hbar \omega_0 (\hat{a}^\dagger \hat{a} + 1/2)\,, \tag{3.28}$$

$\hat{H}^0_{\text{B}}$ is the Hamiltonian of the free bath, and $\hat{V}$ is the interaction (coupling) between the oscillator and the bath. In (3.28) ω_0 is the fundamental frequency of the oscillator; $\hat{a}^\dagger$ and $\hat{a}$ are, respectively, the creation and destruction operators associated with the oscillator and satisfy the Boson commutation relation

$$[\hat{a}, \hat{a}^\dagger] = 1\,. \tag{3.29}$$

In anticipation of solving the equation of motion for $\langle \hat{n}(t) \rangle$ approximately in the van Hove weak-coupling limit, we have multiplied the oscillator-bath interaction by an ordering (or strength) parameter λ.

Instead of dealing directly with $\langle \hat{n}(t) \rangle$, we may consider

$$\langle \Delta \hat{n}(t) \rangle = \text{Tr}\{\hat{\rho}_{\text{S}}(t) \Delta \hat{n}_{\text{S}}\} = \text{Tr}\{\hat{\rho}_{\text{H}}(0) \Delta \hat{n}_{\text{H}}(t)\}\,, \tag{3.30}$$

---

[3] The developments of Subsects. 3.3.1–4 are based on DIESTLER and WILSON [3.85].
[4] Note that the Hamiltonian does not include the radiation field or its coupling with the oscillator. Hence, the lifetime τ' which we calculate is actually not analogous to that measured [see (3.15)]. We shall assume the radiative and non-radiative modes of decay are independent. Hence, our neglecting the radiative mode will not affect the results.

where

$$\Delta \hat{n} \equiv \hat{n} - \bar{n}, \tag{3.31}$$

$$\bar{n} = \mathrm{Tr}\{e^{-\beta \hat{H}^0_{\mathrm{osc}}} \hat{n}\}/\mathrm{Tr}\{e^{-\beta \hat{H}^0_{\mathrm{osc}}}\}, \tag{3.32}$$

and $\beta \equiv (kT)^{-1}$ \hfill (3.33)

as usual[5]; $\langle \Delta \hat{n}(t) \rangle$ is the expected value at time t of the deviation of the occupation number (population) from its equilibrium value $\bar{n}$. We expect that

$$\lim_{t \to \infty} \langle \Delta \hat{n}(t) \rangle = 0. \tag{3.34}$$

We take the initial density operator to be

$$\hat{\rho}_{\mathrm{H}}(0) = \hat{\rho}_{\mathrm{S}}(0) = \hat{\rho}(0) = \hat{\rho}_{\mathrm{osc}}(0)\hat{\rho}_{\mathrm{B}}(0). \tag{3.35}$$

Here $\hat{\rho}_{\mathrm{osc}}(0)$, which describes the initial non-equilibrium condition of the harmonic oscillator in the n-th vibrational level, may be written explicitly as

$$\hat{\rho}_{\mathrm{osc}}(0) = |n\rangle \langle n|, \tag{3.36}$$

where the ket $|n\rangle$ satisfies

$$\hat{H}^0_{\mathrm{osc}} |n\rangle = \hbar \omega_0 (n + 1/2) |n\rangle, \quad n = 0, 1, 2, \ldots; \tag{3.37}$$

$\hat{\rho}_{\mathrm{B}}(0)$, given explicitly by

$$\hat{\rho}_{\mathrm{B}}(0) = e^{-\beta \hat{H}^0_{\mathrm{B}}}/\mathrm{Tr}\{e^{-\beta \hat{H}^0_{\mathrm{B}}}\}, \tag{3.38}$$

describes the initially free bath in thermal equilibrium.

3.3.2 Derivation of Equation of Motion for $\langle \hat{n}(t) \rangle$ by Zwanzig's Projection-Operator Method

To simplify the solution of the equation of motion for $\langle \Delta \hat{n}(t) \rangle$, we introduce a projection operator $\hat{P}$ defined by

$$\hat{P}\hat{X} \equiv \Delta \hat{n}(0) \langle \hat{X} \rangle / \langle \Delta \hat{n}(0) \rangle, \tag{3.39}$$

---

[5] The subscripts S and H denote the Schrödinger and Heisenberg pictures, respectively. As we shall be working henceforth in Subsects. 3.3.1–4 exclusively in the Heisenberg picture, we shall suppress these subscripts where there is no ambiguity.

where $\hat{X}$ is an arbitrary operator and

$$\langle \hat{X} \rangle \equiv \text{Tr} \{ \hat{\rho}(0) \hat{X} \} . \tag{3.40}$$

We also define a normalized initial ensemble average of $\hat{X}$ by

$$\langle\langle \hat{X} \rangle\rangle \equiv \langle \hat{X} \rangle / \langle \Delta \hat{n}(0) \rangle . \tag{3.41}$$

The properties of $\hat{P}$ and $\langle\langle ... \rangle\rangle$ are derived in Appendix A.
Now the Heisenberg equation of motion for $\Delta \hat{n}(t)$ is

$$\Delta \dot{\hat{n}}(t) = i \mathscr{L} \Delta \hat{n}(t) , \tag{3.42}$$

where the Liouville operator $\mathscr{L}$ is defined by

$$\mathscr{L} \hat{X} \equiv \hbar^{-1} [\hat{\mathscr{H}}, \hat{X}] . \tag{3.43}$$

Note that we shall subsequently employ "partial" Liouville operators defined by

$$\hat{\mathscr{L}}^0 \hat{X} \equiv \hbar^{-1} [\hat{H}^0, \hat{X}] , \tag{3.44}$$

$$\hat{\mathscr{L}}^0_{\text{osc}} \hat{X} \equiv \hbar^{-1} [\hat{H}^0_{\text{osc}}, \hat{X}] , \tag{3.45}$$

$$\hat{\mathscr{L}}^0_{\text{B}} \hat{X} \equiv \hbar^{-1} [\hat{H}^0_{\text{B}}, \hat{X}] , \tag{3.46}$$

and $\hat{\mathscr{L}}_{\text{V}} \hat{X} \equiv \hbar^{-1} \lambda [V, \hat{X}] . \tag{3.47}$

The formal solution of (3.42) [or those of the "free" analogs, i. e. $\dot{\hat{X}}(t) = i \hat{\mathscr{L}}^0 \hat{X}(t)$]
is given by

$$\Delta \hat{n}(t) = e^{i \mathscr{L} t} \Delta \hat{n} , \tag{3.48}$$

as may be verified simply by differentiating both members of (3.48) with respect to t.
As (3.48) is of little practical avail, we shall take another approach. We begin by using (3.283) to rewrite (3.42) as

$$\Delta \dot{\hat{n}}(t) = i \mathscr{L} \Delta \hat{n}_{\text{R}}(t) + i \mathscr{L} \Delta \hat{n}_{\text{IRR}}(t) . \tag{3.49}$$

Operating upon both members of (3.49) with $\hat{P}$ and $(\hat{1} - \hat{P})$ yields, respectively,

$$\Delta \dot{\hat{n}}_{\text{R}}(t) = i \hat{P} \mathscr{L} \Delta \hat{n}_{\text{R}}(t) + i \hat{P} \mathscr{L} \Delta \hat{n}_{\text{IRR}}(t) , \tag{3.50}$$

$$\Delta \dot{\hat{n}}_{\text{IRR}}(t) = i (\hat{1} - \hat{P}) \mathscr{L} \Delta \hat{n}_{\text{R}}(t) + i (\hat{1} - \hat{P}) \mathscr{L} \Delta \hat{n}_{\text{IRR}}(t) . \tag{3.51}$$

Solving (3.51) formally for $\Delta \hat{n}_{IRR}(t)$, we obtain

$$\Delta \hat{n}_{IRR}(t) = e^{i(\hat{1}-\hat{P})\hat{\mathscr{L}}t} \Delta \hat{n}_{IRR}(0) + i \int_0^t e^{i(\hat{1}-\hat{P})\hat{\mathscr{L}}\tau}(\hat{1}-\hat{P})\hat{\mathscr{L}}\Delta \hat{n}_R(t-\tau)d\tau . \quad (3.52)$$

But by (3.280) and (3.39)

$$\begin{aligned} \Delta \hat{n}_{IRR}(0) &= (\hat{1}-\hat{P})\Delta \hat{n}(0) \\ &= \Delta \hat{n}(0) - \hat{P}\Delta \hat{n}(0) \\ &= \Delta \hat{n}(0) - \Delta \hat{n}(0) \\ &= 0 . \end{aligned} \quad (3.53)$$

Substituting (3.52) into (3.50) and invoking (3.53), we get

$$\Delta \dot{\hat{n}}_R(t) = i\hat{P}\mathscr{L}\Delta \hat{n}_R(t) - \int_0^t \hat{P}\mathscr{L}e^{i(\hat{1}-\hat{P})\hat{\mathscr{L}}\tau}(\hat{1}-\hat{P})\mathscr{L}\Delta \hat{n}_R(t-\tau)d\tau . \quad (3.54)$$

Operating upon both sides of (3.54) with $\langle\langle ... \rangle\rangle$ and applying (3.279), (3.281), and (3.284), we obtain

$$\langle\langle \Delta \dot{\hat{n}}(t)\rangle\rangle = i\langle\langle \mathscr{L}\hat{P}\Delta \hat{n}(t)\rangle\rangle - \int_0^t \langle\langle \mathscr{L}e^{i(\hat{1}-\hat{P})\hat{\mathscr{L}}\tau}(\hat{1}-\hat{P})\mathscr{L}\hat{P}\Delta \hat{n}(t-\tau)\rangle\rangle d\tau . \quad (3.55)$$

Finally, using (3.41) and (3.285), we may transform (3.55) into

$$\dot{G}(t) = ig\,G(t) - \int_0^t \kappa(\tau)G(t-\tau)d\tau \quad (3.56)$$

where

$$G(t) \equiv \langle \Delta \hat{n}(t)\rangle , \quad (3.57)$$

$$g \equiv \langle\langle \hat{\mathscr{L}}\Delta \hat{n}(0)\rangle\rangle , \quad (3.58)$$

and

$$\kappa(\tau) \equiv \langle\langle \hat{\mathscr{L}}e^{i(\hat{1}-\hat{P})\hat{\mathscr{L}}\tau}(\hat{1}-\hat{P})\hat{\mathscr{L}}\Delta \hat{n}(0)\rangle\rangle . \quad (3.59)$$

Eq. (3.56) is completely general as it stands. However, in order to arrive at a tractable solution in closed form, even in the weak-coupling limit, we must introduce a simplifying approximation.

3.3.3 Linear Coupling: Evaluation of g and $\kappa(\tau)$ Through $O(\lambda^2)$

Let us assume that the oscillator-bath coupling $\hat{V}$ is linear in the coordinate of the oscillator, i. e.

$$\hat{V} = \hat{Q}\hat{\Gamma}, \tag{3.60}$$

where

$$\hat{Q} \equiv \hat{a}^\dagger + \hat{a} \tag{3.61}$$

is proportional to the oscillator coordinate, the constant of proportionality having been assimilated in $\hat{\Gamma}$, which depends only on the bath coordinates, but is otherwise unspecified.

Using the definitions (3.43–47) along with (3.60) and (3.61), we have

$$\begin{aligned}
\hat{\mathscr{L}} \, \Delta \hat{n}(0) &= \hat{\mathscr{L}}_V \hat{n} \\
&= -\lambda \hbar^{-1}[\hat{a}^\dagger \hat{a}, \hat{V}] \\
&= -\lambda \hbar^{-1}(\hat{a}^\dagger - \hat{a})\hat{\Gamma}.
\end{aligned} \tag{3.62}$$

As mentioned above, we are interested in evaluating $G(t)$ in the van Hove weak-coupling limit, so we now proceed to calculate g and $\kappa(\tau)$ through $O(\lambda^2)$. From (3.58) and (3.62) we have

$$g = -\lambda \hbar^{-1}[\langle\langle \hat{a}^\dagger \hat{\Gamma} \rangle\rangle - \langle\langle \hat{a} \hat{\Gamma} \rangle\rangle]. \tag{3.63}$$

To simplify (3.63) further, we consider the quantity $\langle\langle \hat{F}(\hat{a}, \hat{a}^\dagger)\hat{G}(\hat{\Gamma})\rangle\rangle$, where $\hat{F}$ is a function only of $\hat{a}$ and $\hat{a}^\dagger$ and $\hat{G}$ is a function only of bath variables $\hat{\Gamma}$. Then

$$\begin{aligned}
\langle\langle \hat{F}\hat{G} \rangle\rangle &= \langle \hat{F}\hat{G} \rangle / \langle \Delta \hat{n}(0) \rangle \\
&= \mathrm{Tr}\,\{\hat{\rho}(0)\hat{F}\hat{G}\} / \langle \Delta \hat{n}(0) \rangle \\
&= \mathrm{Tr}\,\{\hat{\rho}_{\mathrm{osc}}(0)\hat{F}\}\,\mathrm{Tr}\,\{\hat{\rho}_{\mathrm{B}}(0)\hat{G}\} / \langle \Delta \hat{n}(0) \rangle \\
&= \langle \hat{F} \rangle_{\mathrm{osc}} \langle \hat{G} \rangle_{\mathrm{B}} / \langle \Delta \hat{n}(0) \rangle,
\end{aligned} \tag{3.64}$$

where $\langle\,\rangle_{\mathrm{osc}}$ and $\langle\,\rangle_{\mathrm{B}}$ signify ensemble averages over the initially free oscillator and bath, respectively. The third equality in (3.64) follows from the facts that $\hat{F}$ acts only on oscillator states and G only on bath states and that the sums over oscillator and bath states implied by the operation Tr may be effected independently. We shall use this idea repeatedly in simplifying $\kappa(\tau)$ below.

Using (3.36) we observe that

$$\langle (\hat{a}^\dagger)^s \rangle_{osc} = \sum_{n'} \langle n'|n \rangle \langle n|(\hat{a}^\dagger)^s|n' \rangle$$

$$= \langle n|(\hat{a}^\dagger)^s|n \rangle \qquad\qquad (3.65)$$

$$= 0 \qquad\qquad s=1,2,\ldots$$

and

$$\langle \hat{a}^s \rangle_{osc} = \langle n|\hat{a}^s|n \rangle$$

$$= 0. \qquad\qquad s=1,2,\ldots \qquad\qquad (3.66)$$

Hence, from (3.63–66) it follows that

$$g = 0 \qquad\qquad (3.67)$$

to all orders in λ.

Substituting $\hat{\mathscr{L}} \Delta \hat{n}(0)$ given by (3.62) into (3.59), we obtain

$$\kappa(\tau) = -\lambda\hbar^{-1} \langle\langle \hat{\mathscr{L}} e^{i(\hat{1}-\hat{P})\hat{\mathscr{L}}\tau}(\hat{1}-\hat{P})(\hat{a}^\dagger - \hat{a})\hat{\Gamma} \rangle\rangle . \qquad\qquad (3.68)$$

Since

$$\hat{P}\hat{a}^\dagger\hat{\Gamma} = \Delta\hat{n}(0)\langle\langle \hat{a}^\dagger\hat{\Gamma} \rangle\rangle = 0$$

and

$$\hat{P}\hat{a}\hat{\Gamma} = \Delta\hat{n}(0)\langle\langle \hat{a}\hat{\Gamma} \rangle\rangle = 0 \qquad\qquad (3.69)$$

via (3.39) and (3.64–66), (3.68) simplifies to

$$\kappa(\tau) = -\lambda\hbar^{-1} \langle\langle \hat{\mathscr{L}} e^{i(\hat{1}-\hat{P})\hat{\mathscr{L}}\tau}(\hat{a}^\dagger - \hat{a})\hat{\Gamma} \rangle\rangle . \qquad\qquad (3.70)$$

Using the general relation [3.40]

$$e^{i(\hat{X}+\hat{Y})t} = e^{i\hat{X}t} + i\int_0^t e^{i(\hat{X}+\hat{Y})(t-t')}\hat{Y}e^{i\hat{X}t'}\, dt' , \qquad\qquad (3.71)$$

we may recast (3.70) as

$$\kappa(\tau) = -\lambda\hbar^{-1} \left\langle\left\langle \hat{\mathscr{L}} \left[e^{i(\hat{1}-\hat{P})\mathscr{L}^0\tau} + i\int_0^\tau e^{i(\hat{1}-\hat{P})(\mathscr{L}^0 + \mathscr{L}_V)(\tau-t)} \right.\right.\right.$$

$$\left.\left.\left. \times (\hat{1}-\hat{P})\mathscr{L}_V e^{i(\hat{1}-\hat{P})\mathscr{L}^0 t}\, dt \right](\hat{a}^\dagger - \hat{a})\hat{\Gamma} \right\rangle\right\rangle . \qquad\qquad (3.72)$$

Now since $\mathcal{L}_V$ carries λ implicitly [see (3.47)] and we are considering contributions to $\kappa(\tau)$ only through $O(\lambda^2)$, $\mathcal{L}_V$ can be neglected in the exponent in the integrand in (3.72), which consequently becomes

$$\kappa(\tau) = -\lambda \hbar^{-1} \left\langle\!\!\left\langle [\mathcal{L}^0 + \mathcal{L}_V] \left[e^{i(\hat{1}-\hat{P})\mathcal{L}^0 \tau} + i \int_0^\tau e^{i(\hat{1}-\hat{P})\mathcal{L}^0 (\tau - t)} \right.\right.$$
$$\left.\left. \times (\hat{1}-\hat{P})\mathcal{L}_V e^{i(\hat{1}-\hat{P})\mathcal{L}^0 t} \, dt \right] (\hat{a}^\dagger - \hat{a})\hat{\Gamma} \right\rangle\!\!\right\rangle . \tag{3.73}$$

We consider first the only contribution to $\kappa(\tau)$ of $O(\lambda)$, namely

$$\kappa^{(1)}(\tau) = -\lambda \hbar^{-1} \langle\!\langle \mathcal{L}^0 e^{i(\hat{1}-\hat{P})\mathcal{L}^0 \tau}(\hat{a}^\dagger - \hat{a})\hat{\Gamma} \rangle\!\rangle . \tag{3.74}$$

Note that

$$\mathcal{L}^0 \hat{P} \hat{X} = \langle\!\langle X \rangle\!\rangle \mathcal{L}^0 \Delta n(0) = 0 \tag{3.75}$$

by (3.39), (3.45) and (3.46). We may easily verify that

$$\mathcal{L}^0 e^{i(\hat{1}-\hat{P})\mathcal{L}^0 \tau} = \mathcal{L}^0 e^{i\mathcal{L}^0 \tau} \tag{3.76}$$

by expanding the exponential in the left member of (3.76) and applying (3.75) term by term. Then by (3.76) we have from (3.74)

$$\kappa^{(1)}(\tau) = -\lambda \hbar^{-1} \langle\!\langle \mathcal{L}^0 e^{i\mathcal{L}^0 \tau}(\hat{a}^\dagger \hat{\Gamma} - \hat{a}\hat{\Gamma}) \rangle\!\rangle . \tag{3.77}$$

Now, since $[\hat{H}^0_{osc}, \hat{H}^0_B] = 0$, we see that

$$e^{i\mathcal{L}^0 \tau} \hat{a}^\dagger \hat{\Gamma} = e^{i\mathcal{L}^0_{osc} \tau} \hat{a}^\dagger e^{i\mathcal{L}^0_B \tau} \hat{\Gamma}$$
$$= \hat{a}^\dagger_0(\tau)\hat{\Gamma}_0(\tau) \tag{3.78}$$

and likewise that

$$e^{i\mathcal{L}^0 \tau} \hat{a}\hat{\Gamma} = e^{i\mathcal{L}^0_{osc} \tau} \hat{a} e^{i\mathcal{L}^0_B \tau} \hat{\Gamma}$$
$$= \hat{a}_0(\tau)\hat{\Gamma}_0(\tau) , \tag{3.79}$$

where the zero subscript denotes an unperturbed (freely evolving) Heisenberg operator (see (3.48)). Using (3.28, 29) and (3.45), we may easily demonstrate that

$$\mathcal{L}^0_{osc} \begin{Bmatrix} \hat{a}^{\dagger 2} \\ \hat{a}^\dagger \\ \hat{a}^\dagger \hat{a}, \hat{a}\hat{a}^\dagger \\ \hat{a} \\ \hat{a}^2 \end{Bmatrix} = \begin{Bmatrix} 2\omega_0 \hat{a}^{\dagger 2} \\ \omega_0 \hat{a}^\dagger \\ 0 \\ -\omega_0 \hat{a} \\ -2\omega_0 \hat{a}^2 \end{Bmatrix} . \tag{3.80}$$

Employing (3.78–80) along with (3.64–66), we have from (3.77)

$$\kappa^{(1)}(\tau) = -\lambda \hbar^{-1} [\omega_0 e^{i\omega_0\tau} \langle \hat{a}^\dagger \rangle_{osc} \langle \hat{\Gamma}_0(\tau) \rangle_B + e^{i\omega_0\tau} \langle \hat{a}^\dagger \rangle_{osc} \langle \mathscr{L}_B^0 \hat{\Gamma}_0(\tau) \rangle_B$$
$$+ \omega_0 e^{-i\omega_0\tau} \langle \hat{a} \rangle_{osc} \langle \hat{\Gamma}_0(\tau) \rangle_B - e^{-i\omega_0\tau} \langle \hat{a} \rangle_{osc} \langle \mathscr{L}_B^0 \hat{\Gamma}_0(\tau) \rangle_B] / \langle \Delta \hat{n}(0) \rangle$$
$$= 0. \tag{3.81}$$

Turning now to terms of $O(\lambda^2)$ in (3.73), we consider first

$$\kappa^{(2)\prime}(\tau) \equiv -i\lambda \hbar^{-1} \int_0^\tau \langle\langle \mathscr{L}^0 e^{i(\hat{1}-\hat{P})\mathscr{L}^0(\tau-t)} (\hat{1}-\hat{P}) \mathscr{L}_V e^{i(\hat{1}-\hat{P})\mathscr{L}^0 t} (\hat{a}^\dagger - \hat{a}) \hat{\Gamma} \rangle\rangle \, dt, \tag{3.82}$$

which by virtue of (3.75) simplifies to

$$\kappa^{(2)\prime}(\tau) = -i\lambda \hbar^{-1} \int_0^\tau \langle\langle \mathscr{L}^0 e^{i\mathscr{L}^0(\tau-t)} \mathscr{L}_V e^{i(\hat{1}-\hat{P})\mathscr{L}^0 t} (\hat{a}^\dagger - \hat{a}) \hat{\Gamma} \rangle\rangle \, dt. \tag{3.83}$$

Using the relation (3.71), we may rewrite (3.83) as

$$\kappa^{(2)\prime}(\tau) = -i\lambda \hbar^{-1} \int_0^\tau \left\langle\left\langle \mathscr{L}^0 e^{i\mathscr{L}^0(\tau-t)} \mathscr{L}_V \left[e^{i\mathscr{L}^0 t} - i\int_0^t e^{i(\hat{1}-\hat{P})\mathscr{L}^0(t-t')} \hat{P} \mathscr{L}^0 e^{i\mathscr{L}^0 t'} dt' \right] \right.\right.$$
$$\left.\left. \times (\hat{a}^\dagger - \hat{a}) \hat{\Gamma} \right\rangle\right\rangle \, dt. \tag{3.84}$$

But by (3.78–80) we find

$$\mathscr{L}^0 e^{i\mathscr{L}^0 t'} (\hat{a}^\dagger - \hat{a}) \hat{\Gamma} = (\omega_0 e^{i\omega_0 t'} \hat{a}^\dagger + \omega_0 e^{-i\omega_0 t'} \hat{a}) \hat{\Gamma}_0(t)$$
$$+ (e^{i\omega_0 t'} \hat{a}^\dagger - e^{-i\omega_0 t'} \hat{a}) \mathscr{L}_B^0 \hat{\Gamma}_0(t). \tag{3.85}$$

Furthermore, from (3.39) and (3.64–66) it is clear that application of $\hat{P}$ to each term of the right member of (3.85) yields zero. Hence

$$\hat{P} \mathscr{L}^0 e^{i\mathscr{L}^0 t'} (\hat{a}^\dagger - \hat{a}) \hat{\Gamma} = 0 \tag{3.86}$$

and the second term in brackets in (3.84) vanishes, leaving finally

$$\kappa^{(2)\prime}(\tau) = -i\lambda \hbar^{-1} \int_0^\tau \langle\langle \mathscr{L}^0 e^{i\mathscr{L}^0(\tau-t)} \mathscr{L}_V e^{i\mathscr{L}^0 t} (\hat{a}^\dagger - \hat{a}) \hat{\Gamma} \rangle\rangle \, dt. \tag{3.87}$$

Again by (3.78–80) we have

$$e^{i\mathscr{L}^0 t} (\hat{a}^\dagger - \hat{a}) \hat{\Gamma} = [e^{i\omega_0 t} \hat{a}^\dagger - e^{-i\omega_0 t} \hat{a}] \hat{\Gamma}_0(t), \tag{3.88}$$

which upon substitution into (3.87) yields explicitly

$$
\kappa^{(2)'}(\tau) = -\mathrm{i}\lambda^2\hbar^{-2}\int_0^\tau \langle\!\langle (\mathcal{L}_{\mathrm{osc}}^0 + \mathcal{L}_B^0)\mathrm{e}^{\mathrm{i}\mathcal{L}_{\mathrm{osc}}^0(\tau - t)}\mathrm{e}^{\mathrm{i}\mathcal{L}_B^0(\tau - t)}
$$
$$
\times\,[(\hat{a}+\hat{a}^\dagger)\hat{\Gamma},(\mathrm{e}^{\mathrm{i}\omega_0 t}\hat{a}^\dagger - \mathrm{e}^{-\mathrm{i}\omega_0 t}\hat{a})\hat{\Gamma}_0(t)]\rangle\!\rangle\,dt\,.
$$
(3.89)

Using (3.41) and (3.64), we may recast a "typical" one of the eight terms in (3.89) in the form

$$
-\mathrm{i}\lambda^2\hbar^{-2}\int_0^\tau \langle\mathcal{L}_{\mathrm{osc}}^0\,\mathrm{e}^{\mathrm{i}\mathcal{L}_{\mathrm{osc}}(\tau - t)}(\hat{a}+\hat{a}^\dagger)\hat{a}^\dagger\rangle_{\mathrm{osc}}\langle\mathrm{e}^{\mathrm{i}\mathcal{L}_B^0(\tau - t)}\hat{\Gamma}_0\hat{\Gamma}_0(t)\rangle_B\,\mathrm{e}^{\mathrm{i}\omega_0 t}\,dt/\langle\Delta\hat{n}(0)\rangle\,,
$$
(3.90)

which by (3.65, 66), and (3.80) vanishes. Analogous considerations show that each of the other terms in (3.89) vanishes, so that the sole remaining contribution to $\kappa(\tau)$ of $O(\lambda^2)$ is

$$
\kappa^{(2)}(\tau) = -\lambda\hbar^{-1}\langle\!\langle\mathcal{L}_V\,\mathrm{e}^{\mathrm{i}(\hat{1}-\hat{P})\mathcal{L}^0\tau}(\hat{a}^\dagger - \hat{a})\hat{\Gamma}\rangle\!\rangle\,.
$$
(3.91)

By manipulations precisely parallel to those leading from (3.83) to (3.89) we can convert (3.91) into

$$
\kappa^{(2)}(\tau) = -\lambda^2\hbar^{-2}\langle\!\langle[\hat{Q}\hat{\Gamma},\mathrm{e}^{\mathrm{i}\mathcal{L}^0\tau}(\hat{a}^\dagger - \hat{a})\Gamma]\rangle\!\rangle\,.
$$
(3.92)

Again using (3.64) and (3.80), we may rewrite (3.92) as

$$
\kappa^{(2)}(\tau) = -\lambda^2\hbar^{-2}[\langle\hat{Q}(\mathrm{e}^{\mathrm{i}\omega_0\tau}\hat{a}^\dagger - \mathrm{e}^{-\mathrm{i}\omega_0\tau}\hat{a})\rangle_{\mathrm{osc}}\langle\hat{\Gamma}_0\hat{\Gamma}_0(\tau)\rangle_B
$$
$$
-\langle(\mathrm{e}^{\mathrm{i}\omega_0\tau}\hat{a}^\dagger - \mathrm{e}^{-\mathrm{i}\omega_0\tau}\hat{a})\hat{Q}\rangle_{\mathrm{osc}}\langle\hat{\Gamma}_0(\tau)\hat{\Gamma}_0\rangle_B]/\langle\Delta\hat{n}(0)\rangle\,.
$$
(3.93)

By (3.61), (3.65) and (3.66) and the results

$$
\langle\hat{a}^\dagger\hat{a}\rangle_{\mathrm{osc}} = \sum_{n'}\langle n'|n\rangle\langle n|\hat{a}^\dagger a|n'\rangle
$$
$$
= \langle n|\hat{a}^\dagger\hat{a}|n\rangle = n\,,
$$
(3.94)

$$
\langle\hat{a}\hat{a}^\dagger\rangle_{\mathrm{osc}} = \langle\hat{a}^\dagger\hat{a}\rangle_{\mathrm{osc}} + 1
$$
$$
= n+1\,,
$$
(3.95)

(3.93) becomes *finally*

$$
\kappa^{(2)}(\tau) \equiv \lambda^2\kappa(\tau) = -\lambda^2\hbar^{-2}\{[(1+n)\mathrm{e}^{\mathrm{i}\omega_0\tau} - n\,\mathrm{e}^{-\mathrm{i}\omega_0\tau}]\langle\hat{\Gamma}_0\hat{\Gamma}_0(\tau)\rangle_B
$$
$$
-[n\,\mathrm{e}^{\mathrm{i}\omega_0\tau} - (n+1)\mathrm{e}^{-\mathrm{i}\omega_0\tau}]\langle\hat{\Gamma}_0(\tau)\hat{\Gamma}_0\rangle_B\}/\langle\Delta\hat{n}(0)\rangle\,.
$$
(3.96)

3.3.4 Solution of the Equation of Motion for $\langle \Delta \hat{n}(t) \rangle$ in the Weak-Coupling Limit

For the linear coupling model developed above, the equation of motion (3.56) becomes

$$\dot{G}(t) = -\lambda^2 \int\limits_0^t \kappa(\tau) G(t-\tau) d\tau \tag{3.97}$$

through $O(\lambda^2)$, where we have also employed (3.67) and (3.96). We now wish to solve (3.97) in van Hove's [3.2] weak-coupling limit, which consists of allowing the coupling strength λ to approach zero as the time t approaches infinity in such a manner that the product $\lambda^2 t$ remains constant. Introducing the variable

$$x \equiv \lambda^2 t \tag{3.98}$$

and letting

$$G(t) \equiv F(x), \tag{3.99}$$

we have

$$\dot{G} = \frac{dG}{dt} = \lambda^2 \frac{dF}{dx}. \tag{3.100}$$

Then (3.97) may be rewritten as

$$\frac{dF}{dx} = -\int\limits_0^{x/\lambda^2} \kappa(\tau) F(x - \lambda^2 \tau) d\tau. \tag{3.101}$$

In the simultaneous limits $\lambda \to 0$, $t \to \infty$ and $x = \text{const}$, (3.101) becomes

$$\frac{dF}{dx} = -\hat{\kappa} F \tag{3.102}$$

where

$$\hat{\kappa} = \int\limits_0^\infty \kappa(\tau) d\tau. \tag{3.103}$$

The solution of (3.102) is

$$F(x) = e^{-\hat{\kappa} x} F(0) \tag{3.104}$$

and accordingly

$$G(t) = e^{-\kappa t} G(0).$$ (3.105)

Note that in (3.105) we have replaced λ^2 in the exponential by unity, since we have calculated all quantities involved correctly up to $O(\lambda^2)$.

Now from (3.96) and (3.103), we have

$$\hat{\kappa} = -\hbar^{-2} \left[(1+n) \int_0^\infty e^{i\omega_0\tau} \langle \hat{\Gamma}_0 \hat{\Gamma}_0(\tau) \rangle_{\mathrm{B}} d\tau - n \int_0^\infty e^{-i\omega_0\tau} \langle \hat{\Gamma}_0 \hat{\Gamma}_0(\tau) \rangle_{\mathrm{B}} d\tau \right.$$

$$\left. - n \int_0^\infty e^{i\omega_0\tau} \langle \hat{\Gamma}_0(\tau) \hat{\Gamma}_0 \rangle_{\mathrm{B}} d\tau + (1+n) \int_0^\infty e^{-i\omega_0\tau} \langle \hat{\Gamma}_0(\tau) \hat{\Gamma}_0 \rangle_{\mathrm{B}} d\tau \right] / \langle \Delta\hat{n}(0) \rangle.$$ (3.106)

Letting $\tau' = -\tau$, we may write

$$\int_0^\infty e^{i\omega_0\tau} \langle \hat{\Gamma}_0(\tau) \hat{\Gamma}_0 \rangle_{\mathrm{B}} d\tau = - \int_0^{-\infty} e^{-i\omega_0\tau'} \langle \hat{\Gamma}_0(-\tau') \hat{\Gamma}_0 \rangle_{\mathrm{B}} d\tau'$$

$$= \int_{-\infty}^0 e^{-i\omega_0\tau'} \langle \hat{\Gamma}_0 \hat{\Gamma}_0(\tau') \rangle_{\mathrm{B}} d\tau'.$$ (3.107)

where we have also invoked the invariance of the correlation function under time translation

$$\langle \hat{X}(t) \hat{Y} \rangle = \langle \hat{X} \hat{Y}(-t) \rangle.$$ (3.108)

Using (3.107) and the analogous result

$$\int_0^\infty e^{i\omega_0\tau} \langle \hat{\Gamma}_0 \hat{\Gamma}_0(\tau) \rangle_{\mathrm{B}} d\tau = \int_{-\infty}^0 e^{-i\omega_0\tau} \langle \hat{\Gamma}_0(\tau) \hat{\Gamma}_0 \rangle_{\mathrm{B}} d\tau,$$ (3.109)

we can combine the terms on the right-hand side of (3.106) to obtain

$$\hat{\kappa} = -\hbar^{-2} \left[(1+n) \int_{-\infty}^{+\infty} e^{-i\omega_0\tau} \langle \hat{\Gamma}_0(\tau) \hat{\Gamma}_0 \rangle_{\mathrm{B}} d\tau - n \int_{-\infty}^{+\infty} e^{-i\omega_0\tau} \langle \hat{\Gamma}_0 \hat{\Gamma}_0(\tau) \rangle_{\mathrm{B}} d\tau \right] / \langle \Delta\hat{n}(0) \rangle$$

$$= \hbar^{-2} \left\{ n \int_{-\infty}^{+\infty} e^{-i\omega_0\tau} \langle [\hat{\Gamma}_0, \hat{\Gamma}_0(\tau)] \rangle_{\mathrm{B}} d\tau - \int_{-\infty}^{+\infty} e^{-i\omega_0\tau} \langle \hat{\Gamma}_0(\tau) \hat{\Gamma}_0 \rangle_{\mathrm{B}} d\tau \right\} / \langle \Delta\hat{n}(0) \rangle$$

$$= \hbar^{-2} \left\{ n \int_{-\infty}^{+\infty} e^{-i\omega_0\tau} \langle [\hat{\Gamma}_0, \hat{\Gamma}_0(\tau)] \rangle_{\mathrm{B}} d\tau - \int_{-\infty}^{+\infty} e^{i\omega_0\tau} \langle \hat{\Gamma}_0 \hat{\Gamma}_0(\tau) \rangle_{\mathrm{B}} d\tau \right\} / \langle \Delta\hat{n}(0) \rangle.$$ (3.110)

The last line of (3.110) follows by making the change of variables $\tau' = -\tau$ in the second integral and then applying (3.108). Finally, using (3.293) and (3.303) of Appendix C, we have from (3.110)

$$\hat{\kappa} = \hbar^{-2} [n I(-\omega_0) - (e^{\beta\hbar\omega_0} - 1)^{-1} I(-\omega_0)] / \langle \Delta n(0) \rangle . \tag{3.111}$$

But since

$$\bar{n} = (e^{\beta\hbar\omega_0} - 1)^{-1} \tag{3.112}$$

and

$$\langle \Delta \hat{n}(0) \rangle = \langle \hat{n} \rangle - \bar{n} = n - \bar{n} , \tag{3.113}$$

(3.111) simplifies to

$$\hat{\kappa} = \hbar^{-2} I(-\omega_0) = 2 \operatorname{Re} \{ \gamma(\omega_0) \}$$
$$\equiv \tau'^{-1} , \tag{3.114}$$

where $\gamma(\omega_0)$ is defined by (3.288) of Appendix B. Hence, the final solution for the *linear coupling model* in the *weak-coupling limit* is

$$\langle \Delta \hat{n}(t) \rangle = e^{-t/\tau'} \langle \Delta \hat{n}(0) \rangle \tag{3.115}$$

or by (3.113)

$$\langle \hat{n}(t) \rangle = \bar{n} + e^{-t/\tau'} (n - \bar{n}) . \tag{3.116}$$

Eq. (3.116) states that the expectation value of the occupation number (which, recall, is proportional to the population of vibrationally excited molecules) of an harmonic oscillator coupled linearly to a heat bath (solvent) decays exponentially with a time constant (lifetime) τ' from an initial ($t=0$) value n to a final ($t=\infty$) value $\bar{n}$.

Also using the Zwanzig projection-operator approach, WILSON et al. [3.41] have calculated the correlation functions $\langle \hat{a}^\dagger \hat{a}(t) \rangle$ and $\langle \hat{a} \hat{a}^\dagger(t) \rangle$ for our model, obtaining in the van Hove weak-coupling limit

$$\langle \hat{a} \hat{a}^\dagger(t) \rangle = e^{i\omega't} e^{-\operatorname{Re}\{\gamma(\omega_0)\}|t|} (1 + \bar{n}) \tag{3.117}$$

$$\langle \hat{a}^\dagger \hat{a}(t) \rangle = e^{-i\omega't} e^{-\operatorname{Re}\{\gamma(\omega_0)\}|t|} \bar{n} , \tag{3.118}$$

where

$$\omega' \equiv \omega_0 + \operatorname{Im} \{ \gamma(\omega_0) \} . \tag{3.119}$$

Thus, neglecting the direct terms $\langle \hat{a}\hat{a}(t)\rangle$ and $\langle \hat{a}^\dagger \hat{a}^\dagger(t)\rangle$, which oscillate rapidly compared to the cross terms, we see that the vibrational correlation function

$$G_v(t) \equiv \langle Q(0)Q(t)\rangle \approx [(1+\bar{n})e^{i\omega't} + \bar{n}e^{-i\omega't}]e^{-\mathrm{Re}\{\gamma(\omega_0)\}|t|} \tag{3.120}$$

(or rather its envelope) decays exponentially with a time constant (correlation time)

$$\tau_v = [\mathrm{Re}\{\gamma(\omega_0)\}]^{-1}. \tag{3.121}$$

Comparison of (3.114) and (3.121) leads to the relation

$$\tau_v = 2\tau', \tag{3.122}$$

i. e., for the *linear coupling model* the vibrational correlation time is just twice as long as the vibrational lifetime. It is important to note that (3.122) holds only for the linearly coupled harmonic oscillator. However, the projection-operator method may be extended to more general systems, as KIM and WILSON [3.42] have shown.

3.3.5 Other Approaches to Treating the Dynamics of Vibrational Relaxation

NITZAN and JORTNER [3.43] have also studied the dynamics of relaxation of an harmonic oscillator coupled to a heat bath. They take the bath to be a collection of harmonic oscillators having a frequency distribution ρ_v and assume that the oscillator-bath coupling is linear in the oscillator coordinate. Invoking the random phase approximation, which is equivalent to assuming that the bath remains in thermal equilibrium as the oscillator relaxes, they *linearized* Heisenberg's equations of motion for $\hat{a}^\dagger$ and $\hat{a}$ and then solved these in the Wigner-Weisskopf [3.44] approximation.

Using KUBO's [3.45] cumulant-expansion technique, NITZAN and SILBEY [3.46] have treated precisely the model developed above in Subsects. 3.3.1–4, arriving at an expression for the vibrational lifetime τ' identical to (3.114). They also extend their treatment to the two-state model.

Recently FLEMING et al. [3.47] have applied the Pauli master equation [3.48] to vibrational redistribution in isolated large molecules. They solved the master equation for the relaxation of a single molecular normal mode to second order in the coupling, taken to be linear in the relaxing mode and quadratic in the remaining (effective bath) modes, which are assumed to remain in thermal equilibrium throughout the duration of the relaxation. As we might expect, the results of certain of the limiting cases considered by FLEMING et al. [3.47] are in agreement with those found by NITZAN and JORTNER [3.43].

LIN [3.49] has generalized the master-equation approach to vibrational relaxation obtaining solutions through fourth order in the perturbation. In this work he also derived explicit expressions for the rate constant for vibrational relaxation in some specific models. We shall consider these further in Subsect. 3.6.5.

3.4 Formal Expressions for the Rate of Non-Radiative Vibrational Transitions

Let us reconsider our prototypic system from another viewpoint. Instead of (3.27) we may write for the Hamiltonian

$$\hat{\mathscr{H}} = \hat{H}_M^0 + \hat{T}_L + \hat{V}, \tag{3.123}$$

where $\hat{H}_M^0$ is the "internal" Hamiltonian of the free diatomic solute molecule, $\hat{T}_L$ is the total kinetic energy associated with "external" translation-like degrees of freedom of the centers of mass of all molecules (including the diatomic solute), and $\hat{V}$ is the total potential energy of interaction of all molecules. Note that we make no assumptions about the form of either the internal potential energy or the external interactions.

Let $|\gamma\rangle$ be an eigenket of the isolated-diatomic Hamiltonian, i. e.

$$\hat{H}_M^0 |\gamma\rangle = \varepsilon_\gamma^M |\gamma\rangle, \tag{3.124}$$

where ε_γ^M is the energy associated with vibrational-rotational state γ. Then a many-body stationary state $|\psi\rangle$ of the system may be expanded as a linear combination of direct products of internal eigenkets $|\gamma\rangle$ and external vectors $|\chi^{(\gamma)}\rangle$:

$$|\psi\rangle = \sum_\gamma |\gamma\rangle \otimes |\chi^{(\gamma)}\rangle. \tag{3.125}$$

The $|\chi^{(\gamma)}\rangle$ describe the external motion of the solvent molecules, to which the internal motion is coupled via $\hat{V}$. Now $|\psi\rangle$ must satisfy the Schrödinger equation

$$\hat{\mathscr{H}} |\psi\rangle = \varepsilon |\psi\rangle. \tag{3.126}$$

Substituting the expansion (3.125) into (3.126), projecting both members of the resulting equation onto the eigenket $|\gamma'\rangle$, and invoking the orthonormality condition

$$\langle \gamma' | \gamma \rangle = \delta_{\gamma'\gamma}, \tag{3.127}$$

we arrive at the following result

$$\hat{T}_L |\chi^{(\gamma')}\rangle + \sum_\gamma \hat{V}_{\gamma'\gamma} |\chi^{(\gamma)}\rangle = (\varepsilon - \varepsilon_{\gamma'}^M) |\chi^{(\gamma')}\rangle \ . \tag{3.128}$$

If the off-diagonal coupling matrix elements

$$\hat{V}_{\gamma'\gamma} = \langle \gamma' | \hat{V} | \gamma \rangle \tag{3.129}$$

are neglected, (3.128) simplifies to

$$(\hat{T}_L + \hat{V}_{\gamma\gamma}) |\chi_0^{(\gamma)}\rangle = (\varepsilon - \varepsilon_\gamma^M) |\chi_0^{(\gamma)}\rangle \ , \tag{3.130}$$

where $|\chi_0^{(\gamma)}\rangle$ are zero-order eigenkets of the effective Hamiltonian

$$\hat{H}_{L0}^{(\gamma)} \equiv \hat{T}_L + \hat{V}_{\gamma\gamma} \tag{3.131}$$

governing the external motion of the solvent molecules with the diatomic fixed in internal state γ. Eq. (3.131) has the following physical interpretation. The internal force binding the atoms of the diatomic solute is much stronger than the external forces acting between molecules of the solvent. Thus, the diatomic executes many vibrations during the period of "vibration" of a typical solvent molecule, and in zero-order the effective potential energy governing the external motion of the solvent molecules (in the presence of the diatomic in state γ) is an average of the instantaneous interactions $\hat{V}$ over the internal state γ.

Let us rewrite the eigenket $|\chi_0^{(\gamma)}\rangle$ as

$$|\chi_0^{(\gamma)}\rangle = |n_\gamma\rangle \ , \tag{3.132}$$

where n_γ denotes the set of quantum numbers required to specify the external state of the solvent. Now since the set of kets $\{|\gamma\rangle \otimes |n_\gamma\rangle = |\gamma\, n_\gamma\rangle\}$ is complete, we can write the Hamiltonian (3.123) in the form of a spectral resolution as

$$\hat{\mathscr{H}} = \sum_{\gamma n_\gamma} \sum_{\gamma' n_{\gamma'}} |\gamma\, n_\gamma\rangle \langle \gamma\, n_\gamma | \hat{\mathscr{H}} | \gamma'\, n_{\gamma'}\rangle \langle \gamma'\, n_{\gamma'} | \ , \tag{3.133}$$

which by (3.130) splits into two contributions

$$\hat{\mathscr{H}} = \hat{H}_0 + \hat{H}' \ , \tag{3.134}$$

where

$$\hat{H}_0 \equiv \sum_\gamma \sum_{n_\gamma} |\gamma\, n_\gamma\rangle \, \varepsilon_{\gamma n_\gamma} \langle \gamma\, n_\gamma | \tag{3.135}$$

and

$$\hat{H}' \equiv \sum_{\gamma} \sum_{\boldsymbol{n}_\gamma} \sum_{\gamma'} \sum_{\boldsymbol{n}_{\gamma'}} |\gamma \, \boldsymbol{n}_\gamma\rangle \, \langle \gamma \, \boldsymbol{n}_\gamma| \hat{V} |\gamma' \, \boldsymbol{n}_{\gamma'}\rangle \, \langle \gamma' \, \boldsymbol{n}_{\gamma'}| ,$$
$$\gamma \neq \gamma' .$$
(3.136)

In (3.135) $\varepsilon_{\gamma \boldsymbol{n}_\gamma}$, the total energy of the system in state $\gamma \boldsymbol{n}_\gamma$, is given by

$$\varepsilon_{\gamma \boldsymbol{n}_\gamma} = \varepsilon_\gamma^{\mathrm{M}} + \varepsilon_{\boldsymbol{n}_\gamma}^{\mathrm{L}} ,$$
(3.137)

where $\varepsilon_{\boldsymbol{n}_\gamma}^{\mathrm{L}}$ is the total external energy. If the initial state of the system is well approximated by

$$|\psi(0)\rangle = |\alpha \, \boldsymbol{n}_\alpha\rangle ,$$
(3.138)

then $\hat{H}'$ gives rise to non-radiative transitions of the system to other states $|\beta \, \boldsymbol{n}_\beta\rangle$ and is therefore responsible for vibrational relaxation. We now proceed to calculate the transition rate.

Suppose the system is prepared in such a way that the diatomic solute is initially (at $t=0$) in state α. Then, if we assume that the external degrees of freedom of the solvent are initially in thermal equilibrium, the system is characterized at $t=0$ by the density operator

$$\hat{\rho}(0) = |\alpha\rangle \, \langle \alpha| \otimes \hat{\rho}_{\mathrm{L}}(0) ,$$
(3.139)

where

$$\hat{\rho}_{\mathrm{L}}(0) = \mathrm{e}^{- \beta \hat{H}_{\mathrm{L0}}^{(\alpha)}}/\mathrm{Tr} \, \{\mathrm{e}^{- \beta \hat{H}_{\mathrm{L0}}^{(\alpha)}}\} = \mathrm{e}^{- \beta \hat{H}_{\mathrm{L0}}^{(\alpha)}} \, Q_\alpha^{-1}$$
(3.140)

and

$$\hat{H}_{\mathrm{L0}}^{(\alpha)} = \sum_{\boldsymbol{n}_\alpha} |\boldsymbol{n}_\alpha\rangle \varepsilon_{\boldsymbol{n}_\alpha}^{\mathrm{L}} \langle \boldsymbol{n}_\alpha| .$$
(3.141)

Now the probability that the diatomic is in state β at time t, regardless of the external state of the solvent, may be expressed

$$P_{\alpha \to \beta}(t) = \mathrm{Tr} \, \{\hat{\rho}(t) \hat{P}_\beta\} ,$$
(3.142)

where $\hat{P}_\beta$ is a projection operator defined by

$$\hat{P}_\beta \equiv |\beta\rangle \, \langle \beta|$$
(3.143)

and $\hat{\rho}(t)$ is the density operator at time t, given by [3.44],

$$\hat{\rho}(t) = \hat{U}(t) \hat{\rho}(0) \hat{U}^\dagger(t) .$$
(3.144)

In (3.144) $\hat{U}(t)$ is the time-evolution operator of the complete system

$$\hat{U}(t) = e^{-i\hat{\mathscr{H}}t/\hbar}. \tag{3.145}$$

To first order in the perturbation $\hat{H}'$ we may approximate $\hat{U}(t)$ as [3.50]

$$\hat{U}(t) \simeq \hat{U}_0(t)\left[1 - i\hbar^{-1}\int_0^t \hat{U}_0^\dagger(t')\hat{H}'\hat{U}_0(t')\,dt'\right] \tag{3.146}$$

where

$$\hat{U}_0(t) = e^{-i\hat{H}_0 t/\hbar}. \tag{3.147}$$

Thus,

$$
\begin{aligned}
P_{\alpha\to\beta}^{(1)}(t) = {}& \mathrm{Tr}\left\{\hat{U}_0(t)\left[1 - i\hbar^{-1}\int_0^t \hat{U}_0^\dagger(t')\hat{H}'\hat{U}_0(t')\,dt'\right]\right.\\
& \times \hat{\rho}(0)\left[1 + i\hbar^{-1}\int_0^t \hat{U}_0^\dagger(t'')\hat{H}'\hat{U}_0(t'')\,dt''\right]\left.\hat{U}_0^\dagger(t)\hat{P}_\beta\right\}\\
= {}& \hbar^{-2}\mathrm{Tr}\left\{\hat{U}_0(t)\int_0^t \hat{U}_0^\dagger(t')\hat{H}'\hat{U}_0(t')\,dt'\,\hat{\rho}(0)\right.\\
& \times \left.\int_0^t \hat{U}_0^\dagger(t'')\hat{H}'\hat{U}_0(t'')\,dt''\,\hat{U}_0^\dagger(t)\hat{P}_\beta\right\},
\end{aligned}
\tag{3.148}
$$

where the second line of (3.148) follows by employing the invariance of the trace under cyclic permutation along with relations (3.306–308) of Appendix D. Using (3.139–141) and (3.143), we may recast (3.148) as

$$
\begin{aligned}
P_{\alpha\to\beta}^{(1)}(t) = {}& \hbar^{-2}\sum_\gamma\sum_{n_\gamma}\langle\gamma\,n_\gamma|\,\hat{U}_0(t)\left[\int_0^t \hat{U}_0^\dagger(t')\hat{H}'\hat{U}_0(t')\,dt'\right]|\alpha\rangle\langle\alpha|\\
& \otimes\sum_{n_\alpha}|n_\alpha\rangle\,p_{n_\alpha}\langle n_\alpha|\left[\int_0^t \hat{U}_0^\dagger(t'')\hat{H}'\hat{U}_0(t'')\,dt''\right]\hat{U}_0^\dagger(t)|\beta\rangle\langle\beta|\gamma\,n_\gamma\rangle,
\end{aligned}
\tag{3.149}
$$

where

$$p_{n_\alpha} = e^{-\beta\varepsilon_{n_\alpha}^L}/\sum_{n_\alpha} e^{-\beta\varepsilon_{n_\alpha}^L} \tag{3.150}$$

is the probability that the solvent is initially in state n_α. Carrying out the sum on γ and rearranging the factors in (3.149) yields

$$
\begin{aligned}
P_{\alpha\to\beta}^{(1)}(t) = {}& \hbar^{-2}\sum_{n_\alpha} p_{n_\alpha}\sum_{n_\beta}\langle\alpha\,n_\alpha|\int_0^t \hat{U}_0^\dagger(t'')\hat{H}'\hat{U}_0(t'')\,dt''\,U_0^\dagger(t)|\beta\,n_\beta\rangle\\
& \times\langle\beta\,n_\beta|\,\hat{U}_0(t)\int_0^t \hat{U}_0^\dagger(t')\hat{H}'\hat{U}_0(t')\,dt'\,|\alpha\,n_\alpha\rangle.
\end{aligned}
\tag{3.151}
$$

Now we note that

$$\hat{H}_0 |\alpha\, n_\alpha\rangle = \left(\sum_{n'_\alpha} \varepsilon_{\alpha n'_\alpha} |n'_\alpha\rangle \langle n'_\alpha|\right) |\alpha\, n_\alpha\rangle$$
$$= (\varepsilon^{\mathrm{M}}_\alpha + \hat{H}^{(\alpha)}_{\mathrm{LO}}) |\alpha\, n_\alpha\rangle\,. \tag{3.152}$$

Hence

$$\hat{U}_0(t) |\alpha\, n_\alpha\rangle = \mathrm{e}^{-\mathrm{i}\varepsilon^{\mathrm{M}}_\alpha t/\hbar}\, \mathrm{e}^{-\mathrm{i}\hat{H}^{(\alpha)}_{\mathrm{LO}} t/\hbar} |\alpha\, n_\alpha\rangle$$
$$= \mathrm{e}^{-\mathrm{i}\varepsilon^{\mathrm{M}}_\alpha t/\hbar} |\alpha\rangle \otimes \hat{U}^{(\alpha)}_{\mathrm{LO}}(t) |n_\alpha\rangle\,. \tag{3.153}$$

By (3.153) we may rewrite (3.151) as

$$P^{(1)}_{\alpha\to\beta}(t) = \hbar^{-2} \int_0^t dt'' \int_0^t dt' \sum_{n_\alpha} p_{n_\alpha} \sum_{n_\beta} \mathrm{e}^{\mathrm{i}(\varepsilon^{\mathrm{M}}_\alpha - \varepsilon^{\mathrm{M}}_\beta)(t''-t')/\hbar}$$
$$\times \langle n_\alpha| \hat{U}^{(\alpha)\dagger}_{\mathrm{LO}}(t'') \langle \alpha| \hat{H}' |\beta\rangle \hat{U}^{(\beta)}_{\mathrm{LO}}(t'') |n_\beta\rangle \langle n_\beta| \hat{U}^{(\beta)\dagger}_{\mathrm{LO}}(t') \langle\beta| \hat{H}' |\alpha\rangle \hat{U}^{(\alpha)}_{\mathrm{LO}}(t') |n_\alpha\rangle$$
$$= \hbar^{-2} \int_0^t dt'' \int_0^t dt'\, \mathrm{e}^{\mathrm{i}\omega_{\alpha\beta}(t''-t')} \mathrm{Tr}\{\hat{\rho}_{\mathrm{L}}(0) \hat{V}_{\alpha\beta}(t'') \hat{V}^\dagger_{\alpha\beta}(t')\} \tag{3.154}$$

where

$$\hat{V}_{\alpha\beta}(t) \equiv \hat{U}^{(\alpha)\dagger}_{\mathrm{LO}}(t) \langle\alpha| \hat{H}' |\beta\rangle \hat{U}^{(\beta)}_{\mathrm{LO}}(t)\,. \tag{3.155}$$

and

$$\omega_{\alpha\beta} \equiv (\varepsilon^{\mathrm{M}}_\alpha - \varepsilon^{\mathrm{M}}_\beta)/\hbar\,. \tag{3.156}$$

Invoking the cyclic invariance of the trace in (3.154), we can again rewrite $P^{(1)}_{\alpha\to\beta}(t)$ as

$$P^{(1)}_{\alpha\to\beta}(t) = \hbar^{-2} \int_0^t dt'' \int_0^t dt'\, \mathrm{e}^{\mathrm{i}\omega_{\alpha\beta}(t''-t')} \mathrm{Tr}\{\hat{\rho}_{\mathrm{L}}(0) \langle\alpha| \hat{H}' |\beta\rangle [\hat{U}^{(\beta)}_{\mathrm{LO}}(t') \hat{U}^{(\beta)\dagger}_{\mathrm{LO}}(t'')]^\dagger$$
$$\times \langle\beta| \hat{H}' |\alpha\rangle [\hat{U}^{(\alpha)}_{\mathrm{LO}}(t') \hat{U}^{(\alpha)\dagger}_{\mathrm{LO}}(t'')]\} \tag{3.157}$$
$$= \hbar^{-2} \int_0^t dt'' \int_0^t dt'\, \mathrm{e}^{-\mathrm{i}\omega_{\alpha\beta}(t'-t'')} \mathrm{Tr}\{\hat{\rho}_{\mathrm{L}}(0) \hat{V}_{\alpha\beta}(0) \hat{V}^\dagger_{\alpha\beta}(t'-t'')\}\,.$$

Now the rate of transition from state α to state β is given by

$$W^{(1)}_{\alpha\to\beta}(t) = P^{(1)}_{\alpha\to\beta}(t)/t = \hbar^{-2} t^{-1} \int_0^t dt'' \int_0^t dt'\, \mathrm{e}^{-\mathrm{i}\omega_{\alpha\beta}(t'-t'')}$$
$$\times \mathrm{Tr}\{\hat{\rho}_{\mathrm{L}}(0) \hat{V}_{\alpha\beta}(0) \hat{V}^\dagger_{\alpha\beta}(t'-t'')\}\,. \tag{3.158}$$

Changing the variables of integration in (3.158) to $\tau' = t' - t''$, $\tau'' = t''$, we get

$$W^{(1)}_{\alpha \to \beta}(t) = \hbar^{-2} t^{-1} \left\{ \int_0^t d\tau' \int_0^{t-\tau'} d\tau'' e^{-i\omega_{\alpha\beta}\tau'} \langle \hat{V}_{\alpha\beta}(0) \hat{V}^\dagger_{\alpha\beta}(\tau') \rangle_{0\alpha} \right.$$

$$\left. + \int_{-t}^0 d\tau' \int_{|\tau'|}^t d\tau'' e^{-i\omega_{\alpha\beta}\tau'} \langle \hat{V}_{\alpha\beta}(0) \hat{V}^\dagger_{\alpha\beta}(\tau') \rangle_{0\alpha} \right\} \tag{3.159}$$

$$= \hbar^{-2} t^{-1} \int_{-t}^t d\tau' (t - |\tau'|) e^{-i\omega_{\alpha\beta}\tau'} \langle \hat{V}_{\alpha\beta}(0) \hat{V}^\dagger_{\alpha\beta}(\tau') \rangle_{0\alpha},$$

where $\langle ... \rangle_{0\alpha}$ denotes the ensemble average over the initial density operator $\hat{\rho}_L(0)$. Now in the limit $t \to \infty$

$$W^{(1)}_{\alpha \to \beta} = \hbar^{-2} \int_{-\infty}^{+\infty} d\tau e^{-i\omega_{\alpha\beta}\tau} \langle \hat{V}_{\alpha\beta}(0) \hat{V}^\dagger_{\alpha\beta}(\tau) \rangle_{0\alpha}, \tag{3.160}$$

since the first term (proportional to t) in the integrand in (3.159) dominates the second.

We may easily convert (3.160) to the commonly assumed Golden-Rule expression for $W^{(1)}_{\alpha \to \beta}$. Using (3.140, 141), and (3.155, 156) we can rewrite (3.160)

$$W^{(1)}_{\alpha \to \beta} = \hbar^{-2} \int_{-\infty}^{+\infty} dt e^{-i\varepsilon^M_\alpha t/\hbar} e^{+i\varepsilon^M_\beta t/\hbar} \sum_{n_\alpha} P_{n_\alpha} \langle n_\alpha | \langle \alpha | \hat{H}' | \beta \rangle$$

$$\times \exp[i \sum_{n'_\beta} \varepsilon^L_{n'_\beta} |n'_\beta\rangle \langle n'_\beta| t/\hbar] \langle \beta | \hat{H}' | \alpha \rangle \exp[-i \sum_{n'_\alpha} \varepsilon^L_{n'_\alpha} |n'_\alpha\rangle \langle n'_\alpha| t/\hbar] | n_\alpha \rangle \tag{3.161}$$

Evaluating $\langle \alpha | \hat{H}' | \beta \rangle$ and $\langle \beta | \hat{H}' | \alpha \rangle$ by (3.136) and substituting the resulting expressions into (3.161) gives

$$W^{(1)}_{\alpha \to \beta} = \hbar^{-2} \int_{-\infty}^{+\infty} dt e^{-i\varepsilon^M_\alpha t/\hbar} e^{i\varepsilon^M_\beta t/\hbar} \sum_{n_\alpha} P_{n_\alpha}$$

$$\times \sum_{n''_\alpha} \sum_{n''_\beta} \sum_{n'''_\alpha} \sum_{n'''_\beta} \langle n_\alpha | n''_\alpha \rangle \langle \alpha n''_\alpha | \hat{V} | \beta n''_\beta \rangle \langle n''_\beta | \exp[i \sum_{n'_\beta} \varepsilon^L_{n'_\beta} |n'_\beta\rangle \langle n'_\beta| t/\hbar] | n'''_\beta \rangle$$

$$\times \langle \beta n'''_\beta | \hat{V} | \alpha n'''_\alpha \rangle \langle n'''_\alpha | \exp[-i \sum_{n'_\alpha} \varepsilon^L_{n'_\alpha} |n'_\alpha\rangle \langle n'_\alpha| t/\hbar] | n_\alpha \rangle \tag{3.162}$$

$$= \hbar^{-2} \sum_{n_\alpha} P_{n_\alpha} \sum_{n_\beta} \langle \alpha n_\alpha | \hat{V} | \beta n_\beta \rangle \langle \beta n_\beta | \hat{V} | \alpha n_\alpha \rangle$$

$$\times \int_{-\infty}^{+\infty} dt e^{-i(\varepsilon^M_\alpha + \varepsilon^L_{n_\alpha})t/\hbar} e^{i(\varepsilon^M_\beta + \varepsilon^L_{n_\beta})t/\hbar}$$

$$= \hbar^{-2} \sum_{n_\alpha} P_{n_\alpha} \sum_{n_\beta} |\langle \alpha n_\alpha | \hat{V} | \beta n_\beta \rangle|^2 \int_{-\infty}^{+\infty} dt e^{-i(\varepsilon_{\alpha n_\alpha} - \varepsilon_{\beta n_\beta})t/\hbar}.$$

Finally, making use of the relations [3.50]

$$2\pi\delta(x) = \int\limits_{-\infty}^{+\infty} d\kappa\, e^{i\kappa x} \tag{3.163}$$

and

$$\delta(ax) = \frac{1}{|a|}\delta(x), \qquad a \neq 0 \tag{3.164}$$

in (3.162), we have

$$W_{\alpha\to\beta}^{(1)} = 2\pi\hbar^{-1} \sum_{n_\alpha} p_{n_\alpha} \sum_{n_\beta} |\langle \alpha\, n_\alpha| \hat{V}|\beta\, n_\beta\rangle|^2\, \delta(\varepsilon_{\alpha n_\alpha} - \varepsilon_{\beta n_\beta}), \tag{3.165}$$

which is precisely the usual Golden-Rule rate. Eq. (3.165) is correct to first order in the perturbation.

Before proceeding we wish to establish a relation between $W_{\alpha\to\beta}^{(1)}$ and $W_{\beta\to\alpha}^{(1)}$. By symmetry from (3.165) we observe that

$$W_{\beta\to\alpha}^{(1)} = 2\pi\hbar^{-1} \sum_{n_\beta} p_{n_\beta} \sum_{n_\alpha} |\langle \beta\, n_\beta| \hat{V}|\alpha\, n_\alpha\rangle|^2\, \delta(\varepsilon_{\beta n_\beta} - \varepsilon_{\alpha n_\alpha}). \tag{3.166}$$

Note that the delta function in (3.165) implies that

$$\varepsilon_\beta^M + \varepsilon_{n_\beta}^L = \varepsilon_\alpha^M + \varepsilon_{n_\alpha}^L. \tag{3.167}$$

Thus, using (3.164), (3.167) and the fundamental relation

$$\langle \beta\, n_\beta| \hat{V}|\alpha\, n_\alpha\rangle = \langle \alpha\, n_\alpha| \hat{V}|\beta\, n_\beta\rangle^*, \tag{3.168}$$

we may rewrite (3.166) as

$$\begin{aligned}
W_{\beta\to\alpha}^{(1)} &= 2\pi\hbar^{-1} \sum_{n_\beta} \sum_{n_\beta} e^{-\beta(\varepsilon_\alpha^M - \varepsilon_\beta^M)} e^{-\beta\varepsilon_{n_\alpha}^L} |\langle \alpha\, n_\alpha| \hat{V}|\beta\, n_\beta\rangle|^2\, \delta(\varepsilon_{\alpha n_\alpha} - \varepsilon_{\beta n_\beta}) \\
&= e^{-\beta\hbar\omega_{\alpha\beta}} Q_\alpha Q_\beta^{-1} W_{\alpha\to\beta}^{(1)}, \\
&\equiv \gamma_{\alpha\beta} W_{\alpha\to\beta}^{(1)},
\end{aligned} \tag{3.169}$$

the last line defining $\gamma_{\alpha\beta}$.

3.5 Relation of the Golden-Rule Transition Rate to the Vibrational Lifetime for the Two-State Model

Suppose the diatomic solute has only two internal states—an upper state α and a lower state β. Then the rate of change of the population of state α is given by

$$\dot{N}_\alpha = -W^{(1)}_{\alpha \to \beta} N_\alpha + W^{(1)}_{\beta \to \alpha} N_\beta , \tag{3.170}$$

where N_γ is the population of diatomics in state γ and the $W^{(1)}$'s are first-order transition rates [see (3.160) and (3.165)]. From (3.170) we have

$$\dot{N}_\alpha = [N_\beta \gamma_{\alpha\beta} - N_\alpha] W^{(1)}_{\alpha \to \beta} \tag{3.171}$$

by (3.169). Defining

$$f \equiv N_\alpha / N , \tag{3.172}$$

where

$$N \equiv N_\alpha + N_\beta , \tag{3.173}$$

we may recast (3.171) in the form

$$\dot{f} = [(1-f)\gamma_{\alpha\beta} - f] W^{(1)}_{\alpha \to \beta} . \tag{3.174}$$

When the system is in thermal equilibrium, $f = \bar{f}$, $\dot{f} = 0$ and from (3.174) we deduce that

$$\bar{f}(1-\bar{f})^{-1} = \gamma_{\alpha\beta} . \tag{3.175}$$

Now let us define the deviation of f from its equilibrium value $\bar{f}$ by

$$\Delta f \equiv f - \bar{f}. \tag{3.176}$$

Clearly

$$\dot{f} = \Delta \dot{f} = [(1 - \bar{f} - \Delta f)\gamma_{\alpha\beta} - \bar{f} - \Delta f] W^{(1)}_{\alpha \to \beta}$$
$$= -\Delta f(1 + \gamma_{\alpha\beta}) W^{(1)}_{\alpha \to \beta} , \tag{3.177}$$

the last line of (3.177) following from (3.175). The solution of (3.177) is simply

$$\Delta f(t) = \Delta f(0) e^{-t/\bar{\tau}} , \tag{3.178}$$

where

$$\bar{\tau}^{-1} \equiv (1 + \gamma_{\alpha\beta}) W^{(1)}_{\alpha \to \beta} . \tag{3.179}$$

We note that Δf is identical with $\langle \Delta \hat{n} \rangle$ for the two-state model, so that $\bar{\tau}$ should be identified with the vibrational lifetime τ' [see (3.114)].

Using (3.169), we may rewrite (3.179) as

$$\bar{\tau}^{-1} = (1 + \gamma_{\beta\alpha}) W^{(1)}_{\beta\to\alpha} . \tag{3.180}$$

Then from (3.160) we have

$$W^{(1)}_{\beta\to\alpha} = \hbar^{-2} \int_{-\infty}^{+\infty} d\tau \, e^{i\omega_{\alpha\beta}\tau} \langle \hat{V}_{\beta\alpha}(0) \hat{V}^{\dagger}_{\beta\alpha}(\tau) \rangle_{0\beta} \tag{3.181}$$

by symmetry. According to (3.155),

$$\hat{V}_{\beta\alpha}(\tau) = \hat{V}^{\dagger}_{\alpha\beta}(\tau) . \tag{3.182}$$

Hence, substituting (3.182) into (3.181) and combining the result with (3.180), we obtain

$$\bar{\tau}^{-1} = (1 + \gamma_{\beta\alpha}) \hbar^{-2} \int_{-\infty}^{+\infty} d\tau \, e^{i\omega_{\alpha\beta}\tau} \langle \hat{V}^{\dagger}_{\alpha\beta}(0) \hat{V}_{\alpha\beta}(\tau) \rangle_{0\beta} . \tag{3.183}$$

This expression is a generalization of the two-state result arrived at by NITZAN and SILBEY [3.46]. If the external states of the solvent are independent of the internal state of the diatomic solute, then (3.183) degenerates to the Nitzan-Silbey formula.

Note that in the limit of low temperatures (3.183) applies to the multi-state case of a harmonic diatomic. Furthermore, in this limit it applies equally well to the case of polyatomic solutes for which the relaxation process involves only the lowest-frequency mode.

3.6 Model Calculations and Comparison with Experiment: Diatomics in Monatomic Matrices

In this Section and the following one we consider the problem of obtaining approximate closed expressions for the vibrational transition rate $W^{(1)}_{\alpha\to\beta}$ and related characteristic times, especially the vibrational lifetime τ', in terms of molecular properties and thermodynamic state variables. For the most part, we restrict our attention to systems conforming to our prototype (see the beginning of Sect. 3.3) and to conditions $(\hbar\omega_{\alpha\beta} \gg kT)$ under which the two-state approximation is adequate. Implementing the viewpoint developed in Section 3.4, we adopt specific models for the zero-order many-body states [see (3.130)] of the solvent in order to evaluate $W^{(1)}_{\alpha\to\beta}$ explicitly. We then compare the theoretical predictions with experimental data for specific systems. Finally, we review briefly alternate approaches to obtaining explicit formulas for $W^{(1)}_{\alpha\to\beta}$.

3.6.1 Description of the Model Hamiltonian[6]

In the present case the real system of interest, namely a dilute dispersion of diatomic molecules in a rare-gas matrix, may be represented by a single diatomic substituting as an impurity in an otherwise perfect monatomic crystal. Neglecting rotation of the diatomic, i. e. assuming that its orientation in the matrix is fixed, we may write the Hamiltonian of the complete system as

$$\hat{\mathcal{H}} = \hat{H}_M^0 + \hat{T}_L + \hat{V}, \tag{3.184}$$

where the free-diatomic Hamiltonian is

$$\hat{H}_M^0 = \hat{p}^2/2\mu + \hat{V}_B(\hat{q}) \tag{3.185}$$

and the kinetic energy of the lattice is

$$\hat{T}_L = \sum_{i=1}^{N} \hat{p}_i^2/2m_i \tag{3.186}$$

and $\hat{V}$ is the total potential energy of interaction of all lattice particles (atoms plus the diatomic impurity). In (3.185) and (3.186) $\hat{q}$ is the internuclear separation in the diatomic, $\hat{p}$ is the conjugate momentum, μ is the reduced mass of the diatomic, and $\hat{V}_B$ is the potential energy binding its atoms; $\hat{p}_i$ is the momentum of the i-th lattice particle of mass m_i, there being a total of N such particles.

Since the diatomic does not rotate, a single quantum number (γ) is sufficient to characterize each of its energy eigenstates [see (3.124)]. Let us assume that the zero-order many-body states of the lattice are adequately described within the harmonic approximation. Then (3.130) becomes explicitly

$$\hat{H}_{L0}^{(\alpha)}|\chi_0^{(\alpha)}\rangle = \sum_k (\hat{P}_k^2/2 + \omega_k^{(\alpha)2}\hat{Q}_k^2)|\chi_0^{(\alpha)}\rangle = \varepsilon^L|\chi_0^{(\alpha)}\rangle, \tag{3.187}$$

where $\hat{Q}_k$, $\hat{P}_k$, $\omega_k^{(\alpha)}$ are, respectively, the normal coordinate, its conjugate momentum, and the fundamental frequency of the k-th normal mode. More explicitly we may write

$$|\chi_0^{(\alpha)}\rangle = \prod_{k=1}^{3N} |n_k^{(\alpha)}\rangle = |n_\alpha\rangle \tag{3.188}$$

and

$$\varepsilon^L = \varepsilon_{n_\alpha}^L = \sum_{k=1}^{3N} \hbar\omega_k^{(\alpha)}(n_k^{(\alpha)} + 1/2), \tag{3.189}$$

---

[6] The developments of Subsects. 3.6.1–4 are based on DIESTLER [3.86].

where $\boldsymbol{n}_\alpha = \{n_1^{(\alpha)}, n_2^{(\alpha)}, \ldots, n_k^{(\alpha)}, \ldots, n_{3N}^{(\alpha)}\}$ denotes the set of occupation numbers of the lattice normal modes with the diatomic in vibrational state α. To simplify the evaluation of $W_{\alpha \to \beta}^{(1)}$ below, we further assume that the lattices modes k are "parallel"[7] with the diatomic impurity in various vibrational states $\gamma = 0, 1, \ldots$, [see Fig. 3.4]. Taking α as a reference state, we have then

$$\hat{H}_{L0}^{(\gamma)} = \sum_k [\hat{P}_k^2/2 + \omega_k^{(\gamma)2}(\hat{Q}_k - \Delta Q_k^{(\gamma)})^2/2 + V_{\gamma k}^0] \ . \tag{3.190}$$

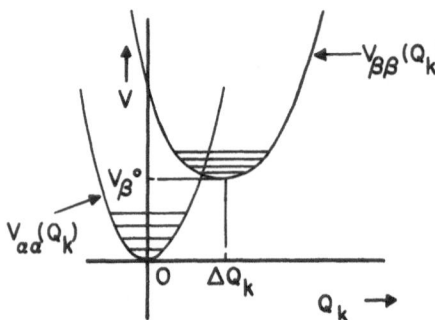

Fig. 3.4. Schematic diagram of harmonic potentials of the k-th normal mode of the lattice corresponding to two different internal states α and β of the diatomic impurity. The equilibrium position and force constant corresponding to state β are simply shifted from their values associated with reference state α

In evaluating the expression for $W_{\alpha \to \beta}^{(1)}$ we find it convenient to work in the number-operator representation, in which we define Bose creation and destruction operators $\hat{b}_k^\dagger$ and $\hat{b}_k$, respectively, for the k-th normal mode of the lattice (with the diatomic in state α) by

$$\hat{Q}_k \equiv (\hbar/2\omega_k^{(\alpha)})^{1/2}(\hat{b}_k^\dagger + \hat{b}_k) \ , \tag{3.191}$$

$$\hat{P}_k \equiv i(\hbar\omega_k^{(\alpha)}/2)^{1/2}(\hat{b}_k^\dagger - \hat{b}_k) \ . \tag{3.192}$$

Using (3.191) and (3.192), we may rewrite (3.187) and (3.190) as

$$\hat{H}_{L0}^{(\alpha)} = \sum_k \hbar\omega_k^{(\alpha)}(\hat{b}_k^\dagger \hat{b}_k + 1/2) \ , \tag{3.193}$$

$$\hat{H}_{L0}^{(\gamma)} = \sum_k [\hbar\omega_k^{(\gamma)}(\hat{b}_k^\dagger \hat{b}_k + 1/2) - \Delta Q_k^{(\gamma)}(\hbar\omega_k^{(\gamma)3}/2)^{1/2}(\hat{b}_k^\dagger + \hat{b}_k)$$
$$\omega_k^{(\gamma)2} \Delta Q_k^{(\gamma)2}/2 + V_{\gamma k}^0] \ ; \tag{3.194}$$

---

[7] The term "parallel" implies that the equilibrium positions and frequencies of the lattice normal modes in state β are simply shifted from their values in state α.

where in (3.194) we have neglected terms of order $\Delta\omega_k \equiv (\omega_k^{(\alpha)} - \omega_k^{(\gamma)})$ and higher. Defining a reduced displacement $g_k^{(\gamma)}$ by

$$\hbar\omega_k^{(\gamma)}g_k^{(\gamma)} = -\Delta Q_k^{(\gamma)}(\hbar\omega_k^{(\gamma)^3}/2)^{1/2}, \tag{3.195}$$

we may transform (3.194) into

$$\hat{H}_{L0}^{(\gamma)} = \sum_k \left[\hbar\omega_k^{(\gamma)}(\hat{b}_k^\dagger\hat{b}_k + 1/2) + \hbar\omega_k^{(\gamma)}g_k^{(\gamma)}(\hat{b}_k^\dagger + \hat{b}_k) + g_k^{(\gamma)^2}\hbar\omega_k^{(\gamma)} + V_{\gamma k}^0\right]. \tag{3.196}$$

Clearly

$$\varepsilon_{n_\gamma}^L = \sum_k \left[\hbar\omega_k^{(\gamma)}(n_k^{(\gamma)} + 1/2) + V_{\gamma k}^0\right] \tag{3.197}$$

and

$$g_k^{(\alpha)} = V_{\alpha k}^0 = 0. \tag{3.198}$$

3.6.2 Formal Expression for the Non-Radiative Rate of Vibrational Relaxation

In order to evaluate $W_{\alpha\to\beta}^{(1)}$ using (3.160), we need an expression for $\hat{V}_{\alpha\beta}$ and hence for $\langle\alpha|\hat{H}'|\beta\rangle$, given by

$$\langle\alpha|\hat{H}'|\beta\rangle = \sum_{n_\alpha'}\sum_{n_\beta'}|n_\alpha'\rangle\langle\alpha n_\alpha'|\hat{V}|\beta n_\beta'\rangle\langle n_\beta'|. \tag{3.199}$$

The matrix element $\langle\alpha n_\alpha'|\hat{V}|\beta n_\beta'\rangle$ may be written in the coordinate representation as

$$\begin{aligned}\langle\alpha n_\alpha'|\hat{V}|\beta n_\beta'\rangle &= \int d\hat{Q}\langle n_\alpha'|\hat{Q}\rangle\left[\int dq\langle\alpha|\hat{q}\rangle\hat{V}(\hat{q},\hat{Q})\langle\hat{q}|\beta\rangle\right]\langle\hat{Q}|n_\beta'\rangle \\ &= \int d\hat{Q}\langle n_\alpha'|\hat{Q}\rangle\hat{V}_{\alpha\beta}(\hat{Q})\langle\hat{Q}|n_\beta'\rangle.\end{aligned} \tag{3.200}$$

Neglecting the dependence of the coupling potential $V_{\alpha\beta}(\hat{Q})$ upon the normal coordinates $(\hat{Q})$, we have from (3.199) and (3.200)

$$\langle\alpha|\hat{H}'|\beta\rangle \simeq V_{\alpha\beta}(0), \tag{3.201}$$

in which we have evaluated $V_{\alpha\beta}(Q)$ at the equilibrium configuration of the lattice in the reference state α. Introducing the approximation (3.201) into (3.160), we get

$$W_{\alpha\to\beta}^{(1)} = \hbar^{-2}|V_{\alpha\beta}(0)|^2 \sum_{n_\alpha} p_{n_\alpha} \int\limits_{-\infty}^{+\infty} dt\, e^{-i\omega_{\alpha\beta}t}\langle n_\alpha|e^{i\hat{H}_{L0}^{(\beta)}t/\hbar}e^{-i\hat{H}_{L0}^{(\alpha)}t/\hbar}|n_\alpha\rangle. \tag{3.202}$$

After considerable straightforward Boson operator algebra [3.44] we transform (3.202) into

$$W^{(1)}_{\alpha \to \beta} = \hbar^{-2} |V_{\alpha\beta}(0)|^2 \exp \left[-\sum_k g_k^{(\beta)^2} (2\bar{n}_k + 1) \right] \int\limits_{-\infty}^{+\infty} dt \exp [G(t)] , \qquad (3.203)$$

where

$$G(t) \equiv -i\omega_{\alpha\beta} t + \sum_k g_k^{(\beta)^2} (\bar{n}_k + 1) \exp(i\omega_k^{(\beta)} t) + \sum_k g_k^{(\beta)^2} \bar{n}_k \exp(-i\omega_k^{(\beta)} t) \qquad (3.204)$$

and $\bar{n}_k$, the expected occupation number of the k-th lattice mode, is given by

$$\bar{n}_k = [\exp(\beta\hbar\omega_k^{(\beta)}) - 1]^{-1} . \qquad (3.205)$$

To reach (3.203) it is necessary to impose the restriction $\Delta\omega_k = 0$, which is consistent with our neglecting terms of order $\Delta\omega_k$ and higher in the zero-order Hamiltonian (3.194). We note in passing that the formal procedure of converting (3.202) into (3.203) is precisely analogous to that employed in a host of previous treatments of electronic relaxation in large molecules [3.51–54] (see Ch. 2) and in solids [3.55].

We now wish to evaluate the time integral in (3.203) by the saddle-point method. Assuming that only a narrow range of frequencies about $\omega_m^{(\beta)}$ effectively contributes to the sums in (3.204), we may express $G(t)$ as

$$G(t) = -i\omega_{\alpha\beta} t + N_m g_m^2 [(\bar{n}_m + 1) e^{i\omega_m t} + \bar{n}_m e^{-i\omega_m t}] , \qquad (3.206)$$

where N_m is the effective number of modes having frequency ω_m. The superscript β has been dropped for simplicity. After a little algebraic manipulation, (3.206) becomes

$$G(t) = -i\omega_{\alpha\beta} t + N_m g_m^2 \cos\omega_m (t - \tfrac{1}{2} i\beta\hbar) (\sinh\tfrac{1}{2}\beta\hbar\omega_m)^{-1} . \qquad (3.207)$$

By setting the derivative of $G(t)$ equal to zero, we find the saddle-point

$$t_s = \tfrac{1}{2} i\hbar\beta - i\omega_m^{-1} \ln [\delta + (1 + \delta^2)^{1/2}] , \qquad (3.208)$$

where

$$\delta \equiv \omega_{\alpha\beta} \sinh(\tfrac{1}{2}\beta\hbar\omega_m) (N_m g_m^2 \omega_m)^{-1} . \qquad (3.209)$$

Thus,

$$G(t_s) = \tfrac{1}{2}\beta\hbar\omega_{\alpha\beta} - \omega_{\alpha\beta} \ln [\delta + (1 + \delta^2)^{1/2}] (\omega_m)^{-1} \\ + N_m g_m^2 (1 + \delta^2)^{1/2} (\sinh\tfrac{1}{2}\beta\hbar\omega_m)^{-1} , \qquad (3.210)$$

$$G''(t_s) = - N_m g_m^2 \omega_m^2 (1 + \delta^2)^{1/2} (\sinh\tfrac{1}{2}\beta\hbar\omega_m)^{-1} , \qquad (3.211)$$

and

$$G'''(t_s) = -i\omega_{\alpha\beta}\,\omega_m^2\,. \tag{3.212}$$

Then the general expression for the transition rate becomes finally

$$\begin{aligned}
W^{(1)}_{\alpha\to\beta} &= \hbar^{-2}|V_{\alpha\beta}(0)|^2 \{2\pi\sinh(\tfrac{1}{2}\beta\hbar\omega_m)\,[N_m g_m^2\omega_m^2(1+\delta^2)^{1/2}]^{-1}\}^{1/2}\\
&\quad\times\exp[-N_m g_m^2(2\bar{n}_m+1)]\exp\{\tfrac{1}{2}\beta\hbar\omega_{\alpha\beta}-\omega_{\alpha\beta}\ln[\delta+(1+\delta^2)^{1/2}]/\omega_m\\
&\quad+N_m g_m^2(1+\delta^2)^{1/2}(\sinh\tfrac{1}{2}\beta\hbar\omega_m)^{-1}\}\,,
\end{aligned} \tag{3.213}$$

which will hold provided [3.56]

$$\begin{aligned}
\rho &\equiv |G'''(t_s)|\,|G''(t_s)|^{-3/2}\\
&= [N_m g_m^2(1+\delta^2)^{1/2}(\sinh\tfrac{1}{2}\beta\hbar\omega_m)^{-1}]^{-3/2}\,\omega_{\alpha\beta}/\omega_m \ll 1\,.
\end{aligned} \tag{3.214}$$

Eq. (3.213) has also been derived by LAUER and FONG [3.57] in another context. If conditions are such that $\delta\gg1$, then $W^{(1)}_{\alpha\to\beta}$ simplifies to

$$\begin{aligned}
W^{(1)}_{\alpha\to\beta} &= \hbar^{-2}|V_{\alpha\beta}(0)|^2\,[2\pi(\omega_m\omega_{\alpha\beta})^{-1}]^{1/2}\exp[-N_m g_m^2(2\bar{n}_m+1)]\\
&\quad\times\exp\left[-\omega_{\alpha\beta}\left(\ln\left[\frac{\omega_{\alpha\beta}}{N_m g_m\omega_m(\bar{n}_m+1)}\right]-1\right)\omega_m^{-1}\right]\,,
\end{aligned} \tag{3.215}$$

wich is identical in form to the so-called "weak-coupling" expression obtained by ENGLMAN and JORTNER [3.51].

In the expression (3.251) we see clearly the dependence of the non-radiative relaxation rate upon microscopic parameters such as the coupling strength $V_{\alpha\beta}(0)$, transition energy $\omega_{\alpha\beta}$ of the diatomic, the effective frequency ω_m of the lattice modes accepting the diatomic's energy, and the effective shift g_m of the equilibrium positions of these modes. The temperature dependence of $W_{\alpha\beta}$ is also manifested through $\bar{n}_m$.

3.6.3 Comparison of Theory with Experiment for Carbon Monoxide in an Argon Matrix

The non-radiative transition of interest here is from the first excited vibrational level ($\alpha=1$) to the ground vibrational level ($\beta=0$) of CO. The experimental relaxation rates τ'^{-1} measured by DUBOST et al. [3.15] are listed in the second column of Table 3.1. The τ'^{-1} were determined in the manner discussed in Subsect. 3.2.4, the radiative lifetime of CO ($\alpha=1$) being taken equal to the gas-phase value. The theoretical expressions (3.213) and (3.215) both involve the unknown parameters ω_{10}, $V_{10}(0)$, ω_m, and $N_m g_m^2$. From the absorption spectrum for CO [3.14] we find $\omega_{10}=4.03\times10^{14}\,\text{s}^{-1}$. Note that since $\hbar\omega_{10}\gg kT$, $W^{(1)}_{\alpha\beta}\simeq\tau'^{-1}$ [see (3.179)]. Hence, we have determined the remaining parameters $[V_{10}(0), \omega_m, N_m g_m^2]$ by fitting the experimental data directly to the expressions (3.213) or (3.215) using the non-linear least squares treatment of Marquardt [3.58].

The relaxation rates predicted by the weak-coupling formula (3.215) for the "best" set of parameters (given in Table 3.3) are listed in the third column

Table 3.3. "Best" parameters determined by Marquardt's nonlinear least-squares fit of theoretical rate expressions (3.213) and (3.215) to the experimental data for CO relaxing in an Ar matrix. The fundamental frequency of CO was taken to be $\omega = 4.03 \times 10^{14} \text{s}^{-1}$

Parameter[a]	Weak-coupling limit, (3.215)	General expression, (3.213)
V_{10} [erg]	7.11×10^{-15}	1.02×10^{-15}
$N_m g_m^2$	18.68	18.98
ω_m [s^{-1}]	7.21×10^{12}	7.65×10^{12}
χ^2 [b]	48.6	57.8

[a] Defined in text.
[b] The sum of the squares of the differences between the experimental rates and those calculated by the "best" fit at temperatures 6, 8, 10, 13, and 18 K.

of Table 3.1. When these weak-coupling parameters are used in the general expression (3.213), the predicted values disagree at higher temperatures with the experimental values. Hence, we have also fitted the more general expression (3.213) to the experimental data over the entire range of temperatures, obtaining the relaxation rates listed in the fourth column of Table 3.1. The

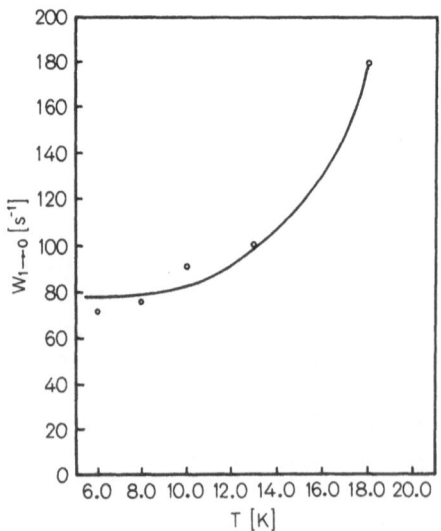

Fig. 3.5. Non-radiative vibrational decay rate versus absolute temperature for CO relaxing in an Ar matrix. The open circles are based on the experimental measurements of DUBOST and LEGAY [3.15]. The solid line is the best fit of these experimental data to the general theoretical rate expression (3.213)

"best" parameters found in this case are also given in Table 3.3. Plots of the experimental and theoretical [(3.213)] relaxation rates are shown in Fig. 3.5. We see that the fit is quite good, as indicated also by the very reasonable values of χ^2 in Table 3.3.

It is interesting to observe the behavior of the relaxation rate at much higher temperature, beyond the range of the experiment of DUBOST et al. [3.14]. Unfortunately, our experiment is imaginary, but it can perhaps give us some qualitative notion of the expected behavior of the relaxation rate in solid matrices which melt at higher temperatures. In Table 3.1 we list relaxation rates predicted by the "best" fit to (3.213) for temperatures up to 400 K. These results are also plotted in Fig. 3.6 to show more clearly how the

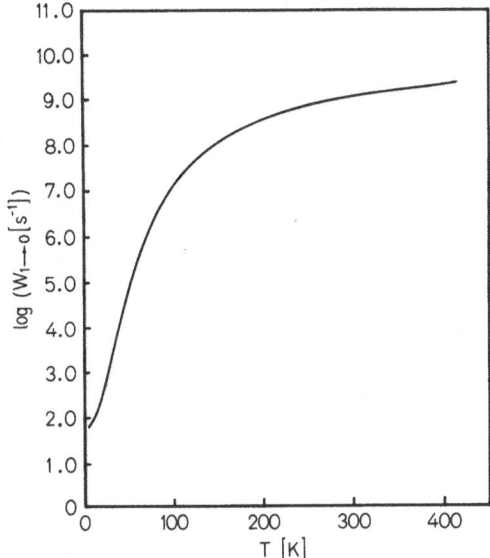

Fig. 3.6. Logarithm of the non-radiative vibrational decay rate versus temperature predicted from the general rate expression (3.213) using the "best" parameters of Table 3.3

relaxation rate increases very rapidly at intermediate temperatures ($T < 100$ K) but then levels out at higher temperatures. At room temperature (~ 300 K), the non-radiative lifetime is about 8.3×10^{-10} s, apparently much longer than those measured for normal liquids (see Subsect. 3.2.7) under similar conditions.

We might hope to test the theory further by making use of isotopic effects. However, introduction of an isotopic diatomic impurity affects the frequency ω_m and shift g_m of the equilibrium position of the effective lattice mode in addition to the transition frequency ω_{10} of the diatomic. If we *assume* that the lattice modes are disturbed negligibly, then we may estimate the expected relaxation rates from expression (3.213) or (3.215) using the previously determined parameters $V_{10}(0)$, ω_m, and $N_m g_m^2$ along with the new value of the transition frequency ω'_{10}. If we further assume that the diatomic is harmonic, ω'_{10} is related to the original frequency ω_{10} by

$$\omega'_{10} = (\mu/\mu')^{1/2} \omega_{10} ,$$

where μ and μ' are the original and isotopic reduced masses, respectively. For CO^{18} we find

$$\omega'_{10} = 3.93 \times 10^{14} \, s^{-1} .$$

Using this value of ω'_{10} along with the values of the remaining parameters listed in Table 3.3, we have generated the relaxation rates for CO^{18} in Ar. These are given in Table 3.4.

Table 3.4. Non-radiative vibrational relaxation rates predicted for CO^{18} in an Ar matrix using the general rate expression (3.213). The fundamental frequency of CO^{18} is taken to be $\omega' = 3.93 \times 10^{14} \, s^{-1}$; the parameters V_{10}, $N_m g_m^2$ and ω_m are taken from the third column of Table 3.3

Temperature [K]	Relaxation rate [s^{-1}]	ρ [a]
6	2.95×10^2	0.140
8	2.99×10^2	0.140
10	3.13×10^2	0.139
13	3.70×10^2	0.139
18	6.50×10^2	0.139

[a] Ratio defined by (3.214).

3.6.4 Discussion

The central equation of this section, (3.213), exhibits the explicit dependence of the relaxation rate upon parameters characterizing the system, namely T, $\omega_{\alpha\beta}$, $V_{\alpha\beta}(0)$, ω_m, N_m, and g_m. Unfortunately, it is not possible experimentally to vary these parameters (except for T) independently. Nor is it feasible to calculate them from additional theoretical considerations. Hence, the theory is incapable of predicting relaxation rates *a priori*. Even so, our comparison of theory with experiment for CO relaxing in Ar does afford some insight into the dependence of the rate upon two of these parameters, namely, the temperature T and the diatomic transition frequency $\omega_{\alpha\beta}$. Although the precise dependence of $W^{(1)}_{\alpha\to\beta}$ on T is not obvious from either (3.213) or (3.215), the experimental and theoretical results (see Tables 3.1 and 3.4 and Figs. 3.5 and 3.6) seem to be in accord with one's physical intuition. Increasing T results in an enhanced rate on account of increased thermal agitation of the lattice.

However, the very rapid increase of rate with rising temperature is due to a delicate counterbalancing of several factors involving T in the rate expression.

Assuming that $V_{10}(0)$ and the lattice parameters ω_m, N_m, and g_m are not affected significantly by the introduction of an isotopic diatomic impurity, we have arrived at the results of Table 3.4. We conclude that the vast increase in relaxation rate for CO^{18} over CO^{16} in Ar is due to the decrease in $\omega_{\alpha\beta}$ alone. This increase, which can be attributed to the negative contribution of the logarithmic term in the argument of the exponential, see (3.213) and (3.215), is also physically reasonable. The smaller $\omega_{\alpha\beta}$ is, the closer the diatomic impurity mode is to "resonance" with the effective lattice mode and hence the less is the "impedance mismatch". A similar phenomenon is observed in the inelastic scattering of diatomic molecules in gases [3.59].

Even though it is not practical to calculate the parameters $V_{10}(0)$, ω_m, and $N_m g_m^2$ from first principles, we can check the empirically determined values in Table 3.3 to see whether they are reasonable and consistent with other independently determined properties of the system. Thus, the value for $V_{10}(0) \simeq 10^{-14}\text{--}10^{-15}$ erg, which is of the order of magnitude of the potential energy of interaction of two argon (lattice) atoms at their equilibrium separation in the crystal [3.60], seems reasonable. Because of the orthogonality of the diatomic states $|0\rangle$ and $|1\rangle$, the only terms in $\hat{V}$ which contribute to $V_{10}(0)$ contain the internal coordinate $\hat{q}$ and the most significant of these terms arise from the interaction of the diatomic with its nearest neighbors. Hence we expect $V_{10}(0)$ to be of the order of magnitude of the interaction between nearest neighbors in the lattice. We also note that ω_m (7.65×10^{12} s^{-1}) is significantly less than the phonon cutoff frequency ($\sim 12.7 \times 10^{12}$ s^{-1}) of solid Ar [3.61], as it must be (see Fig. 3.7). More interesting is the fact that ω_m falls approxi-

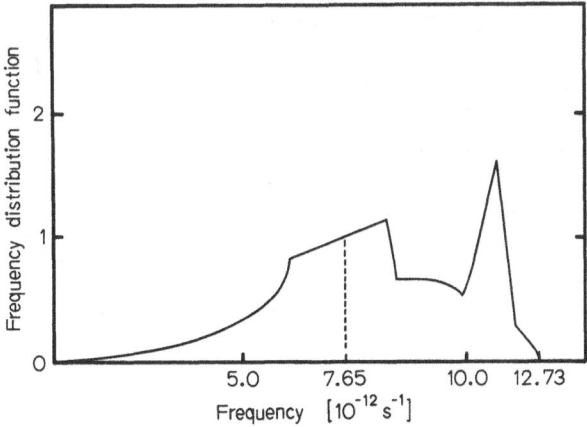

Fig. 3.7. Frequency distribution of lattice normal modes of Ar as a function of circular frequency ω. Curve is determined by "best" two-neighbor fit of observed phonon frequencies to Born-Von Kármán model [3.61]

mately in the middle of the broad peak of the phonon frequency distribution function. This is in accord with one's notion that the effective accepting mode should have a reasonably large degeneracy N_m. By implication the equilibrium displacement of modes associated with the narrow peak around 1.1×10^{13} s^{-1} is negligible, since these modes do not accept energy. Finally, we note that it is not possible to factor $N_m g_m^2$ uniquely in order to obtain N_m and g_m separately.

3.6.5 Review and Discussion of Other Models

LIEBERMANN [3.62] has derived an expression for the acoustical absorption constant (see Sect. 3.2.1) of a pure molecular crystal, calculating the rate of vibrational relaxation by a semi-classical method analogous to the semi-classical treatment of the interaction of radiation and matter. GOUTERMAN [3.63] has independently developed a general semi-classical theory of non-radiative transitions, also precisely analogous to the semi-classical theory of radiative transitions. For the interaction Hamiltonian he assumed a form linear in both the internal coordinates of the relaxing molecule and the lattice coordinates. In a more recent investigation of a one-dimensional model for vibrational relaxation of an electronically excited diatomic molecule in a monatomic crystal, SCHURR [3.64] has explicitly derived an interaction term of the form assumed by GOUTERMAN. Using the HEITLER and MA [3.65] perturbation theory, SCHURR has obtained expressions for the rates of both "cascade damping" and "lattice relaxation"[8].

In an interesting early contribution to the theory of VR, SUN and RICE [3.66] also studied a one-dimensional model for vibrational relaxation of a ground-state diatomic molecule embedded in a monatomic lattice. They employed a semi-classical binary-collision model, calculating the collision frequency between adjacent lattice atoms by SLATER's [3.67] technique and the transition probability per collision by the distorted-wave method [3.68] to obtain a formal expression for the relaxation rate, from which they estimated the relaxation time of N_2 ($v = 1$) in Ar to be about 10^{-2} s.

A major shortcoming of the models (except that of SUN and RICE) discussed above is that they must resort to high-order perturbation theory to account for multi-phonon decay processes, which certainly function in the vast majority of "interesting" systems in which the phonon cutoff (Debye) frequency of the lattice is less than the transition frequency of the diatomic. However, SCHURR [3.64] does give an *upper bound* on the multi-phonon cascade damping rate, but he does not derive a tractable explicit expression.

---

[8] In addition to the bilinear term, which gives rise to "cascade damping", i.e. creation or destruction of phonons with a simultaneous vibrational transition of the diatomic, SCHURR's model Hamiltonian contains another interaction term leading to "lattice relaxation", the process by which the diatomic undergoes an electronic transition while remaining in the same vibrational state.

The treatment of NITZAN and JORTNER [3.43] (see Subsect. 3.3.5) also takes account of multi-phonon processes. Recently NITZAN et al. [3.69], using an approximate version of NITZAN and JORTNER's [3.43] general formula for the vibrational relaxation width, have fitted the experimental VR data on CO in Ar [3.14] reasonably well.

To avoid using high-order perturbation theory, FISCHER [3.70] has proposed an interesting model for VR of diatomics in rare-gas lattices based upon a Born-Oppenheimer type separation of the high frequency (fast) diatomic vibrational mode from the low frequency (slow) lattice modes. His formal development of the model parallels the usual treatment of non-radiative transitions between electronic manifolds (see Ch. 2). FISCHER qualitatively compared his predictions with experimental data for CO in Ar [3.14] and C_2^- in Ar [3.16].

In applying the master-equation approach to vibrational relaxation in condensed media (see Subsect. 3.3.5) LIN [3.49] has evaluated the first-order relaxation rate constant [identical to expression (3.165)] for a specific model in which the interaction potential between the diatomic and lattice atoms is a repulsive exponential. For this special case a closed expression for the rate constant may be obtained using procedures analogous to that employed above in Subsect. 3.6.2.

3.7 Model Calculations and Comparison with Experiment: Diatomics in Monatomic Liquids

Few studies of VR in simple liquids and solutions which one might hope to be amenable to theoretical treatment have been reported (see, however, [3.35–37]). In any event, the theory of VR in liquids is not well developed. Solids and matrices, understandably, have received the most attention, for the many-body vibrational states of the solid lattice can be conveniently and adequately handled in the well-known harmonic approximation (see Subsect. 3.6.1). Adequate description of the many-body states of a liquid, on the other hand, poses a formidable task. To circumvent this difficulty and arrive at useful expressions for the relaxation rate constant, one must invoke some severe approximations. In an early attempt to explain the role of VR in ultrasonic absorption in liquids, HERZFELD [3.71] employed the cell model of LENNARD-JONES and DEVONSHIRE (LJD) [3.72, 73]. He expanded the cell potential (acting on the relaxing molecule) in a Taylor series about the center of the cell, retaining second- and fourth-degree terms only. In zero-order the second-degree term leads to harmonic motion of the center-of-mass of the molecule; the fourth-degree term is a perturbation coupling the internal and external vibrations and thereby giving rise to VR. HERZFELD calculated a VR time for liquid benzene which is in remarkably good agreement with experiment. Later LITOVITZ [3.74] proposed an isolated-binary-collision (IBC) theory of VR in which the relaxation rate constant is proportional to the product of the probability of de-excitation per collision and the average frequency of collisions. Estimating the collision frequency using a cell model, LITOVITZ obtained good agreement of his theory with experimental data on VR in liquid CS_2.

Although the theories of HERZFELD and LITOVITZ both have the virtue of yielding explicit expressions for the relaxation rate constant in terms of microscopic properties of the system, neither is without flaw. An especially important limitation of Herzfeld's model is that the first-order perturbation treatment dictates that the internal vibrational excitation must decay into three external "phonons". Thus, if the transition frequency of the diatomic is more than three times greater than the "Debye" frequency of the liquid, one must account for VR using two alternative approaches: 1) carry the perturbation expansion to higher order; 2) carry the Taylor series expansion of the cell potential to higher degree. Neither of these alternatives is very satisfactory in practice; the first leads to intractable expressions and the second to a rate constant which is extremely sensitive to the details of the cell potential.

ZWANZIG [3.75] has argued that the IBC assumption of Litovitz's theory cannot hold in liquids. However, by a rigorous statistical-mechanical treatment DAVIS and OPPENHEIM [3.76] have demonstrated that an IBC theory of liquids can be derived under rather stringent assumptions about the nature of the system.

The purpose of the remainder of this section is to derive and apply an approximate quantum-mechanical expression for $W_{\alpha\beta}^{(1)}$ for a diatomic molecule in a simple (i.e. unassociated) monatomic liquid. As in Section 3.6, a guiding principle in the development here is the desirability of arriving finally at a closed (and reasonably tractable) expression for the relaxation rate in terms of microscopic parameters of the system. Achievement of this goal, as pointed out above, necessitates the imposition of a number of apparently drastic simplifying assumptions, which we set down carefully below.

3.7.1 The Basic Model[9]

We assume that a dilute solution of diatomic solute molecules in a monatomic liquid solvent can be represented adequately by a single diatomic and a large number, say $N - 1 \simeq 6.02 \times 10^{23}$, of atoms. Let us neglect the rotation of the diatomic and assume that its orientation is fixed throughout the duration of the relaxation process. Furthermore, let us assume that the mechanism of VR involves only head-on collisions. Under these assumptions, the model may be

---

[9] The developments of Subsects. 3.7.1–4 are based on DIESTLER [3.87].

represented schematically as in Fig. 3.8. The Hamiltonian of the entire system, identical in form to that given in (3.184), is

$$\mathcal{H} = \hat{H}_s^0 + \hat{T}_L + \hat{V}, \tag{3.216}$$

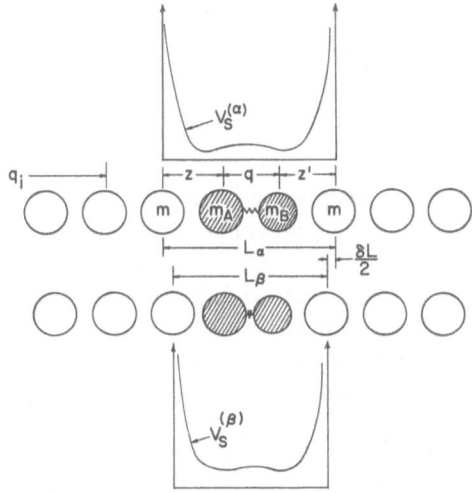

Fig. 3.8. Schematic diagram of diatomic in monatomic liquid, defining the various coordinates and depicting the contraction of the diatomic's cell upon the transition $\alpha \to \beta$

where the Hamiltonian of the isolated diatomic is

$$\mathcal{H}_s^0 = \hat{p}^2/2\mu + \hat{V}_s(\hat{q}), \tag{3.217}$$

the kinetic energy of the liquid is

$$T_L = \sum_{i=1}^{N} \hat{p}_i^2/2m_i = \sum_{i \neq s}^{N} \hat{p}_i^2/2m + \hat{p}_s^2/2m_s. \tag{3.218}$$

and $\hat{V} = \hat{V}(\hat{q}, \hat{q}_i)$ is the total potential energy of interaction of all particles in the liquid. $V_s(\hat{q})$ is the potential energy binding the atoms A and B of the diatomic; $\hat{q}$ and $\hat{p}$ are the vibrational coordinate and conjugate momentum, respectively, of the diatomic; $\mu = m_A m_B/(m_A + m_B)$ is its reduced mass. $\hat{q}_i$ and $\hat{p}_i$ are, respectively, the coordinate and conjugate momentum of the i-th particle and m_i is its mass. Note that $i = s$ corresponds to the diatomic solute.

Again, since the diatomic is assumed not rotate, a single quantum number γ is sufficient to label each of its energy eigenstates $|\gamma\rangle$. As implied by the introductory remarks above, however, a rigorous description of the many-body external states $|\chi_0^{(\gamma)}\rangle$ of a liquid is hopeless. Furthermore, the harmonic approximation is clearly inadequate. Consistent with the guiding principle stressed there, let us assume that each liquid particle moves in the potential of its nearest neighbors fixed at their equilibrium positions. This assumption is just the one-dimensional analogue of the LJD cell model (see Fig. 3.8). The zero-order external states of the liquid may now be expressed as

$$|\chi_0^{(\gamma)}\rangle = \prod_{k=1}^{N} |n_k^{(\gamma)}\rangle = |\boldsymbol{n}_k^{(\gamma)}\rangle , \tag{3.219}$$

where $|n_k^{(\gamma)}\rangle$ satisfies

$$[\hat{p}_k^2/2m + \hat{V}_k^{(\gamma)}(\hat{q}_k)] |n_k^{(\gamma)}\rangle = \varepsilon_{n_k}^{(\gamma)} |n_k^{(\gamma)}\rangle , \tag{3.220}$$

i. e. it is the eigenvector of the external Hamiltonian of the k-th molecule moving in its effective cell potential $\hat{V}_k^{(\gamma)}$, which presumably depends in no simple way upon the state γ of the diatomic. In Fig. 3.8 are shown schematic plots of effective cell potentials (arising from assumed LJ(12,6) interactions between nearest neighbors) for the diatomic solute corresponding to two different internal states α and β. Key features of these diagrams are 1) the steep rise in $\hat{V}_s^{(\gamma)}$ at the ends of cell and 2) the dependence of the effective cell length on γ. The steep rise is easily seen to be due to the r^{-12} repulsive term in the LJ(12,6) interaction. Actually, the γ-dependence of the effective cell length is an implicit assumption, albeit not a necessary one. As will be seen from the derivation in Subsect. 3.7.2 below, it is necessary only that

$$\langle n_k^{(\gamma')} | n_k^{(\gamma)} \rangle \neq 0 . \tag{3.221}$$

In view of these two features of the cell potential, it seems reasonable to approximate $\hat{V}_k^{(\gamma)}$ by an infinite square well with a γ-dependent length. Note that there is a precedent for this approximation in the work of Hamann [3.77], who sought to estimate quantum corrections to the LJD equation of state for light gases.

3.7.2 The Vibrational Transition Rate

We start with the Golden-Rule expression (3.165) for the rate, which may be rewritten in our present notation as

$$W_{\gamma_i \to \gamma_f}^{(1)} = 2\pi\hbar^{-1} \sum_{n_{\gamma_i}} P_{n_{\gamma_i}} \sum_{n_{\gamma_f}} |\langle n_{\gamma_i} | \hat{V}_{\gamma_i \gamma_f} | n_{\gamma_f} \rangle|^2 \delta(\varepsilon_{\gamma_i n_{\gamma_i}} - \varepsilon_{\gamma_f n_{\gamma_f}}) , \tag{3.222}$$

where

$$\hat{V}_{\gamma_i \gamma_f} \equiv \langle \gamma_i | \hat{V} | \gamma_f \rangle , \tag{3.223}$$

$n_\gamma = \{n_1^{(\gamma)}, n_2^{(\gamma)}, \ldots, n_N^{(\gamma)}\}$ denotes the set of quantum numbers specifying the external many-body zero-order state of the liquid, $\varepsilon_{\gamma n_\gamma}$ is the total energy of the liquid in state $|\gamma\, n_\gamma\rangle = |\gamma\rangle \otimes |n_\gamma\rangle$

$$\varepsilon_{\gamma n_\gamma} = \varepsilon_\gamma^{s} + \varepsilon_{n_\gamma} = \varepsilon_\gamma^{s} + \sum_{k=1}^{N} \varepsilon_{n_k^{(\gamma)}} , \tag{3.224}$$

and ε_γ^{s} is the energy of the diatomic solute in state γ.

In order to simplify the evaluation of the matrix element in (3.222) let us assume that the coupling matrix element $\hat{V}_{\gamma_i \gamma_f}$ is not strongly dependent upon the external coordinates $\hat{q}$. Upon expanding $\hat{V}_{\gamma_i \gamma_f}$ in a Taylor series about the midpoint (q_i^0) of each cell, one obtains

$$\hat{V}_{\gamma_i \gamma_f}(\hat{q}) = V_{\gamma_i \gamma_f}(\hat{q}^0) + \sum_{i=1}^{N} (\partial \hat{V}_{\gamma_i \gamma_f} / \partial \hat{q}_i)_{\hat{q} = q^0} (\hat{q}_i - q_i^0) + \cdots . \tag{3.225}$$

Retaining only the first term in (3.225) leads to

$$|\langle n_{\gamma_i} | \hat{V}_{\gamma_i \gamma_f} | n_{\gamma_f} \rangle|^2 \simeq |V_{\gamma_i \gamma_f}(q^0)|^2 |\langle n_{\gamma_i} | n_{\gamma_f} \rangle|^2$$

$$= |V_{\gamma_i \gamma_f}^0|^2 \prod_{k=1}^{N} |\langle n_k^{(\gamma_i)} | n_k^{(\gamma_f)} \rangle|^2 . \tag{3.226}$$

Finally, with the additional simplifying assumption that only the cell length of the diatomic solute changes upon the transition $\gamma_i \to \gamma_f$,

$$\langle n_k^{(\gamma_i)} | n_k^{(\gamma_f)} \rangle = \delta_{n_k^{(\gamma_i)} n_k^{(\gamma_f)}} , \quad k \neq s . \tag{3.227}$$

Then using (3.226) and (3.227) in (3.222), we find

$$W_{\gamma_i \to \gamma_f}^{(1)} = 2\pi \hbar^{-1} |V_{\gamma_i \gamma_f}^0|^2 Q_{\gamma_i}^{-1} \sum_{n_s^{(\gamma_i)}} e^{-\beta \varepsilon_{n_s^{(\gamma_i)}}} \sum_{n_s^{(\gamma_f)}} |\langle n_s^{(\gamma_i)} | n_s^{(\gamma_f)} \rangle|^2$$

$$\times \delta(\Delta \varepsilon^{s} + \varepsilon_{n_s^{(\gamma_i)}} - \varepsilon_{n_s^{(\gamma_f)}}) , \tag{3.228}$$

where

$$Q_{\gamma_i} \equiv \sum_{n_s^{(\gamma_i)}} e^{-\beta \varepsilon_{n_s^{(\gamma_i)}}} , \tag{3.229}$$

and

$$\Delta \varepsilon^{s} \equiv \varepsilon_{\gamma_i}^{s} - \varepsilon_{\gamma_f}^{s} . \tag{3.230}$$

To evaluate the overlap matrix element in (3.228), let us refer again to Fig. 3.8, which indicates that the diatomic's cell contracts (or expands) symmetrically about its center under the transitions $\alpha \leftrightarrow \beta$. With this assumption the relevant matrix element is given by

$$\langle n_\alpha | n_\beta \rangle = \int_0^{L_\beta} \phi_{n_\alpha}(q_s) \phi_{n_\beta}(q_s) dq_s , \tag{3.231}$$

where the normalized wavefunctions are

$$\phi_{n_\alpha}(q_s) \equiv (2/L_\alpha)^{1/2} \sin\left[n_\alpha \pi (q_s + \delta L/2) L_\alpha^{-1} \right]$$
$$\phi_{n_\beta}(q_s) \equiv (2/L_\beta)^{1/2} \sin\left[n_\beta \pi q_s L_\beta^{-1} \right] . \tag{3.232}$$

In (3.231) and (3.232) the designation s on the quantum number $n_s^{(\alpha)}$ and $n_s^{(\beta)}$ has been dropped for simplicity; q_s is measured from the left end of the cell corresponding to internal state β; L_α and L_β are, of course, the solute cell lengths corresponding to diatomic internal states α and β, respectively; δL is the decrement (increment) in the cell length accompanying the transitions $\alpha \leftrightarrow \beta$. The integral in (3.231) is easily done to give

$$\langle n_\alpha | n_\beta \rangle = -2\pi^{-1} \Delta^{1/2} n_\beta [1 + (-1)^{n_\alpha + n_\beta}] \sin(n_\alpha \pi \delta/2)/(n_\alpha^2 \Delta^2 - n_\beta^2) , \tag{3.233}$$

in which

$$\delta \equiv \delta L/L_\alpha \equiv 1 - \Delta . \tag{3.234}$$

Clearly $\langle n_\alpha | n_\beta \rangle = 0$, unless n_α and n_β are both even or odd. This fact, which also follows from elementary considerations of symmetry, allows one to separate the rate expression (3.228) into two parts

$$W_{\gamma_i \rightarrow \gamma_f}^{(1)} = 2\pi\hbar^{-1} |V_{\gamma_i \gamma_f}^0|^2 Q_{\gamma_i}^{-1} \left[\sum_{\substack{n_{\gamma_i} \\ \text{even}}} e^{-\beta \varepsilon_{n_{\gamma_i}}} \sum_{\substack{n_{\gamma_f} \\ \text{even}}} |\langle n_{\gamma_i} | n_{\gamma_f} \rangle|^2 \delta(\Delta\varepsilon^s + \varepsilon_{n_{\gamma_i}} - \varepsilon_{n_{\gamma_f}}) \right.$$
$$\left. + \sum_{\substack{n_{\gamma_i} \\ \text{odd}}} e^{-\beta \varepsilon_{n_{\gamma_i}}} \sum_{\substack{n_{\gamma_f} \\ \text{odd}}} |\langle n_{\gamma_i} | n_{\gamma_f} \rangle|^2 \delta(\Delta\varepsilon^s + \varepsilon_{n_{\gamma_i}} - \varepsilon_{n_{\gamma_f}}) \right]. \tag{3.235}$$

In the further simplification of (3.235) there are two cases to be considered. If the solute cell contracts upon the transition $\gamma_i \rightarrow \gamma_f$, then one makes the identifications $\gamma_i = \alpha$ and $\gamma_f = \beta$, [see Fig. 3.8]. On the other hand, if the cell expands, then $\gamma_i = \beta$ and $\gamma_f = \alpha$. We consider the former case first.

Let us evaluate the "odd" contribution to $W_{\alpha \rightarrow \beta}$ first. It may be represented as

$$W_{\alpha \rightarrow \beta}^{(1)}(\text{odd}) = 2\pi\hbar^{-1} |V_{\alpha\beta}^0|^2 Q_\alpha^{-1} \sum_{n_\alpha' = 0}^{\infty} e^{-\beta \varepsilon_{2n_\alpha' + 1}}$$
$$\times \sum_{n_\beta' = 0}^{\infty} |\langle 2n_\alpha' + 1 | 2n_\beta' + 1 \rangle|^2 \delta(\Delta\varepsilon^s + \varepsilon_{2n_\alpha' + 1} - \varepsilon_{2n_\beta' + 1}) . \tag{3.236}$$

The energy levels of the cells are

$$\varepsilon_{2n'_\alpha+1} = (2n'_\alpha+1)^2 \pi^2 \hbar^2/(2m_s L_\alpha^2) = (2n'_\alpha+1)^2 \varepsilon_1^\alpha$$
$$\varepsilon_{2n'_\beta+1} = (2n'_\beta+1)^2 \varepsilon_1^\alpha \Delta^{-2} ,$$

(3.237)

with Δ given by (3.234). If the mass m_s of the diatomic solute is sufficiently large and/or its cell is sufficiently long, then the energy levels ε_{n_β} will be so closely spaced that one can approximate the sum on n'_β in (3.236) by an integral, i. e.

$$\sum_{n'_\beta} \to \int dn'_\beta = \int d\varepsilon (dn'_\beta/d\varepsilon) ,$$

(3.238)

where $\varepsilon_{2n'_\beta+1}$, now continuous, has been replaced by ε. From (3.237)

$$dn'_\beta/d\varepsilon = \Delta \varepsilon_1^{\alpha-1/2} \varepsilon^{-1/2}/4 .$$

(3.239)

Introducing (3.233), (3.237), (3.238), and (3.239) into (3.236) and performing the integration over ε yields

$$W^{(1)}_{\alpha\to\beta}(\text{odd}) \simeq 8\pi^{-1}\hbar^{-1}|V^0_{\alpha\beta}|^2 Q_\alpha^{-1} \varepsilon_1^\alpha (\Delta\varepsilon^s)^{-2} \sum_{n'_\alpha} e^{-\beta(2n'_\alpha+1)^2\varepsilon_1^\alpha}$$
$$\times \sin^2\left[(2n'_\alpha+1)\pi\delta/2\right] \left[\Delta\varepsilon^s/\varepsilon_1^\alpha + (2n'_\alpha+1)^2\right]^{1/2} .$$

(3.240)

Under most experimental conditions, $kT \gg \varepsilon_1^\alpha$ and δ, the fractional change in the cell length, is small. Then the summand in (3.240) will be a slowly varying function of n'_α and the sum may be replaced by an integral. The rate expression (3.240) then becomes

$$W^{(1)}_{\alpha\to\beta}(\text{odd}) \simeq 4\pi^{-1}\hbar^{-1}|V^0_{\alpha\beta}|^2 Q_\alpha^{-1} \varepsilon_1^\alpha (\Delta\varepsilon^s)^{-2}$$
$$\times \int_0^\infty dn\, e^{-\beta n^2\varepsilon_1^\alpha} \sin^2(n\pi\delta/2)(\Delta\varepsilon^s/\varepsilon_1^\alpha + n^2)^{1/2} .$$

(3.241)

By exactly analogous considerations the "even" contribution to $W^{(1)}_{\alpha\to\beta}$ in (3.235) is found to be identical to the odd. Thus, the expression for the total rate is

$$W^{(1)}_{\alpha\to\beta} = 8\hbar^{-1}\pi^{-1}\varepsilon_1^\alpha(\Delta\varepsilon^s)^{-2} Q_\alpha^{-1}|V^0_{\alpha\beta}|^2 \int_0^\infty dn\, e^{-\beta n^2\varepsilon_1^\alpha} \sin^2(n\pi\delta/2)$$
$$\times (\Delta\varepsilon^s/\varepsilon_1^\alpha + n^2)^{1/2} .$$

Under the assumption made above, namely that $kT \gg \varepsilon_1^\alpha$, one can obtain Q_α in closed form as

$$Q_\alpha = (2\pi m_s \beta^{-1})^{1/2} L_\alpha h^{-1} .$$

(3.243)

Defining $x \equiv (\beta \varepsilon_1^z)^{1/2} n$ and substituting (3.243) into (3.242) yields finally

$$W_{\alpha \to \beta}^{(1)} = 16(2\pi m_s)^{-1/2} L_\alpha^{-1} (\Delta \varepsilon^s)^{-3/2} |V_{\alpha\beta}^0|^2 I \tag{3.244}$$

where

$$I \equiv \int_0^\infty dx\, e^{-x^2} \sin^2 \eta \, x (1 + \zeta x^2)^{1/2} \tag{3.245}$$

and

$$\eta \equiv \pi \delta (\varepsilon_1^z / kT)^{-1/2} / 2 \tag{3.246}$$

$$\zeta \equiv kT / \Delta \varepsilon^s . \tag{3.247}$$

In case the cell *expands* upon the transition $\gamma_i \to \gamma_f$, considerations precisely parallel to those above lead to the following expression for the transition rate

$$W_{\beta \to \alpha}^{(1)} = 16(2\pi m_s)^{-1/2} L_\beta^{-1} (\Delta \varepsilon^s)^{-5/2} |V_{\alpha\beta}^0|^2 kT I' , \tag{3.248}$$

where

$$I' \equiv \int_0^\infty dx\, e^{-x^2} x^2 (1 + \zeta x^2)^{-1/2} \sin^2 \eta' (1 + \zeta x^2)^{1/2} \tag{3.249}$$

and

$$\eta' \equiv \pi \delta \Delta^{-1} (\Delta \varepsilon^s / \varepsilon_1^\beta)^{1/2} / 2 . \tag{3.250}$$

Both expressions (3.244) and (3.248) exhibit specific dependences upon the temperature and also upon microscopic parameters, namely the coupling strength $V_{\alpha\beta}^0$ and the diatomic solute's transition energy $\Delta \varepsilon^s$, mass m_s, cell length L_α, and fractional change in cell length δ. These dependences may be rendered more explicit by invoking additional approximations which permit the evaluation of I and/or I' in closed form.

3.7.3 Approximate Closed Expressions for the Transition Rate

Anticipating the application of (3.244) in Subsect. 3.7.4 to VR of diatomics in liquid rare gases, let us restrict our attention to approximating I, given by (3.245).

If $\Delta \varepsilon^s \gg kT$, which will be so for "stiff" diatomics in low-temperature liquid rare gases, then $\zeta \ll 1$ and one may set

$$(1 + \zeta x^2)^{1/2} \sim 1 + \zeta x^2 / 2 + \cdots . \tag{3.251}$$

Introducing (3.251) into (3.245) gives

$$I \simeq \lim_{\lambda \to 1} \left[I_1(\lambda) + \zeta I_2(\lambda)/2 \right],$$ (3.252)

where

$$I_1(\lambda) \equiv \int_0^\infty dx \exp(-\lambda x^2) \sin^2 \eta x = \pi^{1/2} \lambda^{-1/2} \left[1 - \exp(-\eta^2 \lambda^{-1}) \right]/4$$

and

$$I_2(\lambda) \equiv -\partial I_1/\partial \lambda .$$ (3.253)

Taking the limit indicated in (3.252), one finds

$$I \simeq \pi^{1/2} \{ 1 - e^{-\eta^2} + \zeta/2 \left[1/2 + (\eta^2 - 1/2) e^{-\eta^2} \right] \}/4 .$$ (3.254)

If the fractional change δ in the cell length is very small, then $\eta \ll 1$ and (3.254) simplifies to

$$I \simeq \pi^{1/2} \eta^2 (1 + 3\zeta/4)/4 .$$ (3.255)

In the instance that $\Delta \varepsilon^s \lesssim kT$, yet δ is sufficiently small, one may introduce the approximation

$$\sin^2 \eta x \simeq \eta^2 x^2$$

into (3.245) to obtain

$$I \simeq \eta^2 \int_0^\infty dx e^{-x^2} x^2 (1 + \zeta x^2)^{1/2} .$$ (3.256)

With the substitution

$$u = \zeta^{-1} + x^2$$

(3.256) becomes [3.78]

$$I \simeq \frac{1}{2} \eta^2 \zeta^{1/2} \exp(\zeta^{-1}) \int_{\zeta^{-1}}^\infty du \, u^{1/2} (u - \zeta^{-1})^{1/2} e^{-u}$$

$$= -\pi \eta^2 \zeta^{-1/2} \exp(\zeta^{-1}/2) H_1^{(1)}(i\zeta^{-1}/2)/8 ,$$ (3.257)

where $H_1^{(1)}(z)$ is Bessel's function of the third kind.

Note that in the limit $\zeta \to 0$ [3.79]

$$H_1^{(1)}(i\zeta^{-1}/2) \to -(4\zeta^{-1}/\pi)^{1/2} \exp(-\zeta^{-1}/2)$$

and from (3.257)

$$I \to \pi^{1/2}\eta^2/4 \,,$$

in precise agreement with (3.255). Hence the two approximations (3.254) and (3.257) are consistent when both ζ and η are small and under these conditions the transition rate (3.244) assumes the explicit form

$$W_{\alpha \to \beta}^{(1)} = 2^{1/2} \hbar^{-2} m_s^{1/2} L_\alpha (\Delta \varepsilon^s)^{-3/2} |V_{\alpha\beta}^0|^2 \delta^2 k T [1 + 3(\Delta \varepsilon^s)^{-1} k T/4] \,. \tag{3.258}$$

Let us now apply (3.258).

3.7.4 Prediction of the Rates of Vibrational Relaxation of Selected Diatomics in Liquid Rare Gases

In order to predict *a priori* the transition rate $W_{\alpha \to \beta}^{(1)}$, we must estimate the parameters L_α, $V_{\alpha\beta}^0$, and δ. The mass m_s of the diatomic is known and its transition energy $\Delta \varepsilon^s$ can be calculated from its transition frequency ω_s (also known) by $\Delta \varepsilon^s = \hbar \omega_s$.

The cell length L_α may be reckoned from the liquid density ρ. If one assumes that the atoms of the real liquid are cubic close-packed, then the distance between nearest neighbors is given by

$$a = (\sqrt{2}M/\rho N)^{1/3} \,, \tag{3.259}$$

where M is the gram-molecular mass of the solvent and N is Avogadro's number. The sketch in Fig. 3.8 implies that one should take

$$L_\alpha = 2a. \tag{3.260}$$

Next let us assume that *each* of the atoms of the diatomic solute interacts only with its nearest neighbor via the LJ(12,6) potential

$$V_{LJ}(z) = 4\hat{\varepsilon} [(\hat{\sigma}/z)^{12} - (\hat{\sigma}/z)^6] \,, \tag{3.261}$$

where

$$\hat{\varepsilon} = (\varepsilon_s \varepsilon_{s'})^{1/2}$$

and

$$\hat{\sigma} = (\sigma_s + \sigma_{s'})/2$$

are the LJ parameters appropriate for the interaction of two dissimilar molecules [3.80], the subscript s referring to the diatomic solute and s' to the
monatomic solvent. The distance z (see Fig. 3.8) may be expressed as

$$z = q_s - \hat{\gamma} q, \tag{3.262}$$

where

$$\hat{\gamma} = m_B/(m_A + m_B). \tag{3.263}$$

In order to evaluate the matrix elements $V_{\gamma'\gamma}(q_s)$ in closed form, it is convenient
to fit the LJ interaction (3.261) to a Morse potential. Arbitrarily requiring
that the minima and the zeroes of the two potentials coincide, we obtain for
the Morse potential

$$V_M(z) = \hat{\varepsilon}(e^{-2\hat{a}(z-z_0)} - 2e^{-\hat{a}(z-z_0)}), \tag{3.264}$$

where

$$\hat{a} = \ln 2\hat{\sigma}^{-1}/(2^{1/6} - 1) \tag{3.265}$$

and

$$z_0 = 2^{1/6}\hat{\sigma}. \tag{3.266}$$

Now, if the binding potential $\hat{V}_s(\hat{q})$ [see (3.217)] is taken to be harmonic, the
matrix elements of $V_M(z)$ are given by [3.81]

$$\begin{aligned}
\hat{V}_{nm}(\hat{q}_s) = \hat{\varepsilon}\,[&e^{-2\hat{a}(\hat{q}_s-z_0)}\,S(n,m;2\hat{a}\hat{\gamma}/\!\sqrt{\hat{\beta}}) \\
&- 2e^{-\hat{a}(\hat{q}_s-z_0)}\,S(n,m;\hat{a}\hat{\gamma}/\!\sqrt{\hat{\beta}})],
\end{aligned} \tag{3.267}$$

where

$$\begin{aligned}
S(n,m;\tau) = (n!\,m!)^{-1/2}\exp(\tau^2/4) \\
\times \sum_{j=0}^{\min(n,m)} (\tau/\!\sqrt{2})^{n+m-2j}\,n!\,m!/[(n-j)!\,(m-j)!\,j!],
\end{aligned} \tag{3.268}$$

$$\hat{\beta} = \mu\omega_s\hbar^{-1}, \tag{3.269}$$

and n and m are integers replacing the generalized indices γ' and γ. Now
$V_{\alpha\beta}^0 = \hat{V}_{\alpha\beta}(\hat{q}_s = a)$ may be evaluated using (3.267).

Finally, let us estimate δ by the following procedure. Assume that all the
liquid particles are at rest and the potential energy is at a minimum. Then,
if the diatomic is in state n, the condition of equilibrium is

$$\partial V_{nn}(\hat{q}_s)/\partial\hat{q}_s = 0 \tag{3.270}$$

and from (3.267) one deduces that

$$X_{nn} = S(n,n; \hat{\alpha}\hat{\gamma}/\sqrt{\bar{\beta}})/S(n,n; 2\hat{\alpha}\hat{\gamma}/\sqrt{\bar{\beta}}), \tag{3.271}$$

where

$$X_{nn} \equiv \exp\left[-\hat{\alpha}(q_s^{(n)} - z_0)\right]. \tag{3.272}$$

The distance $q_s^{(n)}$ of the center of mass of the diatomic (in state n) from its left-hand neighbor (see Fig. 3.8) is found from (3.272) to be

$$q_s^{(n)} = z_0 - \ln X_{nn}/\hat{\alpha}. \tag{3.273}$$

For a homonuclear diatomic[10], the cell length with the diatomic in state n is by (3.260)

$$L_n = 2z_0 - 2\ln X_{nn}/\hat{\alpha}, \tag{3.274}$$

and hence the fractional change δ in the cell length associated with the transition $n \rightarrow m$ is

$$d \sim (L_m - L_n)/L_n = 2\ln(X_{nn}/X_{mm})/(L_n\hat{\alpha}). \tag{3.275}$$

Using the physical constants given in Table 3.5 and (3.259), (3.260), (3.267), and (3.275), we compute and parameters L_1, δ, and V_{10}^0 for the systems listed

Table 3.5. Physical constants of liquid rare gases and diatomic solutes

	mass,[a] m, [amu]	fundamental frequency [3.90] ω, [$\times 10^{-14} s^{-1}$]	density,[a] ρ [gcm^{-3}]	LJ(12,6) σ [Å]	Parameters,[b] ε/k [K]
Ne	20.18	–	1.21	2.77	35.3
Ar	39.94	–	1.40	3.40	121.0
H$_2$	2.02	7.84	0.07	2.93	37.0
D$_2$	4.03	5.64	0.17	2.93	37.0
N$_2$	28.02	4.39	0.81	3.71	96.0
O$_2$	32.00	2.93	1.15	3.52	118.0
CO	28.01	4.04	0.79	3.76	100.2

[a] WEAST [3.89]. Densities are given at the normal boiling point.
[b] HIRSCHFELDER et al. [3.80]. Force constants are based on second virial coefficients. If more than one set of data is available, the average is taken.

---

[10] If the diatomic is heteronuclear, then its equilibrium position in the rigid, minimum-energy configuration is not precisely midway between its nearest neighbors. One must then calculate $q_s^{(n)}$ for both right- and lefthand interactions and sum them to obtain L_n.

Table 3.6. Estimates of cell length L_1, fractional decrement δ in cell length, and coupling matrix element V_{10} for the $1 \rightarrow 0$ transition of diatomic for selected systems

System	L_1 [Å]	$-\delta \times 10^3$	$V_{10}^0 \times 10^{16}$ [erg]
H_2/Ne	6.80	3.65	1.21
D_2/Ne	6.80	2.55	1.02
H_2/Ar	8.12	2.97	2.25
D_2/Ar	8.12	2.08	1.88
N_2/Ar	8.12	0.31	1.41
O_2/Ar	8.12	0.43	3.71
CO/Ar	8.12	0.34	0.95
$H_2(l)$	8.15	3.45	0.30
$D_2(l)$	7.65	2.42	0.34
$N_2(l)$	8.67	0.28	2.28
$O_2(l)$	8.06	0.41	1.98
CO(l)	8.72	0.31	2.16

in Table 3.6. The approximate range of temperatures over which these systems are liquids is 10–150 K. From this fact and the data in Tables 3.5 and 3.6 one can easily show that both η and ζ, given by (3.246) and (3.247), are sufficiently small that (3.258) is valid. The transition rates $W_{1 \rightarrow 0}^{(1)}$ calculated from (3.258) are listed as a function of temperature in Table 3.7. The rates computed for the pure diatomic liquids are included merely for purposes of comparison. The effects of *intermolecular* resonant transfer of vibrational energy are neglected, the relaxing diatomic's nearest neighbors being assumed rigid. It is

Table 3.7. Predicted rates (in s^{-1}) of vibrational relaxation of diatomic molecules in rare-gas solutions and in the neat diatomic liquid as a function of temperature

Temperature [K]	Solutions							Pure Liquids				
	Ne		Ar									
	H_2	D_2	H_2	D_2	N_2	O_2	CO	H_2	D_2	N_2	O_2	CO
25	142	114	—	—	—	—	—	9.55	12.7	—	—	—
30	171	137	—	—	—	—	—	11.5	15.2	—	—	—
35	199	160	—	—	—	—	—	13.4	17.8	—	—	—
40	228	183	—	—	—	—	—	15.3	20.4	—	—	—
50	—	—	—	—	—	—	—	—	—	—	—	—
60	—	—	—	—	—	—	—	—	—	83.6	245	—
80	—	—	1240	993	—	—	—	—	—	112	329	139
100	—	—	1560	1240	59.7	1560	37.9	—	—	141	413	175
120	—	—	1880	1500	72.3	1890	45.6	—	—	170	500	211
140	—	—	2200	1750	84.2	2220	53.5	—	—	—	585	—
150	—	—	2360	1880	90.5	2380	57.4	—	—	—	628	—

important to note that the results in Table 3.7 are at *constant density*, since L_1 and V_{10}^0 are in fact implicit functions of ρ [see (3.259), (3.260), and (3.267)].

3.7.5 Discussion

For the systems under consideration in Table 3.5, the diatomic transition energies $\Delta\varepsilon^s$ are sufficiently large and the temperature sufficiently low that $kT/\Delta\varepsilon \ll 1$ and from (3.258) the VR rates should be essentially linear in T. The results in Table 3.7 bear out this conclusion. However, the relative rates for any two systems at the same temperature are determined by a counterbalancing of several factors in (3.258), namely m_s, L_α, $\Delta\varepsilon^s$, $V_{\alpha\beta}^0$, and δ. Consider, for example, the isotopic systems H_2/Ne and D_2/Ne.

The ratio of the VR rates at a fixed temperature is

$$W(H_2/Ne)/W(D_2/Ne) = (m_s/m_s')^{1/2}\,(\Delta\varepsilon^s/\Delta\varepsilon^{s'})^{-3/2}(\delta/\delta')^2\,(V_{10}^0/V_{10}^{0'})^2$$

$$= (2.02/4.03)^{1/2} \times (7.84/5.64)^{-3/2}$$

$$\times (1.21/1.02)^2 \times (2.97/2.08)^2 \tag{3.276}$$

$$= 0.70 \times 0.610 \times 1.40 \times 2.03 = 1.22\,.$$

The critical factors in this case are clearly δ and V_{10}^0. The reason that both δ and V_{10}^0 are larger for H_2/Ne than for D_2/Ne is that the vibrational wavefunction for H_2 is spatially more extensive, i. e. from a classical viewpoint, the amplitude of vibration of H_2 is greater since its reduced mass is less. Consequently the overlap of the vibrational wavefunction with the interaction potential $V_M(q_s)$ [see (3.264)] (at a fixed separation q_s of the center of mass of the diatomic and its nearest neighbors) is greater for H_2 than for D_2. The larger overlap leads to larger values of both δ and V_{10}^0. Exactly analogous considerations hold for the isotopic systems H_2/Ar and D_2/Ar.

We observe that the order of the VR rates in the systems N_2/Ar, O_2/Ar, CO/Ar is

$$W(O_2/Ar) \gg W(N_2/Ar) > W(CO/Ar)\,.$$

This order is easily understood by reference to Tables 3.5 and 3.6. For all three systems L_1 is the same and m_s and δ are approximately equal. Thus, the determining factors are $\Delta\varepsilon^s$ and V_{10}^0 [see (3.276)]. The system O_2/Ar relaxes substantially faster than either N_2/Ar or CO/Ar since both $\Delta\varepsilon^s$ is smaller and V_{10}^0 is larger for the first system than for either of the latter two. Comparing just N_2/Ar with CO/Ar, one sees that $\Delta\varepsilon^s$ does not differ greatly between the two systems yet V_{10}^0 for N_2/Ar is about 50% greater than that for CO/Ar. Hence, the order observed. Note that the VR rates of O_2, N_2, and CO in Ar are predicted to be about an order of magnitude less than those of H_2 or D_2 in Ar. This decrease is again ascribable to the kinematic effects of the reduced mass, which is much larger for O_2, N_2, or CO than for H_2 or D_2. From (3.268), (3.269) and (3.271) it is clear that increasing μ decreases $S(n,n; \hat{\alpha}\hat{\gamma}/\sqrt{\hat{\beta}})$ and consequently δ. The coupling matrix elements are not substantially changed because the increase in $\hat{\varepsilon}$ counterbalances the decrease in the overlap factor. Therefore, the decrease in δ alone is primarily responsible for the overall decrease in the VR rate.

It is curious that the relative rates of VR for N_2 and CO reverse in going from Ar to the pure liquids. From Table 3.6 this reversal is seen to be a consequence of the relative increase in the coupling matrix element V_{10}^0 for CO, which, in turn, is due to the increase in strength of the interaction of CO with its nearest neighbors (relative to that of N_2) in going from dilute solutions in Ar to the pure liquid. This relative increase in strength of interaction is also reflected directly by the $\hat{\varepsilon}$ values calculated from Table 3.5.

It is somewhat gratifying that the SRS experimental data on liquid O_2 and N_2 (see Table 3.2) are in accord (in a limited sense) with our predictions. The measured vibrational lifetimes correspond to transition rates of $2 \times 10^4 \, s^{-1}$ for O_2 and $0.67–10^3 \, s^{-1}$ for N_2, whereas from Table 3.7 we estimate the respective theoretical rates to be about $45 \, s^{-1}$ and $15 \, s^{-1}$. Although the predicted rates are in the right order, at least, the absolute magnitudes and the ratio are in poor agreement. Part of this discrepancy may be due to experimental error. As noted in Subsect. 3.2.7, the values of τ' obtained by RENNER and MAIER [3.36] are probably too short.

Recall that the density ρ has been assumed constant in the applications of (3.258) considered above in Subsect. 3.7.4. This assumption, however, is not a limitation inherent in the model. In fact, if ρ is known as a function of pressure p and temperature, then L_1, δ, and V_{10}^0 can easily be calculated as functions of p and T. Note also that the model is not limited merely to the description of the simple vibrational-to-external "translational" (V–T) energy-transfer mechanism operating in solutions of diatomic solutes in monatomic solvents. It may be extended to treat the following more complex mechanisms of VR: 1) V–T transfer from polyatomic solutes to monatomic and polyatomic solvents; 2) externally assisted *intermolecular* V–V transfer between polyatomics, in which any excess energy is converted into excitations of the external modes of the liquid; 3) externally assisted *intramolecular* V–V transfer.

3.8 Summary and Conclusion

The principal goal of this chapter is to develop a theoretical framework within which experiments on vibrational relaxation in condensed media can be interpreted. The approach consists of the following sequence. First we derive an exact equation of motion for the quantity (e. g. the population of vibrationally excited molecules) being measured experimentally. Next, we solve the equation approximately, obtaining a characteristic relaxation time expressed in terms of the fundamental interactions responsible for the relaxation. Finally, by adopting a model for the specific system of interest, we obtain an approximate closed expression for the relaxation time in terms of molecular properties and thermodynamic state variables.

In order to simplify the theoretical treatment and also the interpretation of the experimental data, we have restricted our considerations to vibrational relaxation of diatomic molecules in their ground electronic states in dilute solutions in monatomic solvents. In such systems only the relatively simple mechanism of vibrational-to-"translational" (V–T) energy transfer operates. Of course, the mechanism applies equally well to VR of polyatomics in the lowest vibrationally excited level of the lowest-frequency normal mode.

The assumption of dilute solutions is introduced to avoid possible complications arising from resonant V–V energy transfer between neighboring diatomics, i. e. from vibrational excitons. In fact, the apparent lifetime of CO in Ar decreases markedly with concentration [3.15]. One would not expect this, since resonant V–V energy transfer does not deplete the population of vibrationally excited diatomics. A possible explanation is that the interaction between pairs of CO molecules causes an excitonic splitting of the doubly degenerate first vibrationally excited state of the pair into a *lower* (symmetric) state and an *upper* (antisymmetric) state. The resulting depression of the lower level reduces the effective transition frequency $\omega_{\alpha\beta}$, and hence, according to (3.213), the decay rate increases.

A very important question is: What, precisely, are the relationships among the characteristic VR times determined by various experimental techniques, in particular stimulated Raman scattering (Subsect. 3.2.7) and IR absorption and spontaneous Raman scattering (Subsect. 3.2.6)? The vibrational lifetime τ' (from SRS) is a measure of the rate of decay of the population of vibrationally excited molecules and, therefore, of the rate of loss of vibrational energy. In contrast, both the dephasing time τ (from SRS) and the vibrational correlation time τ_v (from IR and spontaneous Raman) characterize the rate of loss of phase coherence in the vibrational coordinate and should be essentially the same, yet different, in general, from τ'. BRATOS et al. [3.20] have argued that τ and τ_v are essentially equal, at least for dilute solutions. Our calculation in Section 3.3 showed that $\tau_v = 2\tau'$ for a harmonic oscillator coupled linearly to a heat bath. However, it is interesting to note (see Table 3.2) that $\tau' > \tau$ for most liquids. Hence, if the identification of τ with τ_v is valid, our simple model is inadequate. Apparently, we are neglecting important additional interactions in the Hamiltonian which lead to observed dephasing times considerably shorter than corresponding vibrational lifetimes. We should note that the experimental data in Table 3.2 refer to pure liquids, in which vibrational excitons could play a significant role in dephasing the molecular vibrations.

Our restriction to the prototypic system has permitted a reasonably perspicuous theoretical analysis of VR. However, most of the available experimental data refer to neat polyatomic liquids, in which a variety of modes (channels) participate in the overall relaxation process. A particularly interesting example is provided by the SRS investigations [3.31, 32] of ethanol (see Subsect. 3.2.7). The involvement of more than a single normal mode of the relaxing molecule complicates the theoretical treatment somewhat. In order to gain some initial insight into the process, it would be helpful to study the dynamics of the following model: *two* harmonic oscillators coupled to each other and also to a heat bath. One should calculate $\langle \hat{n}_1(t) \rangle$ and $\langle \hat{n}_2(t) \rangle$ by an extension of the Zwanzig projection-operator formalism applied in Section 3.3.

A significant problem that we have not touched upon here is that of molecular rearrangements (chemical reactions) in condensed phases. In general, of course, such processes are accompanied by vibrational relaxation. From a theoretical standpoint they are particularly difficult to manage, since the integrity of the normal modes is not preserved during the rearrangement. It is common to partition the multi-dimensional configuration space into a reaction coordinate (RC) and the remaining degrees of freedom "normal" to the RC. The rearrangement occurs as the phase point of the system traverses the RC from one arrangement (reactants) α to the other (products), β. It is usually assumed that the potential energy is separable into a contribution along the RC plus a remainder (perhaps depending upon the RC). In condensed phases the potential energy along the RC has, in general, the shape (i. e. a double minimum) depicted in Fig. 3.9a. The remaining contribution is frequently taken to be harmonic in the "normal" coordinates.

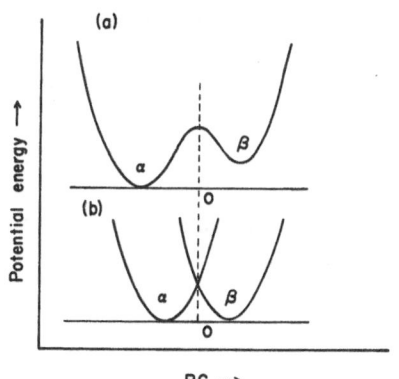

Fig. 3.9a and b. Potential energy profiles along the reaction coordinate (RC) associated with rearrangements in condensed phases. (a) Adiabatic, double-minimum surface for a typical unsymmetric rearrangement; (b) "localized" adiabatic states defined by the method of FONG and DIESTLER [3.82]

FONG and DIESTLER [3.82] have proposed a model for thermally activated symmetric rearrangements. The rearrangement is viewed as a radiationless transition between two different localized adiabatic electronic states α and β. The localized states are calculated in the Born-Oppenheimer approximation and the adiabatic potential energy curves associated with the RC (see Fig. 3.9b) are taken to be displaced harmonic potentials. Transitions between states α and β arise from the usual non-Born-Oppenheimer electronic-nuclear coupling, the transition rate being calculated by Kubo's [3.83] linear response theory. The model was applied to a quantitative analysis of the orientational relaxation of dipolar impurities in ionic crystals and to a qualitative discussion of dielectric relaxation in solutions of polar molecules.

In a recent article FREED and FONG [3.84] have argued that a more realistic view is that the rearrangement takes place on a single adiabatic double-minimum surface (Fig. 3.9a). In this view the electronic-nuclear non-adiabatic coupling cannot be responsible for transitions between the two arrangements. FREED and FONG regard the thermally activated process as a *nuclear* non-adiabatic transition, considering two alternative approaches: (i) the transition arises from potential-energy coupling between zero-order harmonic minima; (ii) an adiabatic separation of the RC from the "normal" degrees of freedom, in which case non-adiabatic nuclear-phonon couplings analogous to the electron-phonon (non-Born-Oppenheimer) coupling becomes responsible for the transition. We note that the latter approach is similar to that taken by FISCHER [3.70] in his treatment of VR of diatomics in rare-gas matrices.

In closing, let us take note of a very important point, namely that all of the above treatments have been based upon first-order perturbation theory, in which it is implicitly assumed that the interaction between the active (absorbing, scattering) solute mode(s) and solvent modes is weak. This is not always true (e. g., in the case of hydrogen-bonded liquids or chemical reactions). In such cases where the solute mode(s) interact strongly with one or more solvent modes, it is necessary to partition the full Hamiltonian into "prediagonalized", multidimensional, strongly coupled solute and solvent portions plus weak residual interactions, which can be handled by perturbation theory. Such a procedure, of course, requires the solution of a *set* of coupled equations of motion for the correlation (or population) functions of the strongly coupled solute modes.

Appendix A

To verify the fact that $\hat{P}$, defined by (3.39), is indeed a projection operator we simply consider the effect of $\hat{P}^2$ on an *arbitrary* operator $\hat{X}$

$$
\begin{aligned}
\hat{P}^2 \hat{X} = \hat{P}(\hat{P}\hat{X}) &= (\hat{P}\,\varDelta\hat{n}(0))\langle\hat{X}\rangle/\langle\varDelta\hat{n}(0)\rangle \\
&= \langle\hat{X}\rangle\,\varDelta\hat{n}(0)\langle\varDelta\hat{n}(0)\rangle/\langle\varDelta\hat{n}(0)\rangle^2 \\
&= \varDelta\hat{n}(0)\langle\hat{X}\rangle/\langle\varDelta\hat{n}(0)\rangle \\
&= \hat{P}\hat{X}.
\end{aligned}
\tag{3.277}
$$

Hence, we have

$$
\hat{P}^2 = \hat{P},
\tag{3.278}
$$

which condition is both necessary and sufficient for $\hat{P}$ to be a projection operator.

We can define the *relevant* and *irrelevant* parts of an arbitrary operator $\hat{X}$ by the following equations:

$$\hat{P}\hat{X} = \hat{X}_R \tag{3.279}$$

$$(\hat{1} - \hat{P})\hat{X} = \hat{X}_{IRR}. \tag{3.280}$$

The rationale for the nomenclature is easily seen by noting that

$$\langle \hat{P}\hat{X} \rangle = \langle [\Delta\hat{n}(0)\langle\hat{X}\rangle/\langle\Delta\hat{n}(0)\rangle] \rangle$$
$$= \langle \hat{X} \rangle, \tag{3.281}$$

whereas

$$\langle (\hat{1} - \hat{P})\hat{X} \rangle = \langle \hat{X} \rangle - \langle \hat{P}\hat{X} \rangle = 0. \tag{3.282}$$

We see that $\hat{P}$ projects out of $\hat{X}$ only that part which contributes to the ensemble average defined by (3.40). From (3.279) and (3.280) it is clear that

$$\hat{X} = \hat{X}_R + \hat{X}_{IRR}. \tag{3.283}$$

Using (3.39) and (3.41), we find

$$\langle\langle \hat{P}\hat{X} \rangle\rangle = \langle [\Delta\hat{n}(0)\langle\hat{X}\rangle/\langle\Delta\hat{n}(0)\rangle] \rangle/\langle\Delta\hat{n}(0)\rangle$$
$$= \langle\hat{X}\rangle/\langle\Delta\hat{n}(0)\rangle = \langle\langle \hat{X} \rangle\rangle. \tag{3.284}$$

Similarly,

$$\langle\langle \hat{Y}\hat{P}\hat{X} \rangle\rangle = \langle \hat{Y}[\Delta\hat{n}(0)\langle\hat{X}\rangle/\langle\Delta\hat{n}(0)\rangle] \rangle/\langle\Delta\hat{n}(0)\rangle$$
$$= \langle \hat{Y}\Delta\hat{n}(0)\rangle\langle\hat{X}\rangle/\langle\Delta\hat{n}(0)\rangle^2$$
$$= \langle\langle \hat{Y}\Delta\hat{n}(0) \rangle\rangle\langle\langle\hat{X}\rangle\rangle. \tag{3.285}$$

Appendix B

Note that

$$\langle [\hat{\Gamma}_0, \hat{\Gamma}_0(t)] \rangle_B^* = \langle [\hat{\Gamma}_0, \hat{\Gamma}_0(t)]^\dagger \rangle_B$$
$$= \langle [\hat{\Gamma}_0(t), \hat{\Gamma}_0] \rangle_B \tag{3.286}$$
$$= -\langle [\hat{\Gamma}_0, \hat{\Gamma}_0(t)] \rangle_B,$$

which shows that $\langle [\hat{\Gamma}_0, \hat{\Gamma}_0(t)] \rangle_B$ is pure imaginary and consequently may be written as

$$\langle[\hat{\Gamma}_0,\hat{\Gamma}_0(t)]\rangle_{\mathrm{B}} = \mathrm{i}\,|\langle[\hat{\Gamma}_0,\hat{\Gamma}_0(t)]\rangle_{\mathrm{B}}|\,. \tag{3.287}$$

Then we have

$$\gamma(\omega) \equiv \hbar^{-2} \int_0^\infty \mathrm{e}^{-\mathrm{i}\omega t}\langle[\hat{\Gamma}_0,\hat{\Gamma}_0(t)]\rangle_{\mathrm{B}}\,dt$$

$$= \hbar^{-2} \int_0^\infty (\cos\omega t - \mathrm{i}\sin\omega t)\,\mathrm{i}\,|\langle[\hat{\Gamma}_0,\hat{\Gamma}_0(t)]\rangle_{\mathrm{B}}|\,dt \tag{3.288}$$

$$= \gamma_1(\omega) + \mathrm{i}\gamma_2(\omega)\,,$$

where

$$\gamma_1(\omega) = \mathrm{Re}\,\{\gamma(\omega)\} = -\mathrm{i}\hbar^{-2} \int_0^\infty \sin\omega t\,\langle[\hat{\Gamma}_0,\hat{\Gamma}_0(t)]\rangle_{\mathrm{B}}\,dt \tag{3.289}$$

and

$$\gamma_2(\omega) = \mathrm{Im}\,\{\gamma(\omega)\} = -\mathrm{i}\hbar^{-2} \int_0^\infty \cos\omega t\,\langle[\hat{\Gamma}_0,\hat{\Gamma}_0(t)]\rangle_{\mathrm{B}}\,dt \tag{3.290}$$

Hence, from (3.289) and (3.290) we conclude that

$$\mathrm{Re}\,\{\gamma(-\omega)\} = -\mathrm{Re}\,\{\gamma(\omega)\} \tag{3.291}$$

and

$$\mathrm{Im}\,\{\gamma(-\omega)\} = \mathrm{Im}\,\{\gamma(\omega)\}\,. \tag{3.292}$$

Appendix C

We define $I(\omega)$ by

$$I(\omega) \equiv \int_{-\infty}^{+\infty} \mathrm{e}^{\mathrm{i}\omega t}\langle[\hat{\Gamma}_0,\hat{\Gamma}_0(t)]\rangle_{\mathrm{B}}\,dt\,. \tag{3.293}$$

Eq. (3.293) may be rewritten as

$$I(\omega) = \int_0^\infty \mathrm{e}^{\mathrm{i}\omega t}\langle[\hat{\Gamma}_0,\hat{\Gamma}_0(t)]\rangle_{\mathrm{B}}\,dt + \int_{-\infty}^0 \mathrm{e}^{\mathrm{i}\omega t}\langle[\hat{\Gamma}_0,\hat{\Gamma}_0(t)]\rangle_{\mathrm{B}}\,dt\,,$$

which by (3.288) may be recast as

$$I(\omega) = \hbar^2\gamma(-\omega) + \int_{-\infty}^0 \mathrm{e}^{\mathrm{i}\omega t}\langle[\hat{\Gamma}_0,\hat{\Gamma}_0(t)]\rangle_{\mathrm{B}}\,dt\,. \tag{3.294}$$

Making the substitution $t' = -t$ in the integral in (3.294) and then invoking (3.108), we obtain

$$I(\omega) = \hbar^2 \gamma(-\omega) - \int_0^\infty e^{-i\omega t} \langle [\hat{\Gamma}_0, \hat{\Gamma}_0(t)] \rangle_B \, dt$$

$$= \hbar^2 \gamma(-\omega) - \hbar^2 \gamma(\omega), \tag{3.295}$$

the second equality following via (3.288). Finally from (3.291) and (3.292) we have

$$I(\omega) = 2\hbar^2 \, \mathrm{Re}\,\{\gamma(-\omega)\}. \tag{3.296}$$

Using the general theorem [3.83]

$$\int_{-\infty}^{+\infty} e^{-i\omega t} \langle \hat{X}\,\hat{Y}(t) \rangle \, dt = e^{\beta\hbar\omega} \int_{-\infty}^{+\infty} e^{i\omega t} \langle \hat{Y}\,\hat{X}(t) \rangle \, dt, \tag{3.297}$$

we can write

$$\int_{-\infty}^{+\infty} e^{-i\omega t} \langle \hat{\Gamma}_0 \hat{\Gamma}_0(t) \rangle_B \, dt = e^{\beta\hbar\omega} \int_{-\infty}^{+\infty} e^{i\omega t} \langle \hat{\Gamma}_0 \hat{\Gamma}_0(t) \rangle_B \, dt. \tag{3.298}$$

Again making a change of variables from t to $t' = -t$ in the integral on the left-hand side of (3.298) and applying (3.108), we may transform (3.298) into

$$\int_{-\infty}^{+\infty} e^{i\omega t} \langle \hat{\Gamma}_0(t) \hat{\Gamma}_0 \rangle_B \, dt = e^{\beta\hbar\omega} \int_{-\infty}^{+\infty} e^{i\omega t} \langle \hat{\Gamma}_0 \hat{\Gamma}_0(t) \rangle_B \, dt. \tag{3.299}$$

Making the substitution

$$\hat{\Gamma}_0(t) \hat{\Gamma}_0 = -[\hat{\Gamma}_0, \hat{\Gamma}_0(t)] + \hat{\Gamma}_0 \hat{\Gamma}_0(t) \tag{3.300}$$

into (3.299) and rearranging terms, we obtain

$$-\int_{-\infty}^{+\infty} e^{i\omega t} \langle [\hat{\Gamma}_0, \hat{\Gamma}_0(t)] \rangle_B \, dt = (e^{\beta\hbar\omega} - 1) \int_{-\infty}^{+\infty} e^{i\omega t} \langle \hat{\Gamma}_0 \hat{\Gamma}_0(t) \rangle_B \, dt. \tag{3.301}$$

Once more letting $t' = -t$ in the integral on the left-hand side of (3.301) and using (3.108), we get

$$\int_{-\infty}^{+\infty} e^{-i\omega t} \langle [\hat{\Gamma}_0, \hat{\Gamma}_0(t)] \rangle_B \, dt = (e^{\beta\hbar\omega} - 1) \int_{-\infty}^{+\infty} e^{i\omega t} \langle \hat{\Gamma}_0 \hat{\Gamma}_0(t) \rangle_B \, dt. \tag{3.302}$$

Finally, comparing (3.293) and (3.296) with (3.302), we observe that

$$I(\omega) = 2\hbar^2 \, \mathrm{Re}\,\{\gamma(-\omega)\} = (e^{-\beta\hbar\omega} - 1) \int_{-\infty}^{+\infty} e^{-i\omega t} \langle \hat{\Gamma}_0 \hat{\Gamma}_0(t) \rangle_B \, dt. \tag{3.303}$$

Appendix D

From (3.135), (3.139) and (3.140) we have

$$
\begin{aligned}
\hat{H}_0\hat{\rho}(0) &= \sum_\gamma \sum_{n_\gamma} \varepsilon_{\gamma n_\gamma} |\gamma\rangle\langle\gamma| \otimes |n_\gamma\rangle\langle n_\gamma| \, (|\alpha\rangle\langle\alpha| \otimes e^{-\beta\hat{H}_0^{(\alpha)}})/\mathrm{Tr}\,\{e^{-\beta\hat{H}_0^{(\alpha)}}\} \\
&= |\alpha\rangle\langle\alpha| \sum_{n_\alpha} \varepsilon_{\alpha n_\alpha} |n_\alpha\rangle\langle n_\alpha| \, e^{-\beta\hat{H}_0^{(\alpha)}}/\mathrm{Tr}\,\{e^{-\beta\hat{H}_0^{(\alpha)}}\}\,.
\end{aligned}
\tag{3.304}
$$

Using (3.137) and (3.141), we can rewrite (3.304)

$$
\begin{aligned}
\hat{H}_0\hat{\rho}(0) &= |\alpha\rangle\langle\alpha| \Big[\varepsilon_\mathrm{M}^\alpha \sum_{n_\alpha} |n_\alpha\rangle\langle n_\alpha| + \hat{H}_{\mathrm{L}0}^{(\alpha)}\Big] e^{-\beta\hat{H}_{\mathrm{L}0}^{(\alpha)}}/\mathrm{Tr}\,\{e^{-\beta\hat{H}_{\mathrm{L}0}^{(\alpha)}}\} \\
&= |\alpha\rangle\langle\alpha| \, e^{-\beta\hat{H}_{\mathrm{L}0}^{(\alpha)}} \Big[\varepsilon_\mathrm{M}^\alpha \sum_{n_\alpha} |n_\alpha\rangle\langle n_\alpha| + H_{\mathrm{L}0}^{(\alpha)}\Big]/\mathrm{Tr}\,\{e^{-\beta\hat{H}_{\mathrm{L}0}^{(\alpha)}}\} \\
&= |\alpha\rangle\langle\alpha| \, e^{-\beta\hat{H}_{\mathrm{L}0}^{(\alpha)}} \Big[\sum_\gamma \sum_{n_\gamma} \varepsilon_{\gamma n_\gamma} |\gamma\rangle\langle\gamma| \times |n_\gamma\rangle\langle n_\gamma|\Big]/\mathrm{Tr}\,\{e^{-\beta\hat{H}_{\mathrm{L}0}^{(\alpha)}}\} \\
&= \hat{\rho}(0)\hat{H}_0\,.
\end{aligned}
\tag{3.305}
$$

Thus

$$
[\hat{U}_0(t),\hat{\rho}(0)] = 0\,.
\tag{3.306}
$$

In a similar fashion, we have from (3.135) and (3.143)

$$
\begin{aligned}
\hat{H}_0\hat{P}_\beta &= \sum_\gamma \sum_{n_\gamma} \varepsilon_{\gamma n_\gamma} |\gamma\rangle\langle\gamma| \otimes |n_\gamma\rangle\langle n_\gamma| \, (|\beta\rangle\langle\beta|) \\
&= \sum_{n_\beta} \varepsilon_{\beta n_\beta} |n_\beta\rangle\langle n_\beta| \\
&= \hat{P}_\beta\hat{H}_0\,,
\end{aligned}
\tag{3.307}
$$

so

$$
[\hat{U}_0(t),\hat{P}_\beta] = 0\,.
$$

Finally, we observe

$$
\begin{aligned}
\hat{\rho}(0)\hat{P}_\beta &= |\alpha\rangle\langle\alpha| \otimes \hat{\rho}_L(0) |\beta\rangle\langle\beta| \sum_{n_\beta} |n_\beta\rangle\langle n_\beta| \\
&= 0\,.
\end{aligned}
\tag{3.308}
$$

References

3.1 R. Zwanzig: J. Chem. Phys. **33**, 1338 (1960)
3.2 L. van Hove: Physica **21**, 517 (1955)
3.3 D. Sette: In: *Physics of Simple Liquids*, ed. by H. N. V. Temperley, J. S. Rowlinson, G. S. Rushbrooke (North Holland, Amsterdam 1968) p. 325

3.4   K. F. HERZFELD, T. A. LITOVITZ: *Absorption and Dispersion of Ultrasonic Waves* (Academic Press, New York 1959)

3.5   E. SITTIG: Acustica **10**, 81 (1960)

3.6   R. D. MOUNTAIN: Rev. Mod. Phys. **38**, 205 (1966)

3.7   R. D. MOUNTAIN: J. Res. Natl. Bur. Stand. **70 A**, 207 (1966)

3.8   R. D. MOUNTAIN: J. Res. Natl. Bur. Stand. **72 A**, 95 (1968)

3.9   L. I. KOMAROV, I. Z. FISHER: Sov. Phys. JETP **16**, 1358 (1963)

3.10 L. J. SCHOEN, H. P. BROIDA: J. Chem. Phys. **32**, 1184 (1960)

3.11 D. S. TINTI: J. Chem. Phys. **48**, 1459 (1968)

3.12 J. S. SHIRK, A. M. BASS: J. Chem. Phys. **52**, 1894 (1970)

3.13 D. S. TINTI, G. W. ROBINSON: J. Chem. Phys. **49**, 3229 (1968)

3.14 H. DUBOST, L. ABOUAF-MARGUIN, F. LEGAY: Phys. Rev. Lett. **29**, 145 (1972)

3.15 H. DUBOST, F. LEGAY: Private communication (1973)

3.16 L. J. ALLAMANDOLA, Jr., J. W. NIBLER: Chem. Phys. Lett. **28**, 335 (1974)

3.17 L. A. NAFIE, W. L. PETICOLAS: J. Chem. Phys. **57**, 3145 (1972)

3.18 S. BRATOS, J. RIOS, Y. GUISSANI: J. Chem. Phys. **52**, 439 (1970)

3.19 S. BRATOS, E. MARECHAL: Phys. Rev. **A4**, 1078 (1971)

3.20 S. BRATOS, Y. GUISSANI, J. C. LEICKNAM: In: *Molecular Motions in Liquids*, ed. by J. LASCOMBE (D. Reidel, Dordrecht-Holland 1974) p. 187

3.21 R. B. WRIGHT, M. SCHWARZ, C. H. WANG: J. Chem. Phys. **58**, 5125 (1973)

3.22 J. H. CAMPBELL, J. F. FISHER, J. JONAS: J. Chem. Phys. **61**, 346 (1974)

3.23 G. DÖGE: In *Molecular Motions in Liquids*, ed. by J. LASCOMBE (D. Reidel, Dordrecht-Holland 1974) p. 225

3.24 F. G. BAGLIN, R. B. THOMAS, G. N. FICKES: J. Chem. Phys. **60**, 2475 (1974)

3.25 W. G. ROTHSCHILD: In: *Molecular Motions in Liquids*, ed. by J. LASCOMBE (D. Reidel, Dordrecht-Holland 1974) p. 247

3.26 Y. LEDUFF: J. Chem. Phys. **59**, 1984 (1973)

3.27 M. MAIER, W. KAISER, J. A. GIORDMAINE: Phys. Rev. **177**, 580 (1969)

3.28 C. P. SLICHTER: *Principles of Magnetic Resonance* (Harper and Row, New York 1963)

3.29 A. LAUBEREAU, D. VON DER LINDE, W. KAISER: Phys. Rev. Lett. **27**, 802 (1971)

3.30 A. LAUBEREAU, D. VON DER LINDE, W. KAISER: Phys. Rev. Lett. **28**, 1162 (1972)

3.31 R. R. ALFANO, S. L. SHAPIRO: Phys. Rev. Lett. **29**, 1655 (1972)

3.32 A. LAUBEREAU, G. KEHL, W. KAISER: Opt. Commun. **11**, 74 (1974)

3.33 A. LAUBEREAU, L. KIRSCHNER, W. KAISER: Opt. Commun. **9**, 182 (1973)

3.34 P. R. MONSON, S. PATUMTEVAPIBAL, K. J. KAUFMANN, G. W. ROBINSON: Chem. Phys. Lett. **28**, 312 (1974)

3.35 A. LAUBEREAU: Chem. Phys. Lett. **27**, 600 (1974)

3.36 G. RENNER, M. MAIER: Chem. Phys. Lett. **28**, 614 (1974)

3.37 W. F. CALAWAY, G. E. EWING: Chem. Phys. Lett. **30**, 485 (1975)

3.38 W. KAISER: Private communication (1975)

3.39 B. J. BERNE, G. D. HARP: Advan. Chem. Phys. **17**, 63 (1970)

3.40 T. MATSUBARA: Progr. Theoret. Phys. **14**, 351 (1955)

3.41 R. S. WILSON, W. T. KING, K. S. KIM: Phys. Rev. **175**, 1164 (1968)

3.42 K. S. KIM, R. S. WILSON: Phys. Rev. **A7**, 1396 (1973)

3.43 A. NITZAN, J. JORTNER: Molec. Phys. **25**, 713 (1973)

3.44 W. H. LOUISELL: *Radiation and Noise in Quantum Electronics* (McGraw-Hill, New York 1964)

3.45 R. KUBO: J. Phys. Soc. Japan **17**, 1100 (1962)

3.46 A. NITZAN, R. S. SILBEY: J. Chem. Phys. **60**, 4070 (1974)

3.47 G. R. FLEMING, O. L. J. GIJZEMAN, S. H. LIN: J. Chem. Soc. Faraday Trans. II **70**, 37 (1974)

3.48 R. W. ZWANZIG: Physica **30**, 1109 (1964)

3.49 S. H. LIN: J. Chem. Phys. **61**, 3810 (1974)

3.50 A. MESSIAH: *Quantum Mechanics* (John Wiley, New York 1962)

3.51 R. ENGLMAN, J. JORTNER: Molec. Phys. **18**, 145 (1970)

3.52 K. F. FREED, J. JORTNER: J. Chem. Phys. **52**, 6272 (1970)

3.53 A. D. BRAILSFORD, T. Y. CHANG: J. Chem. Phys. **53**, 3108 (1970)
3.54 S. F. FISCHER: J. Chem. Phys. **53**, 3195 (1970)
3.55 F. K. FONG, S. L. NABERHUIS, M. M. MILLER: J. Chem. Phys. **56**, 4020 (1972)
3.56 R. ERDÉLYI: *Asymptotic Expansions* (Dover, New York 1956)
3.57 H. V. LAUER, F. K. FONG: J. Chem. Phys. **60**, 274 (1974)
3.58 P. R. BEVINGTON: *Data Reduction and Error Analysis for the Physical Sciences* (McGraw-Hill, New York 1969)
3.59 M. E. RILEY, A. KUPPERMANN: Chem. Phys. Lett. **1**, 537 (1968)
3.60 G. L. POLLACK: Rev. Mod. Phys. **36**, 748 (1964)
3.61 D. N. BATCHELDER, M. F. COLLINS, B. C. G. HAYWOOD, G. R. SIDEY: J. Phys. C3, 249 (1970)
3.62 L. N. LIEBERMANN: Phys. Rev. **113**, 1052 (1959)
3.63 M. GOUTERMAN: J. Chem. Phys. **36**, 2846 (1962)
3.64 J. M. SCHURR: Intl. J. Quant. Chem. **5**, 239 (1971)
3.65 W. HEITLER, S. T. MA: Proc. Roy. Irish Acad. **52 A**, 109 (1949)
3.66 H.-Y. SUN, S. A. RICE: J. Chem. Phys. **42**, 3826 (1965)
3.67 N. B. SLATER: *Theory of Unimolecular Reactions* (Cornell Press, Ithaca 1959)
3.68 N. F. MOTT, H. S. W. MASSEY: *The Theory of Atomic Collisions* (Clarendon Press, Oxford 1965)
3.69 A. NITZAN, S. MUKAMEL, J. JORTNER: J. Chem. Phys. **60**, 3929 (1974)
3.70 S. F. FISCHER: Private communication (1974)
3.71 K. F. HERZFELD: J. Chem. Phys. **20**, 288 (1952)
3.72 J. E. LENNARD-JONES, A. F. DEVONSHIRE: Proc. Roy Soc. **163**, 53 (1937)
3.73 J. E. LENNARD-JONES, A. F. DEVONSHIRE: Proc. Roy Soc. **165**, 1 (1938)
3.74 T. A. LITOVITZ: J. Chem. Phys. **26**, 469 (1957)
3.75 R. ZWANZIG: J. Chem. Phys. **34**, 1931 (1961)
3.76 P. K. DAVIS, I. OPPENHEIM: J. Chem. Phys. **57**, 505 (1972)
3.77 S. D. HAMANN: Trans. Faraday Soc. **48**, 303 (1952)
3.78 I. S. GRADSHTEYN, I. M. RYZHIK: *Tables of Integrals, Series and Products* (Academic, New York 1965)
3.79 M. ABRAMOWITZ, I. A. STEGUN: *Handbook of Mathematical Functions* (U.S. Government Printing Office, Washington, D.C. 1964)
3.80 J. O. HIRSCHFELDER, C. F. CURTISS, R. B. BIRD: *Molecular Theory of Gases and Liquids* (Wiley, New York 1966)
3.81 D. J. DIESTLER, P. FEUER: J. Chem. Phys. **54**, 4626 (1971)
3.82 F. K. FONG, D. J. DIESTLER: J. Chem. Phys. **57**, 4953 (1972)
3.83 R. KUBO: J. Phys. Soc. Japan **12**, 570 (1957)
3.84 K. F. FREED, F. K. FONG: J. Chem. Phys. **63**, 2890 (1975)
3.85 D. J. DIESTLER, R. S. WILSON: J. Chem. Phys. **62**, 1572 (1975)
3.86 D. J. DIESTLER: J. Chem. Phys. **60**, 2692 (1974)
3.87 D. J. DIESTLER: Chem. Phys. **7**, 349 (1975)
3.88 D. VON DER LINDE, A. LAUBEREAU, W. KAISER: Phys. Rev. Lett. **26**, 954 (1971)
3.89 R. C. WEAST: *Handbook of Chemistry and Physics* (Chemical Rubber, Cleveland 1972)
3.90 J. C. DAVIS, Jr.: *Advanced Physical Chemistry* (Ronald, New York 1972)
3.91 See, e.g., R. M. MARTIN, L. M. FALICOV, in: *Topics in Applied Physics*, Vol. 8: Light Scattering in Solids, ed. by M. CARDONA (Springer, Berlin, Heidelberg, New York 1975)

4. Up-Conversion and Excited State Energy Transfer in Rare-Earth Doped Materials

J. C. WRIGHT

With 25 Figures

The rare-earth ions are unique spectroscopically because the optically active transitions within the $4f^n$ core are well-shielded from outside influences. This isolation results in the sharp lines that are observed when the ions are doped into crystalline lattices [4.1]. There are, however, important processes such as sensitization and quenching of fluorescence that rely upon the very small interactions between rare-earth ions. Both fluorescence sensitization and quenching involve two ions, one in an excited state and one in the ground state. Recently, new processes have been found that involve interaction between two excited ions. Although such processes were not of importance in early work, they have become very important in recent years with the advent of intense, narrow-band laser sources which are capable of creating high excitation densities. These processes have been used to sensitize laser operation [4.2], develop a new series of infrared to visible phosphors [4.3], increase sensitivity of infrared quantum counters (IRQC) [4.4] (although they can also limit performance as well) [4.5], and create new quenching mechanisms [4.6, 7]. The complexity of levels that are generally present in rare-earth ions makes excited state interactions a very general process in highly excited materials. In this article, we will try to understand the basic mechanisms that are important for these processes. In the first section, we will look at the wide variety of energy transfer phenomena that have been observed and place the excited state interactions in perspective. In the next section, we will examine the basic interactions that occur between ions. In the third section, the general processes of importance in energy transfer will be discussed and the mathematical description for the interactions will be treated. Finally, excited state interactions will be studied specifically and some examples of the important excited state interactions will be presented to understand the mechanisms and experimental procedures available for analyzing them.

4.1 Background Information

BLOEMBERGEN proposed using excited-state absorption for infrared detection following an idea sketched in Fig. 4.1a [4.8]. A strong visible excitation source is tuned to the transition from level 1 to 3 and the fluorescence from level 3 is monitored. If infrared radiation is absorbed at the frequency of the 0 to 1

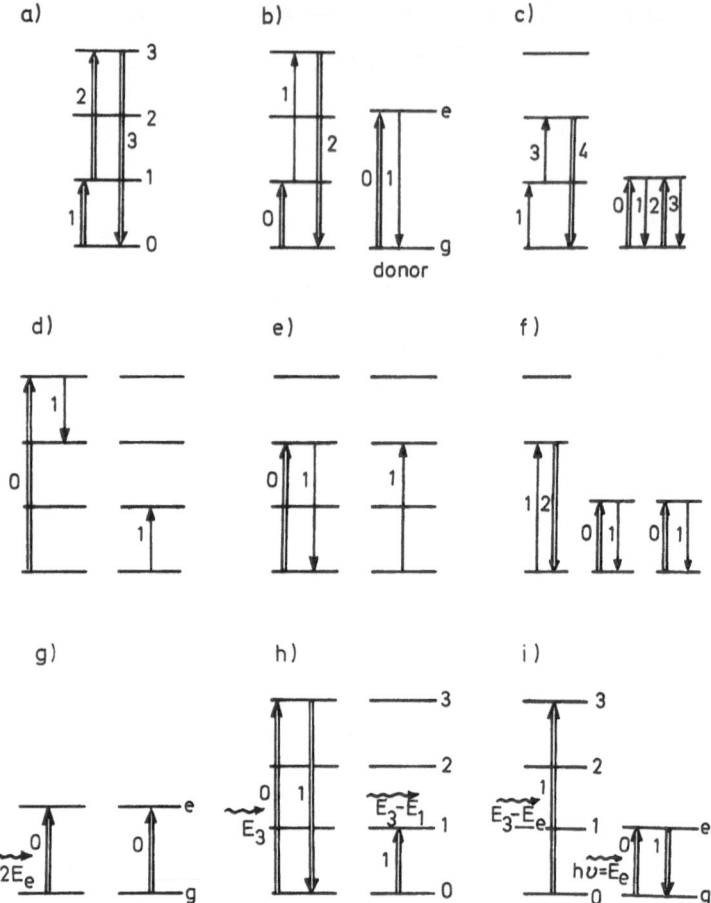

Fig. 4.1a–i. Diagrams of phenomena caused by ion-ion coupling or up-conversion of radiation. The double lined arrows represent transitions induced by the radiation field and the single lines represent transitions caused by inter-ionic coupling. The sequence in time is indicated by the numbers beside each transition with 0 representing the earliest event. Designation of levels is given beside the level. Numbers are used to designate all levels except the simple donor ions where e and g are used to represent excited and ground states

transition, fluorescence will be observed from level 3 and up-conversion of infrared energy to more easily detected visible energy is achieved. The idea was proposed as a noiseless technique for infrared detection but subsequent analysis has shown sources of noise arise from background radiation [4.9, 10]. The field of IRQC was reviewed in an excellent article by BROWN and SHAND [4.11]. ESTEROWITZ et al. [4.4] demonstrated a second scheme in which the transition between level 1 and 3 was provided by energy transfer from an excited donor ion, as shown in Fig. 4.1b. This technique proved to be more efficient in providing the second transition. Carrying this procedure one step further, both the first and second transitions could be the result of successive energy transfers from excited donors (see Fig. 4.1c). This process was independently observed in 1966 by AUZEL [4.12, 13], and OVSYANKIN and FEOFILOV [4.14, 15], and was the probable explanation for the observations of SINGH and GUESIC [4.16]. An excellent review article of this field has been prepared by AUZEL [4.3]. The process can also occur in singly doped systems in which transfer occurs between levels of two identical ions. Examples of this are seen in several different cases. For example, it is responsible for the exciton annihilation that is observed in organic materials when two migrating triplet state excitations combine to form a more excited singlet state excitation [4.17, 18]. The process can also occur in a very troublesome way in some IRQC systems [4.5]. The intense excitation source required for an efficient IRQC can sometimes non-resonantly pump levels from the ground state if the frequency is within several phonon energies of the levels. If the excitation that results can be up-converted by the process of successive transfers, a background fluorescence will result that obscures the infrared signal sought. Although this process may not be efficient, the high intensities that are needed for efficient IRQC detection can sometimes produce appreciable fluorescence. The successive transfer up-conversion mechanism is also responsible for the nonlinear quenching that is observed in highly excited systems as it provides an additional depopulation mechanism for a fluorescent state [4.6, 7, 19].

The three transition schemes presented above are all methods of achieving up-conversion of a lower energy photon to a higher energy. They are very similar to the more traditional down-conversion processes of fluorescence sensitization, quenching, or cross-relaxation. In many cases, they are just the reverse process. A fluorescence quenching process by ion pair decay is shown in Fig. 4.1d. Fluorescence sensitization is identical to the quenching except the ions involved are different (although not necessarily) and the final state is fluorescent [4.20]. The quenching or sensitization can occur directly between ions or an excitation can migrate through the lattice as an exciton (see Fig. 4.1e) before the energy is transferred [4.21, 22].

A very different kind of energy transfer has been observed in doubly doped Yb^{3+}, Tb^{3+} systems [4.23–26]. Two excited donor ions can simultaneously transfer their energy to an acceptor ion, as shown in Fig. 4.1f. The inverse of this process should also occur although it has not been definitively identified experimentally. This cooperative energy transfer results in up-conversion of energy but it is a three-ion process requiring coupling between three ions. Our previous examples required only two-ion interactions. It is the first example of a three-body process in rare-earth doped crystals.

The interaction between ions cause shifts and splittings of individual energy levels as well as causing transitions between them. These effects have been seen as Davydov splittings or dispersive line shapes in concentrated crystals where exciton migration is efficient [4.27, 28] or as anti-resonance in the absorption lineshapes when a continuum of excited states is available [4.29–32].

Since the interaction between ions causes a coupling of energy levels, cooperative radiative transitions are allowed where a single photon can be absorbed or emitted by two ions, as shown in Fig. 4.1g. Cooperative absorption was first found in 1961 by VARSANYI and DIEKE [4.33] and later by many other workers [4.34–36]. The inverse process of cooperative emission was identified by NAKAZAWA and SHINOYA in 1970 [4.37]. A very similar cooperative transition has been studied in a number of systems by VAN DER ZIEL and co-workers [4.38–43]. In this case, an excited ion can emit a photon whose energy is the difference between that of the initial excited level and a level on a neighboring ion. This is shown in Fig. 4.1h.

ALTARELLI and DEXTER [4.44] predicted that a combination of energy transfer and photon absorption should also be observable. An excited donor ion can transfer its energy to an acceptor at the same moment that the acceptor is absorbing a photon. Then an acceptor excitation is created that is equal to the sum of the donor and photon energies. The process is sketched in Fig. 4.1i. It should be noted that it is simply the reverse process of that in Fig. 1.1h. BILAK et al. [4.45] suggest this mechanism to explain a rapid temporal build-up of Tb^{3+} green fluorescence when a Yb^{3+} donor ion is excited at $1.06\,\mu m$. Their experiment, however, was not designed to establish the mechanism and cannot be taken as experimental evidence for the process.

4.2 Ion-Pair Interactions

Inter-ionic interactions produce both static and dynamic effects. The relationships between the diversity of processes can be seen clearly by a description similar to one presented by MELTZER and MOOS [4.28]. The Hamiltonian for the entire crystal is divided into terms which describe the single ions and terms which couple two ions

$$\mathscr{H} = H_0 + \sum_{i>j} H_{ij} \tag{4.1}$$

where the coupling between the m-th electron on the i-th ion and the n-th electron on the j-th ion is written as

$$H_{i,j} = \sum_{m>n} \frac{e^2}{\kappa |r_m - r_n|} \tag{4.2}$$

where κ is the dielectric constant. The eigenfunctions for the single-ion portion of the Hamiltonian are a product of the single-ion eigenstates. We will be primarily concerned with excitations of the entire system. If the ions at r_i and

r_j are in excited states α_r and α_s, respectively, the wavefunction of the system will be designated as

$$|\Psi(r_i,\alpha_r;r_j,\alpha_s)\rangle = |\Psi_{\alpha_0}(r_1),\Psi_{\alpha_0}(r_2)\dots\Psi_{\alpha_r}(r_i)\Psi_{\alpha_s}(r_j)\dots\rangle \qquad (4.3)$$

where $\Psi_{\alpha_0}(r)$ denotes a ground-state wavefunction. Matrix elements will then have the form $\langle\Psi(r_i,\alpha_r;r_j,\alpha_s)|H_{ij}|\Psi(r_i,\alpha_r';r_j,\alpha_s')\rangle$. These matrix elements will have both direct and exchange parts. Consider first of all the direct part of those matrix elements for which $\alpha_s=\alpha_s'=\alpha_0$ (i.e., the ion at r_j is in the ground state). These matrix elements describe the crystal field contribution from the neighboring ion. The exchange part of those matrix elements determines the exchange splittings [4.46, 47] that can lead to magnetic ordering in rare earth materials [4.48]. Similarly, the matrix elements in which $\alpha_s=\alpha_s'$ determine the crystal field contribution and exchange splittings when the neighbor at r_j is in the α_s excited state [4.49]. The matrix elements for which $\alpha_r\neq\alpha_r'$ and $\alpha_s\neq\alpha_s'$ determine the transfer of excitation between ions. Again, there are direct and exchange parts for these matrix elements. The direct part represents the contribution to energy transfer by electric multipole interactions while the exchange part represents the contribution from electron exchange. Many of the different relaxation processes shown in Fig. 4.1 arise as special cases from matrix elements of this form. Some examples of matrix elements are given in Table 4.1. The transition rates can be obtained from the Fermi Golden Rule [4.50]

$$W = \frac{2\pi}{\hbar}|\langle\Psi|H_{ij}|\Psi'\rangle|^2\rho. \qquad (4.4)$$

Cooperative radiative transitions in which a pair of ions are simultaneously excited or de-excited by the radiation field can arise because of the mixing of wavefunctions by the inter-ion perturbation [4.39, 51, 52]. The perturbed wavefunction can be written as

$$|\Psi(r_i,\alpha_r;r_j,\alpha_s)\rangle_{\text{Pert.}} = |\Psi(r_i,\alpha_r;r_j,\alpha_s)\rangle - \sum_{\alpha_r'',\alpha_s''}{}' |\Psi(r_i,\alpha_r'';r_j,\alpha_s'')\rangle$$
$$\times\frac{\langle\Psi(r_i,\alpha_r'';r_j,\alpha_s'')|H_{ij}|\Psi(r_i,\alpha_r;r_j,\alpha_s)\rangle}{(E_{\alpha_r''}-E_{\alpha_r}+E_{\alpha_s''}-E_{\alpha_s})}$$

where the prime on the summation indicates the terms $\alpha_r''=\alpha_r$ and $\alpha_s''=\alpha_s$ are omitted. Optical transitions for the coupled system are now determined by matrix elements such as

$$\langle\Psi(r_i,\alpha_r;r_j\alpha_s)|e_0(R_i+R_j)|\Psi(r_i,\alpha_r';r_j\alpha_s')\rangle \qquad (4.5)$$

Table 4.1

Ion pair relaxation

$$\langle \Psi(r_i,\alpha_2; r_j,\alpha_1) |H_{ij}| \Psi(r_i,\alpha_3; r_j,\alpha_0)\rangle$$

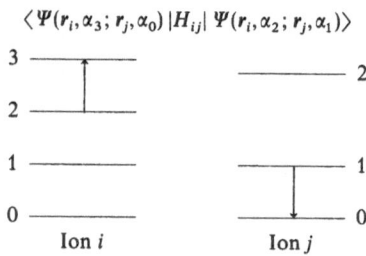

Ion *i*   Ion *j*

Sequential sensitization of up-conversion

$$\langle \Psi(r_i,\alpha_3; r_j,\alpha_0) |H_{ij}| \Psi(r_i,\alpha_2; r_j,\alpha_1)\rangle$$

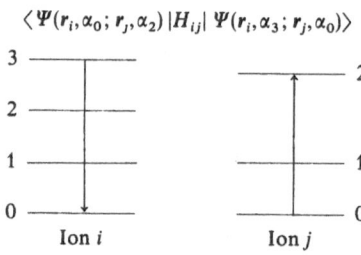

Ion *i*   Ion *j*

Fluorescence sensitization

$$\langle \Psi(r_i,\alpha_0; r_j,\alpha_2) |H_{ij}| \Psi(r_i,\alpha_3; r_j,\alpha_0)\rangle$$

Ion *i*   Ion *j*

Exciton migration

$$\langle \Psi(r_i,\alpha_0; r_j,\alpha_3) |H_{ij}| \Psi(r_i,\alpha_3; r_j,\alpha_0)\rangle$$

Ion *i*   Ion *j*

---

where $R_i=\sum_i r_i$ and $R_j=\sum_j r_j$. The matrix element shown represents a transition from an initial state with ion *i* in the α_r state and ion *j* in the α_s state to the final state with ion *i* in the α_r' state and ion *j* in the α_s' state. The different processes that have been observed are summarized in Table 4.2 along with the matrix elements that characterize them.

Table 4.2

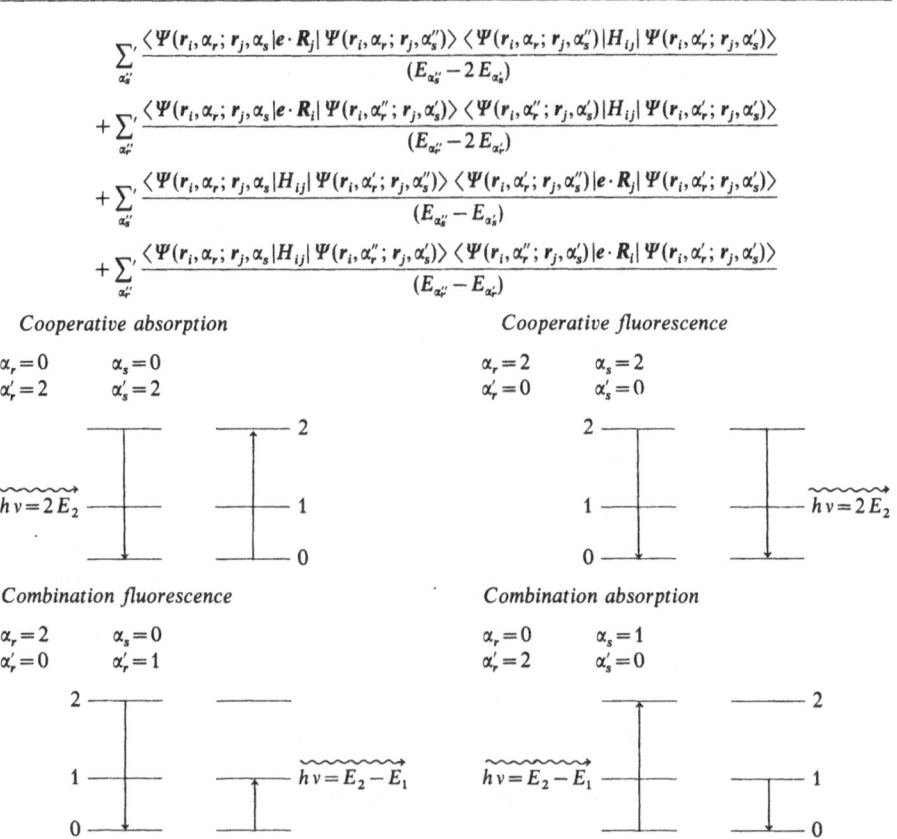

$$\sum_{\alpha_s''}' \frac{\langle \Psi(r_i,\alpha_r; r_j,\alpha_s | e\cdot R_j | \Psi(r_i,\alpha_r; r_j,\alpha_s'')\rangle \langle \Psi(r_i,\alpha_r; r_j,\alpha_s'')|H_{ij}| \Psi(r_i,\alpha_r'; r_j,\alpha_s')\rangle}{(E_{\alpha_s''} - 2E_{\alpha_s'})}$$

$$+\sum_{\alpha_r''}' \frac{\langle \Psi(r_i,\alpha_r; r_j,\alpha_s | e\cdot R_i | \Psi(r_i,\alpha_r''; r_j,\alpha_s')\rangle \langle \Psi(r_i,\alpha_r''; r_j,\alpha_s')|H_{ij}| \Psi(r_i,\alpha_r'; r_j,\alpha_s')\rangle}{(E_{\alpha_r''} - 2E_{\alpha_r'})}$$

$$+\sum_{\alpha_s''}' \frac{\langle \Psi(r_i,\alpha_r; r_j,\alpha_s |H_{ij}| \Psi(r_i,\alpha_r'; r_j,\alpha_s'')\rangle \langle \Psi(r_i,\alpha_r'; r_j,\alpha_s'')|e\cdot R_j | \Psi(r_i,\alpha_r'; r_j,\alpha_s')\rangle}{(E_{\alpha_s''} - E_{\alpha_s'})}$$

$$+\sum_{\alpha_r''}' \frac{\langle \Psi(r_i,\alpha_r; r_j,\alpha_s |H_{ij}| \Psi(r_i,\alpha_r''; r_j,\alpha_s')\rangle \langle \Psi(r_i,\alpha_r''; r_j,\alpha_s')|e\cdot R_i | \Psi(r_i,\alpha_r'; r_j,\alpha_s')\rangle}{(E_{\alpha_r''} - E_{\alpha_r'})}$$

Cooperative absorption   *Cooperative fluorescence*

$\alpha_r = 0$ $\alpha_s = 0$ $\alpha_r = 2$ $\alpha_s = 2$
$\alpha_r' = 2$ $\alpha_s' = 2$ $\alpha_r' = 0$ $\alpha_s' = 0$

$h\nu = 2E_2$ $h\nu = 2E_2$

Combination fluorescence   *Combination absorption*

$\alpha_r = 2$ $\alpha_s = 0$ $\alpha_r = 0$ $\alpha_s = 1$
$\alpha_r' = 0$ $\alpha_s' = 1$ $\alpha_r' = 2$ $\alpha_s' = 0$

$h\nu = E_2 - E_1$ $h\nu = E_2 - E_1$

Interactions between three ions can occur by higher-order terms in the perturbation. Multi-ion interaction has been treated theoretically by several authors [4.53–56]. The coupling between three ions, i, j, and k comes about because the wavefunction for ion j will have some character from ion k mixed with it and this character will be imparted to ion i by the interaction between ions i and j. The perturbation method can be extended to an arbitrary number of interacting particles and results in many-body processes. Such processes have been observed as the cooperative sensitization of up-converted fluorescence [4.25, 26] and have been proposed as explanations for the concentration dependence of ion pair relaxation [4.54].

The electric multiple interactions that couple two ions in the direct part of those matrix elements involved in energy transfer have been treated in detail by DEXTER [4.57]. Earlier, FÖRSTER treated the role of the electric dipole-dipole interaction in energy transfer [4.58, 59].

The Hamiltonian that express the near-zone electric field is conveniently expanded in a series of tensor operators [4.60]

$$H = \sum_{m>n} \frac{e^2}{\kappa |r_m - r_n|}$$

$$= \sum_{\substack{k_1 k_2 \\ q_1 q_2}} \left(\frac{e^2}{R^{k_1+k_2+1}} \right) C_{q_1 q_2}^{k_1 k_2} D_{q_1}^{(k_1)}(r_i) D_{q_2}^{(k_2)}(r_j) \tag{4.6}$$

where

$$D_q^{(k)} = \sum_m \left(\frac{4\pi}{2k+1} \right)^{1/2} r_m^k Y_q^{(k)}(\theta_m, \phi_m),$$

$C_{q_1 q_2}^{k_1 k_2}$ is a coefficient which depends upon the relative orientation of the two ions and the indices of summation, and R is the distance between the ions. For $k_1 = k_2 = 1$, the expression reduces to the familiar dipole-dipole interaction; for $k_1 = 1$ and $k_2 = 2$ it reduces to a dipole-quadrupole interaction; and for $k_1 = 2$ and $k_2 = 2$ it reduces to a quadrupole-quadrupole interaction. Although interactions higher than dipole-dipole are considered negligible in radiative processes, DEXTER and SCHULMAN [4.22] have shown that they can be more important in rare-earth doped materials as one enters the near-field region about an ion (estimated to be within 9 nm.). When this Hamiltonian is used to evaluate appropriate matrix elements in the Golden Rule expression, it is found that the energy transfer rate will vary as $R^{-\theta}$ where $\theta = 6$ for dipole-dipole interaction, $\theta = 8$ for dipole-quadrupole, and $\theta = 10$ for quadrupole-quadrupole. Higher-order multipole contributions have been shown to be negligible.

The exchange part of the matrix element is quite complex for the rare earths. They differ from transition metal ions because they have a non-spherical charge distribution arising from large spin-orbit interactions and smaller crystal field interactions [4.61]. The non-spherical charge distribution results in large departures from the simple isotropic exchange forces that are usually employed with the simple Heisenberg-Dirac-Van Vleck-Néel exchange Hamiltonian $H_{ij} = J S_i \cdot S_j$ in describing magnetic interactions. It has been shown that the complete form of the exchange interaction is highly anisotropic and can be described by the tensor operator formalism [4.61, 62]. The anisotropic nature of the interaction results in a strong angular dependence of the two interacting ions and makes any estimates of the radial interaction very difficult. The anisotropic nature of exchange in rare-earth materials has been well documented in a number of different studies and there is little doubt of its importance [4.60, 63]. The relative importance of exchange and electric multipole processes in causing energy transfer is a matter of considerable debate [4.64–66] and depends primarily upon an accurate experimental determination of the range over which energy transfer can occur. This point will be discussed in more depth in a later section but it is expected that exchange interactions will be of comparable or greater importance than electric multipole interactions for small separations between ions.

If one assumes that one of the electric multipole interaction terms is of primary importance, values for the matrix elements and density of states required in the Golden Rule expression (4.4) can be determined from experimental measurements by a method developed by FÖRSTER [4.58, 59, 67, 68] and DEXTER [4.50, 57, 69]. They have shown, for example, that the transition probability for an electric dipole process in which a donor ion (d) transfers energy to an acceptor (a) can be written in terms of the fluorescence line-shape of the donor (f_d) and the absorption line-shape of the acceptor (F_a)

$$P_{da}(\text{dipole-dipole}) = \frac{3\hbar^4 c^4 Q_a}{4\pi R^6 n^4 \tau_d} \left(\frac{\varepsilon}{\sqrt{\kappa \varepsilon_c}}\right)^4 \int \frac{f_d(E) F_a(E)}{E^4} dE. \tag{4.7}$$

In this expression, Q_a is the integrated absorption cross-section, τ_d is the radiative lifetime of the donor fluorescence, and $(\varepsilon/\sqrt{\kappa \varepsilon_c})^4$ is a factor to correct for the size of the electric field in the crystal, ε_c, relative to the vacuum. It is commonly taken to be unity. The overlap between the donor emission band and the acceptor absorption band is a consequence of the energy conservation requirement.

If there is not appreciable overlap, the transition probability should vanish according to (4.7). However, it has been observed in numerous experimental studies that energy transfer can still be an important process without appreciable overlap. The non-resonant processes require emission or absorption of phonons or photons to conserve energy. Non-resonance processes have been treated by a number of authors [4.65, 70, 71]. MIYAKAWA and DEXTER [4.70] derived an expression for the phonon-assisted energy transfer probabilities in the adiabatic approximation and predicted that the transfer probability should vary exponentially with the number of phonons required in the same way as the multi-phonon relaxation rates described in a separate section of this book. This prediction has been recently confirmed experimentally for large mismatches in energy [4.72]. The dependence for smaller values of non-resonance has not been studied and would be very interesting. On the basis of the theory, one would expect that the transfer rates would be sensitive to the density of states and orbit-lattice coupling parameters of the particular phonon modes involved when the mismatches are less than the phonon cut-off energy. This would account for the observation by IMBUSCH [4.66] that transitions requiring a single, higher-energy optical phonon which is strongly coupled with the ion are more probable than transitions requiring a single acoustical phonon which is coupled more weakly.

4.3 Energy Transfer

4.3.1 Theoretical Treatment of Energy Transfer

Experiments in energy transfer generally do not measure the transfer between isolated pairs of ions. Instead, an experiment represents an ensemble-averaged measurement over all possible pair separations. The fundamental processes that produce the observed behavior must be extrapolated backwards from the ensemble averaged data using least squares fitting procedures to theoretical predictions [4.55, 67, 68]. There are a number of considerations that must be treated in describing such systems. First, the electronic levels involved in the energy transfer processes should be either identified or known to be unique. Secondly, the number of ions involved in the energy transfer should also be known. Most experiments have been interpreted by assuming only pairwise interactions but since it has been shown that three-ion interactions can occur, they must also be considered, at least for non-resonant processes. Thirdly, for a general system there are three types of energy transfer that can occur: donor-donor, acceptor-acceptor, or donor-acceptor transfer. The first two transfers represent exciton migration or excitation diffusion through a crystal while the last is a direct transfer. In order to understand energy transfer, we will first examine the general ensemble averaging in the presence of these various processes and try to understand the assumptions that must be made obtaining a theoretical model. Then, we will examine the theoretical models that describe systems with varying degrees of exciton diffusion and direct transfer.

GRANT [4.55] has derived a general expression from first principles for the kinetics of energy transfer in order to show explicitly the assumptions that are inherent in previous theoretical models. If $P_d(k, \alpha_k)$ expresses the probability of a donor ion located at r_k being in state α_k (similarly, $P_a(j, \beta_j)$ represents the acceptor) the rate of change of $P_d(k, \alpha_k)$ can be written in two parts

$$
\begin{aligned}
\dot{P}_d(k, \alpha_k) = & \sum_j \sum_{\bar{\beta}_j}{}' \sum_{\beta_j} \sum_{\bar{\alpha}_k}{}' \omega_{da}(|r_j - r_k|, \bar{\beta}_j, \bar{\alpha}_k, \beta_j, \alpha_k) \\
& \times [P_d(k, \bar{\alpha}_k) P_a(j, \bar{\beta}_j) - P_d(k, \alpha_k) P_a(j, \beta_j)] \\
& + \sum_i \sum_{\bar{\alpha}_i} \sum_{\alpha_i} \sum_{\bar{\alpha}_k} \omega_{dd}(|r_i - r_k|, \bar{\alpha}_i, \bar{\alpha}_k, \alpha_i, \alpha_k) \\
& \times [P_d(i, \bar{\alpha}_i) P_d(k, \bar{\alpha}_k) - P_d(i, \alpha_i) P_d(k, \alpha_k)] .
\end{aligned}
\tag{4.8}
$$

The first part describes the donor-acceptor transfer while the second part expresses exciton migration among donor ions. The donor ion at r_k initially in state α_k can undergo a transition by transfering its energy to any other donor ion in state α_i resulting in the final states $\bar{\alpha}_k$ and $\bar{\alpha}_i$, respectively, or it can transfer its energy to any acceptor ion labeled as r_j in state β_j resulting in the final states $\bar{\alpha}_k$ and $\bar{\beta}_j$, respectively. The total rate of change of $\dot{P}_d(k, \alpha_k)$ is the sum of the transition probabilities over all possible ions that can be involved and all possible states. The sum is considerably more limited than (4.8) implies because many transition rates will be zero from energy conservation considerations. GRANT then considered the different ways to perform the configuration averaging over all possible configurations. The principle difficulty arises in considering the probability that the site at r_d will be occupied by a donor ion in a dilute system. For the case in which the transition rates for exciton migration are much higher than the donor-acceptor transition rates and the time scale of measurement, the distribution of ions is best taken into account by enlarging the ensemble from which (4.8) was derived to include the probability that the site at r_d contains a donor ion in the state α_k as opposed to the previous case of saying that the donor ion located at r_k was in state α_k. One is now dealing with a joint probability that the site is occupied by the donor ion and it is in state α_k. Since exciton migration is assumed to be very fast, all sites become equivalent and the transition rate observed will follow the donor-acceptor transition rate. Since the transition rates are now independent of the position of the donor, one can define an averaged transition rate

$$\langle W(\bar{\beta}_j, \bar{\alpha}_k, \beta_j, \alpha_k) \rangle = \sum_{r_a} W(|r_d - r_a|, \bar{\beta}_j, \bar{\alpha}_k, \beta_j, \alpha_k) . \tag{4.9}$$

Direct Transfer

The same procedure cannot be applied, however, if the exciton migration rates are small compared with both the donor-acceptor rates and the measurement timescale. The donor excitation then cannot be uniformly distributed over all available sites. The probability of having an excited donor near an acceptor is less than having it more distant since the acceptor rapidly depletes the excited donor population about it. GRANT [4.55] has shown that a configuration average for (4.8) can be performed if probabilities $P(i, \alpha_i)$ are linearized for small displacements from equilibrium values (i.e., the system is not highly excited), the initially excited donor state is fully depleted by the energy transfer, the ground state population of the acceptor ions is not significantly depleted by quenching the excited donor state, and the reverse transfer from the acceptor back to the donor can be neglected. These conditions are required to specify the populations of the interacting states. With these assumptions, the basic equations used in all previous theoretical treatments of energy transfer can be derived. They can be stated and understood intuitively by considering a single donor-acceptor pair separated by a distance R. The total rate of transfer from the excited donor level of this single pair is

$$W_T(R) = \frac{1}{\tau_0} + \frac{C}{f(R)}, \tag{4.10}$$

where τ_0 is the intrinsic lifetime of the excited donor level, and $C/f(R)$ is a function describing the radial dependence of the transition probability (assuming there is no angular dependence). The decay of donor fluorescence from this pair will be an exponential with a time constant of $1/W_T(R)$. If there are N_a total acceptor ions in a volume V around the donor, the observed time dependence will be a multi-exponential because of contributions from all donor-acceptor pairs with different values of R. If $\rho(R)$ expresses the probability distribution of different radial separations, one can write the ensemble averaged time dependence resulting from pulsed excitation as

$$\phi(t) = \exp\left(-\frac{t}{\tau_0}\right) \lim_{\substack{N_a \to \infty \\ V \to \infty}} \left[\int_V e^{-tC/f(R)} \rho(R) dV\right]^{N_a} \tag{4.11}$$

where the limits are taken in such a way that the ratio N_a/V becomes the acceptor concentration c_a. If the experiment is performed by watching the decay after the system has reached a steady state under constant excitation instead of after a pulsed excitation, the time dependence $\phi_s(t)$ observed [4.73] is related to $\phi(t)$ by

$$\phi_s(t) = \frac{\int\limits_t^\infty \phi(t')dt'}{\int\limits_0^\infty \phi(t')dt'} . \tag{4.12}$$

Eq. (4.11) has been evaluated for several different radial dependences assuming a random distribution of acceptor ions. If the energy transfer is governed by a particular electric multipole term, the interaction can be written in the form CR^{-s} where $s=6$ for the dipole-dipole, 8 for the dipole-quadrupole, and 10 for the quadrupole-quadrupole term. For this case, it has been found that [4.74]

$$\phi(t) = \exp\left[-\frac{t}{\tau_0} - \frac{4\pi}{3} c_a \Gamma\left(1 - \frac{3}{s}\right)(Ct)^{3/s}\right] \tag{4.13}$$

where $\Gamma(x)$ is the gamma function. This expression is also written in terms of a "critical transfer distance" R_0 by replacing $C^{3/s}$ by $(R_0^3/\tau_0^{3/s})$. The critical transfer distance is defined as the distance required to make the energy transfer rate equal to the intrinsic decay rate. In addition a "critical concentration" c_{a0} is also in common useage to specify the interaction strength. It enters by replacing $(4\pi/3)C^{3/s}$ by $(c_{a0}\tau_0^{3/s})^{-1}$. Eq. (4.13) is plotted in Fig. 4.2 for various parameters of c_a, and s. The non-exponential character of the decay is most evident at short times and high concentrations. The closest donor-acceptor pairs rapidly undergo energy transfer initially and therefore their population becomes depleted early in the decay. The depletion continues radially outward from the donor ions as time progresses until the only pairs remaining are sufficiently separated that energy transfer does not occur. The decay has then resumed the intrinsic exponential decay.

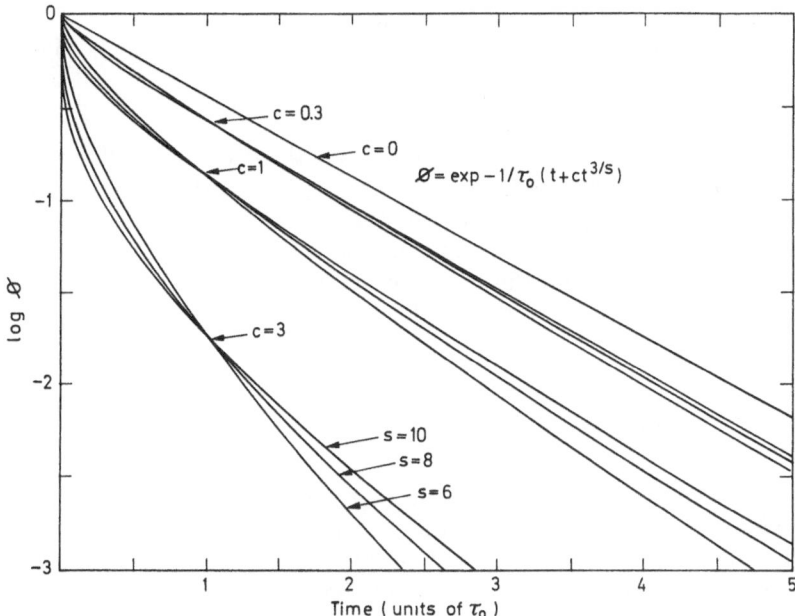

Fig. 4.2. Decay curves of a level with varying values of direct energy transfer to intrinsic decay and different coupling mechanisms. The intrinsic lifetime is τ_0, s represents the electric multipole order and r is the ratio of direct transfer rate to the intrinsic decay rate

If the energy transfer occurs by the exchange interaction, the radial dependence of the interaction is typically described by the expression $1/\tau_0 \exp[\gamma(1 - R/R_{a0})]$ where $\gamma = 2R_{a0}/L$, L is a distance characterizing the spatial extent of an ionic wavefunction, and R_{a0} is the critical transfer distance [4.74]. All angular variations are ignored. With this radial dependence, (4.11) becomes [4.74]

$$\phi(t) = \exp\left[-\frac{t}{\tau_0} - \frac{\pi L^3 c_a}{6} g\left(\frac{e^{\gamma} t}{\tau_0}\right) \right] \tag{4.14}$$

where $g(z)$ is defined by $g(z) = -z \int_0^1 \exp(-zy)(\ln y)^3 dy$.

The Perrin model and the Stern-Volmer model use two other radial dependences that have been used to evaluate (4.11) [4.74]. Although these models may be over-simplifications of reality, they are nevertheless instructive.

The Perrin model states that there is no energy transfer between ions once they are separated by a distance greater than R_0. Within that separation, energy transfer occurs immediately. For this model,

$$\phi(t) = \exp\left(-\frac{t}{\tau_0} - \frac{4\pi R_0^3}{3} c_a\right) \quad \text{if} \quad t > 0,$$

$$= 1 \qquad\qquad\qquad \text{if} \quad t = 0. \tag{4.15}$$

It predicts an immediate drop in population after the excitation pulse as the ions within the critical radius relax followed by an intrinsic exponential decay of all the ions remaining. The Stern-Volmer model represents a simple kinetic picture in which the interaction has no radial dependence but is a constant, κ. In solving (4.11) for this case, it is necessary to let $\kappa \to 0$ as $N_a \to \infty$ such that $N_a \kappa$ is a constant. For this model, (4.11) becomes

$$\phi(t) = \exp\left(-\frac{t}{\tau_0} - \frac{c_a}{c_{a0}} \frac{t}{\tau_0}\right), \tag{4.16}$$

where c_{a0} is a reference concentration which specifies the interaction strength. In this model, there is no depletion of the closest pairs initially because there is no radial dependence. The decay is a simple exponential. The Perrin and Stern-Volmer models are the limiting cases of energy transfer models.

Many experiments have been performed to obtain the interaction mechanism by measuring the radiative quantum efficiency of donor fluorescence.

The quantum efficiency under pulsed excitation can be obtained from an arbitrary time dependence by performing the following integration

$$\frac{\eta_d}{\eta_{d0}} = \frac{\int_0^\infty \phi(t)\,dt}{\int_0^\infty \phi_0(t)\,dt} = \frac{1}{\tau_0}\int_0^\infty \phi(t)\,dt, \tag{4.17}$$

where $\phi_0(t)$ is the intrinsic time decay without energy transfer present. This expression can be easily evaluated for the Stern-Volmer model giving

$$\frac{\eta_d}{\eta_{d0}} = \frac{1}{1 + c_a/c_{a0}}. \tag{4.18}$$

The quantum efficiency predicted by the multipole and exchange interaction models is generally evaluated from (4.17) by numerical integration [4.69]. If the acceptor concentration is sufficiently small that the energy transfer rate is much lower than the intrinsic decay rate, all of the models can be simply evaluated. Under this approximation, the models reduce to a form identical to (4.18) except for a constant. This is expected at low concentrations because a correct model based on a simple pair interaction between donor and acceptor must depend linear upon their concentrations by the law of mass action. The model must therefore reduce to a form similar to (4.18) in the limit of low concentration [4.54]. However, there has been considerable experimental work in fluorescence quenching which utilized the following expression to obtain the form of the interaction

$$\frac{\eta_d}{\eta_{d0}} = \frac{1}{1 + \beta c_a^{\theta/3}} \tag{4.19}$$

where θ was taken as 6, 8, or 10 for electric dipole-dipole, dipole-quadrupole, or quadrupole-quadrupole interactions, respectively [4.75–84]. This expression is inconsistent with the low concentration expansion form of the quantum efficiency. GRANT [4.55] has traced this inconsistency to the ensemble averaging procedure used to obtain (4.19). If the ensemble average is restricted to the closest available acceptor ion, an additional concentration dependence is artifically introduced into the average transfer rate which causes the deviation of (4.19) from the simple limiting dependence expected.

Diffusion Limited Transfer

The donor fluorescence time dependence described by any model of the interaction is controlled by the more rapid depletion of the closely spaced pairs. The radial dependence of the interaction determines the relative rates. If excitation diffusion can occur, the gradient associated with the depletion will be lowered until it reaches zero in the limit of very rapid diffusion. Diffusion will therefore cause departures from the behavior described by (4.13–16) [4.84–86]. The energy transfer between a donor ion and any of the acceptor ions in the presence of diffusion is described by the equation

$$-\frac{\partial \psi_d(r,t)}{\partial t} = -D_d \nabla^2 \psi_d(r,t) + \frac{1}{\tau_0} \psi_d(r,t) + \sum_i V(|r - R_{ia}(t)|) \psi_d(r,t) \tag{4.20}$$

where $\psi_d(r,t)$ is the donor distribution function, D_d is the donor diffusion constant, τ_0 is the intrinsic lifetime, and $v(|r - R_{ia}(t)|)$ is the interaction of a donor ion with the i-th acceptor ion. YOKOTA and TANIMOTO [4.87] have solved this expression approximately for the dipole-dipole interaction, C/R_{ia}^6, using the method of Padé approximants and found the expression

$$\phi_d(t) = e^{-t/\tau_0} \exp\left[-\frac{4}{3}\pi^{3/2} c_a (Ct)^{1/2} \left(\frac{1 + 10.87x + 15.50x^2}{1 + 8.743x} \right)^{3/4} \right] \tag{4.21}$$

where $x = DC^{-1/3}t^{2/3}$. If diffusion is negligible, $x = 0$ and this expression reduces to (4.13). The asymptotic behavior of this expression at large t or $x \gg 1$ is a simple exponential described by

$$\phi_d(t) = e^{-\left(\frac{1}{\tau_0} + 11.404 c_a D^{3/4} C^{1/4}\right)t}.$$

(4.22)

In this region, the directly relaxing pairs have been depleted and relaxation depends upon the diffusion of additional excitation into the spatial region where relaxation occurs. The expression is different from that predicted directly from (4.20) under the same conditions. The correct behavior for these conditions is

$$\phi_d(t) = e^{-\left(\frac{1}{\tau_0} + 8.54 c_a D^{3/4} C^{1/4}\right)t}.$$

(4.23)

The difference between (4.22) and (4.23) arises from approximations needed in deriving (4.21) [4.88].

Rapid Diffusion

If the diffusion constant becomes very large compared with the direct relaxation, all spatial inhomogeneities are removed and the rate limiting step of direct transfer will determine the systems' time dependence. The situation was described earlier in the discussion associated with (4.9). Experimentally, the limit of fast diffusion is observed as a saturation of the diffusion constant at higher concentration.

4.3.2 Experiments in Energy Transfer

There has been an enormous of experimental work performed on energy transfer between rare-earth ions. [4.55, 89–104]. The majority of the work has not been concerned with elucidating the fundamental processes that governed the transfer but with simply demonstrating the presence of energy transfer or evaluating the importance of energy transfer in optimizing particular applications.

The earliest work attempted to determine fundamental information on the interaction mechanism by measuring the concentration dependence of the fluorescence quantum efficiency but there were many problems that were not considered [4.75]. Since the energy levels involved in the transfer steps were not determined experimentally, the number of ions involved in the transfer was unknown. Many of the studies involved host lattices requiring charge-compensation which destroys the random arrangement of rare-earth ions because of dipolar interactions between rare-earth compensation pairs [4.105, 106]. The quantum efficiency concentration dependence was interpreted using (4.19) which has been shown by several authors to be incorrect, see previous discussion of (4.19) [4.54, 55]. In addition, the effects of donor diffusion were not considered in the analysis although the concentrations were large enough that diffusion would be very important [4.107, 108]. Finally when the data was fit, the value obtained for θ in (4.19) was consistently larger than the value of 3 expected for pairwise interactions and usually was 6 [4.75]. Both GRANT [4.55], and FONG and DIESTLER [4.54] have pointed out that this observation is inconsistent with a simple direct transfer between two ions but it would be consistent with a transfer involving one donor and two or more acceptors. This process requires a multi-body interaction. Such processes have been conclusively observed by up-conversion experiments in mixed $Yb^{3+}-Tb^{3+}$ system (see later discussion) [4.25, 26]. However, it should also be pointed out that the experimental data used to obtain values of θ include donor concentrations up to 10–15 mole % where donor excitation diffusion will certainly be important. The value determined for θ therefore cannot be considered reliable.

A number of experiments have been performed which do provide direct information on the phenomena of energy transfer. The experiments typically measure the donor fluorescence time dependence and quantum efficiency as a function of donor and acceptor concentration. The results are analyzed allowing for both direct relaxation and donor migration. Although several excellent experiments have appeared using this approach [4.88, 108–113], only one will be described in detail as an example.

KRASUTSKY and MOOS [4.108] have studied energy transfer in a mixed $Pr^{3+}-Nd^{3+}:LaCl_3$ system. The relevant energy levels are shown in Fig. 4.3. Fluorescence from the Pr^{3+} $^3F_3$ manifold was monitored after excitation to the $^3F_4$ manifold. Nd^{3+} could quench the $^3F_3$ fluorescence by one of the processes shown in the figure but the particular scheme involved in the quenching could not be determined. Pr^{3+} acted as the donor and Nd^{3+} acted as the acceptor. The time dependence of the $^3F_3$ fluorescence is shown in Fig. 4.4

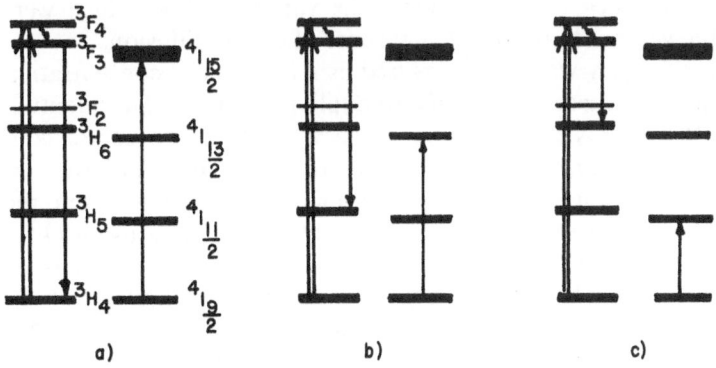

Fig. 4.3a–c. Pair decay processes in $LaCl_3:Nd^{3+}, Pr^{3+}$ [4.108]

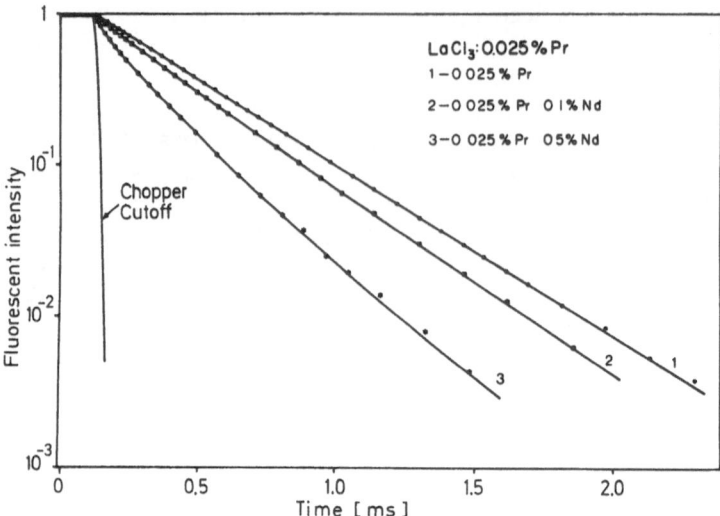

Fig. 4.4. Decay curves of the $^3F_3$ Pr^{3+} fluorescence for varying concentrations of Nd^{3+}. The solid curves are theoretical fits to the data for direct energy transfer. The agreement shows the absence of donor excitation migration [4.108]

for a donor concentration of 0.025 mole % and varying acceptor concentrations. The transient behavior was obtained after the system had reached a steady state equilibrium thus requiring application of (4.12). The solid curves are the theoretical fit to the steady-state equivalent of expression (4.13) for the dipole-dipole interaction. The excellent agreement shows that donor diffusion is not important at low donor concentrations and yields a value for C of 6×10^{-38} cm^6/s. The same expression could be used to fit the 0.25 mole % Pr^{3+} concentration time dependence. At concentrations of 2.0 mole %, however, the effects of donor diffusion become appreciable, and values of the diffusion constant and C had to be obtained by fitting (4.21) to the data (modified for the system having reached a steady state). The values of C obtained were consistent with the values obtained in the absence of donor diffusion.

The quantum efficiency was also measured as a function of concentration to obtain an independent measurement of the coupling constants. The quantum efficiency of the $^3F_3$ level is shown in Fig. 4.5 as a function of acceptor concentration for those donor concentrations where donor diffusion was insignificant. The data point scatter was attributed to variations from the nominal dopant concentrations. The theoretical fit to this data using (4.18) and assuming a dipole-dipole interaction with a $C = 6 \times 10^{-38}$ cm^6/s is shown as the solid line. The authors pointed out that although all of the experimental data could be fit well with the dipole-dipole interaction model, the other forms for the interaction predict results sufficiently close to each other that they cannot be distinguished within experimental error. The im-

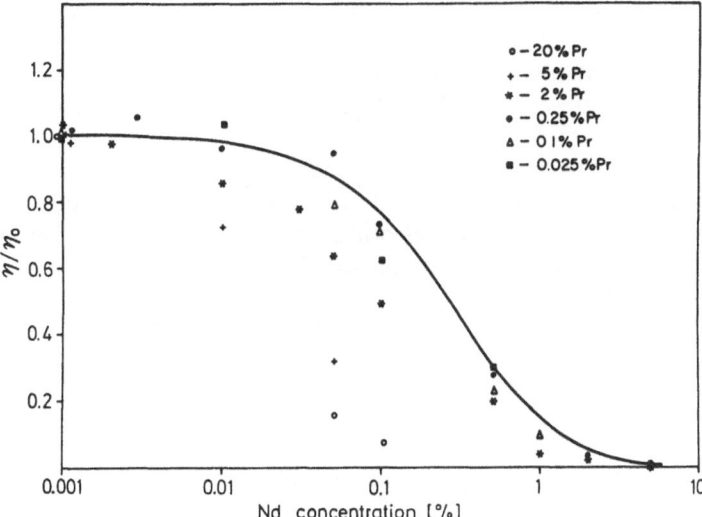

Fig. 4.5. Quantum efficiency of $^3F_3$ fluorescence as a function of the Nd^{3+} acceptor concentration. Only those Pr^{3+} concentrations were negligible Pr^{3+} exciton migration occurs are shown. The solid curve is a theoretical fit which assumes only direct transfer occurs between Pr^{3+} and Nd^{3+} ions [4.108]

portant contributions from work of this type are the elucidation of the relative importances of diffusion and direct transfer and the values of the diffusion constant and the dipole-dipole interaction constant. Once these are known, the system can be described by a model.

Other experimental work of this type has produced additional values for the coupling constants. The values obtained are summarized in Table 4.3. It should be noted from this table that one group of workers conclude the direct transfer in their system follows a dipole-quadrupole interaction [4.110].

The values obtained above raise several additional questions. It is clear that the particular values for the interaction coefficients can vary from one another by several orders of magnitude. It is important to understand the reasons for this variation and to predict how the variation depends upon system parameters. Recently, several experiments have been performed that begin to probe these questions. Three effects have been identified and will be discussed in turn: assistance of non-resonant transfer by phonons, nature of the electronic levels involved, and changing Boltzmann populations.

Most of the experiments in energy transfer have involved some degree of non-resonant transfer requiring a contribution from lattice phonons. The effects are of particular importance in amorphous materials where the equivalency of sites needed for resonant energy migration has been removed. Thus, non-resonant transfer has been used to explain the low diffusion coefficients observed in glass matrices [4.113]. Some authors have suggested that since the linewidths observed in most crystalline lattice hosts are inhomogeneously broadened and the real homogeneous linewidths cannot always be observed, truly resonant energy transfer may be a very rare event [4.66]. The first experimental work to show the dependence of non-resonance was performed by YAMADA et al. [4.72]. They examined the fluorescence time dependence in doubly doped samples of Y_2O_3 which had different degrees of non-resonance between the donor and acceptor rare earths. The donor and acceptor concentrations were held at 1 and 4 mole %, respectively. The time dependence was fit to (4.13) neglecting donor diffusion. They found that the energy transfer rate depended exponentially on the energy difference in the non-resonance, ΔE,

$$W = W_0 e^{-\beta \Delta E}, \tag{4.24}$$

Table 4.3.

a) Diffusion constants [cm^2/s]

   1) KRASUTSKY and MOOS [4.108]
                         LaCl$_3$ Host   4.2 K

			Donor-Concentration [Mole %]				
Acceptor	Nd	Pr	0.025	0.25	2.0	5.0	20
Concen-	0.01		$<10^{-11}$	$<10^{-11}$	1.3×10^{-10}	1.1×10^{-9}	—
tration	0.03		—	—	1.0×10^{-10}	—	—
	0.05		$<10^{-11}$	$<10^{-11}$	1.2×10^{-10}	1.4×10^{-9}	5×10^{-9}
	0.10		$<10^{-11}$	$<10^{-11}$	4.8×10^{-11}	—	6×10^{-9}
	0.50		$<10^{-11}$	$<10^{-11}$	1.5×10^{-11}	—	—
	1.0		—	—	8×10^{-11}	—	—

   2) WATTS and RICHTER [4.112]
                        YF$_3$ Host   300 K

	Yb^{3+} / Ho^{3+}	Donor-Yb^{3+}-Concentration [Mole %]				
		0.10	0.30	1.0	3.0	10.0
Acceptor Ho^{3+}	1.0	$<3 \times 10^{-14}$	2.7×10^{-14}	2.9×10^{-12}	1.1×10^{-11}	1.6×10^{-11}
Concen-	3.0	—	$<8 \times 10^{-14}$	1.1×10^{-12}	—	1.6×10^{-11}
tration	6.0	$<3 \times 10^{-14}$	$<8 \times 10^{-14}$	—	1.8×10^{-11}	2.2×10^{-11}
	10.0	—	—	—	—	—

   3) WEBER [4.113]
      Eu(PO$_3$)$_3$ glass
                   donor = Eu^{3+} − ~100% concentration
               acceptor = Cr^{3+} −0.05, 0.1, 0.3% concentrations
     D: 6×10^{-10}

   4) VAN DER ZIEL [4.88]
      Tb$_3$Al$_5$O$_{12}$ host
              donor = Tb^{3+} − ~100% concentration
              acceptor = Unintentional transition metal impurities ~0.05%
     D estimated as 0.58×10^{-10} at 4.2 K
                      3.76×10^{-10} at 77 K
                      12.5×10^{-10} at 297 K

Table 4.3 (continued)

b) Coupling constants in C/R^s (units of $[cm^6/s]$ for $s=6$ and $[cm^8/s]$ for $s=8$)

Donor	Level	Acceptor	s Value	C	Reference
Pr^{3+}	$^3F_3$	Nd^{3+}	6	6×10^{-38}	4.108
Pr^{3+}	$^3H_6$	Nd^{3+}	6	2×10^{-38}	4.108
Eu^{3+}	$^5D_0$	Cr^{3+}	6	6.3×10^{-38}	4.113
Tb^{3+}	$^5D_4$	? [a]	6 [b]	1.2×10^{-35} [c]	4.88
Yb^{3+}	$^2F_{5/2}$	Ho^{3+}	6	$(3.9 + 0.43) \times 10^{-41}$ [d]	4.112
Tb^{3+}	$^5D_4$	Dy^{3+}	8	$(0.37–0.08) \times 10^{-55}$ [d]	4.110
Tb^{3+}	$^5D_4$	Eu^{3+}	8	$(2.2–1.2) \times 10^{-55}$ [d]	4.110
Tb^{3+}	$^5D_4$	Sm^{3+}	8	$(2.5–1.8) \times 10^{-55}$ [d]	4.110
Tb^{3+}	$^5D_4$	Tm^{3+}	8	$(4.3–3.5) \times 10^{-55}$ [d]	4.110
Tb^{3+}	$^5D_4$	Pr^{3+}	8	$(29–26) \times 10^{-55}$ [d]	4.110
Tb^{3+}	$^5D_5$	Er^{3+}	8	$(80–55) \times 10^{-55}$ [d]	4.110
Tb^{3+}	$^5D_4$	Ho^{3+}	8	$(140–110) \times 10^{-55}$ [d]	4.110
Tb^{3+}	$^5D_4$	Nd^{3+}	8	$(240–230) \times 10^{-55}$ [d]	4.110
Eu^{3+}	$^5D_0$	Sm^{3+}	8	0.016×10^{-55}	4.110
Eu^{3+}	$^5D_0$	Yb^{3+}	8	$(0.21–0.006) \times 10^{-55}$ [d]	4.110
Eu^{3+}	$^5D_0$	Dy^{3+}	8	$(3.3–1.3) \times 10^{-55}$ [d]	4.110
Eu^{3+}	$^5D_0$	Ho^{3+}	8	$(5.0–10) \times 10^{-55}$ [d]	4.110
Eu^{3+}	$^5D_0$	Er^{3+}	8	$(12.4–6.9) \times 10^{-55}$ [d]	4.110
Eu^{3+}	$^5D_0$	Pr^{3+}	8	$(19–7.6) \times 10^{-55}$ [d]	4.110
Eu^{3+}	$^5D_0$	Tm^{3+}	8	$(27–16) \times 10^{-55}$ [d]	4.110
Eu^{3+}	$^5D_0$	Nd^{3+}	8	$(330–140) \times 10^{-55}$ [d]	4.110

[a] Acceptors were unintentional impurities.
[b] Electric dipole mechanism assumed.
[c] Estimated.
[d] Range of values reported.

where β is a parameter characterizing the electron-phonon interaction strength. This relationship had been predicted earlier by MIYAKAWA and DEXTER [4.70]. The experimental results obtained are shown in Fig. 4.6. The result is very similar to that found for multi-phonon relaxation. The details of the electronic interaction become averaged out and the transition rate depends primarily upon the host lattice effective phonon energy and the number of phonons required for energy conservation. The number of phonons involved in a given

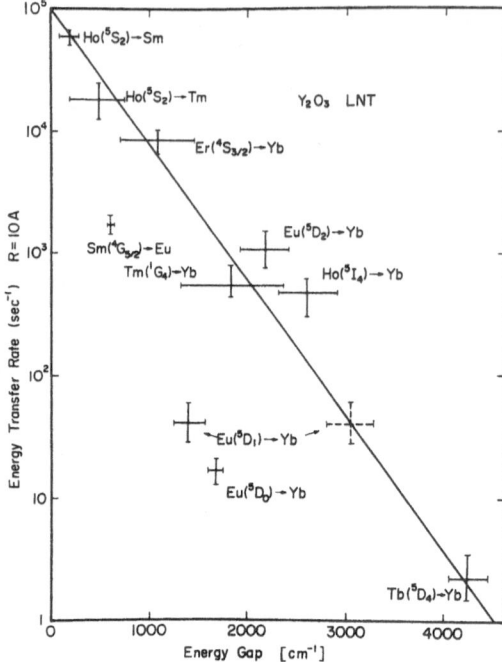

Fig. 4.6. Direct energy transfer rate between various rare earth ions as a function of the energy gap mismatch between levels that are involved. This figure shows how the rate varies for non-resonant energy transfer [4.72]

transition can be obtained from the temperature dependence of the energy transfer rate since the temperature dependence is determined by the phonon population. If the number of phonons associated with a particular energy mismatch is known, the effective energy of each phonon is determined. In Y_2O_3, they find the effective phonon energy to be 420 cm^{-1}. Studies such as these will allow one to estimate energy transfer rates as the parameters appropriate to different crystal classes become available.

The particular nature of the electronic levels involved does not have much influence upon the transfer rates shown in Fig. 4.6 except for special cases such as the $^5D_1$ and $^5D_0$ manifolds of Eu^{3+} where the energy transfer is apparently forbidden by a ΔJ selection rule. KUSHIDA has recently developed a theoretical treatment which uses tensor operator methods to explicitly include the particular electronic levels within a calculation [4.114]. He used this theory to calculate relative efficiencies for up-conversion by successive transfers and cooperative transfers and to obtain order-of-magnitude estimates for cooperative radiative transitions [4.115, 116].

Several studies of donor excitation diffusion have found that the diffusion coefficient is temperature dependent [4.88, 113, 117]. It was found that the temperature dependence arose from participation of thermally excited levels near the ground level which had high radiative oscillator strengths to the level involved in the excitation migration. If the inter-ionic coupling is caused by electric multi-pole interactions, the probability of excitation transfer to a neighbor ion depends quadratically upon the oscillator strengths. By measurement of the oscillator strengths and energies of the levels involved, a temperature dependence of the diffusion constant can be predicted which agrees very closely with experiment [4.113].

Information on donor migration can be obtained in a completely different manner from that described in preceding sections. When inter-ionic coupling occurs, not only is excitation transfer allowed but also small electronic level shifts are produced depending upon the size of the coupling. The shifts will appear experimentally as line-splittings or changes in line-shape. If there is more than one rare-earth ion in a unit cell, a splitting of lines occurs known as a Davydov splitting. Exciton migration between unit cells causes changes in the line-shape [4.27]. It had previously been thought that such effects were too small to observe but very careful measurements of the line-shapes in $GdCl_3$ and $Gd(OH)_3$ by MELTZER and MOOS have shown that the asymmetry of the lines is that predicted by exciton-magnon dispersion theory [4.28]. The exchange parameters obtained by the fitting ranged from about 0.40 to 0.01 cm^{-1} depending upon the transition observed. The average value was about 0.05 cm^{-1}. These parameters can be related to the time required for a single resonant transfer [4.118]. Although this has not yet been done for these crystals, one can estimate that they correspond to times of about 10^{-10} to 3×10^{-9} s. The diffusion constant predicted from these values would be about 2×10^{-5} to 6×10^{-7} cm^2/s. These values are much larger than those observed in the systems of Table 4.3. The discrepancy is probably caused because the values given in this table are for dilute samples whereas the exciton dispersion was observed in concentrated materials. As the concentration is increased, short range interactions such as the exchange interaction can predominate over longer range interactions causing a further increase in transfer rate. The diffusion constant is related to the average time per transfer T_0, and the step length, R, by the equation [4.108]

$$D = \frac{8}{3\pi} \frac{R^2}{T_0}.$$
(4.25)

If the transfer is by the dipole-dipole interaction, this equation can be written as

$$D = \frac{8}{3\pi} \frac{C}{R^4}.$$
(4.26)

If a typical value of C is 6×10^{-38}, a diffusion constant of 10^{-7} cm^2/s is predicted. Since the exciton migration in $GdCl_3$ or $Gd(OH)_3$ is at least a factor of 6 higher, this suggests that exchange dominates energy transfer in very high concentration crystals. In crystals such as $GdCl_3$ or $Gd(OH)_3$, it is known that the exchange interaction is sufficiently large to produce magnetic phase transitions at low temperatures.

4.4 Kinetics of Up-Conversion

We would now like to apply this fundamental information to the problem of up-conversion. The most efficient type of up-conversion mechanism involves the sequential transfer of two or more excitations from the donor ions to an acceptor ion and is the one we will consider in the greatest detail. This problem would appear to be considerably more difficult to describe than transfer of one excitation. If a sample is excited by a pulse of light, up-conversion can occur by two sequential transfers at an acceptor ion if one or both transfers takes place directly between two close neighbors or if exciton diffusion is rapid enough to bring excitations within range for a direct transfer. Both the concentration and fluorescence time dependence will be quite complex for this general case. The primary difficulty is in describing the depletion rate of excited donors radially outward from an acceptor ion after the initial excitation. If exciton diffusion is sufficiently large, we have seen that any spatial depletion regions around acceptor ions are minimized as the donor (or acceptor) excitation is delocalized over the entire donor (or acceptor) system. One can then describe the process with an average transition rate and a simple rate equation analysis can be used. Such an analysis will become inadequate at low donor and acceptor concentrations where exciton diffusion canot occur. The experiments of KRASUTSKY and MOOS described earlier show that donor diffusion

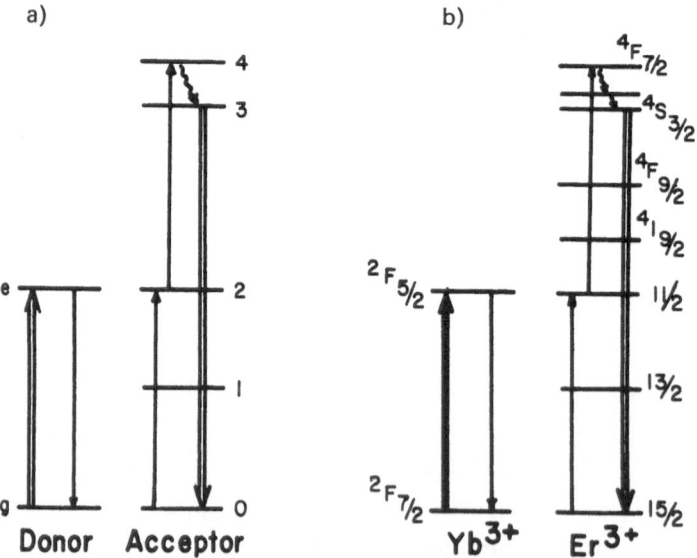

Fig. 4.7. (a) Energy levels used for theoretical rate equation model. (b) Corresponding levels in the Yb^{3+}, Er^{3+} system. The double arrows indicate transitions induced by the radiation field, the wavy arrows indicate non-radiative transitions induced by the phonon field, and the straight single arrows represent transitions induced by inter-ionic coupling

was appreciable at donor concentrations of 2.0 mole % [4.108]. WATTS and RICHTER have performed analogous experiments in YF_3: Yb, Ho [4.112]. Since the Yb^{3+} sensitized systems have been extensively studied for up-conversion applications, their results are of direct applicability. They find that Yb^{3+} exciton diffusion can be noticed at concentrations of 0.3 mole % and that it dominates direct transfer at concentrations of 3 mole %. The rate equations can therefore be expected to describe a sequential transfer up-conversion system at concentrations of 3 mole % or higher.

The majority of the experimental work in up-conversion has been concerned with development of efficient up-conversion phosphors using Yb^{3+} sensitized systems containing Ho^{3+}, Er^{3+}, or Tm^{3+} ions. There are a number of different level schemes that can be used in the different systems and in fact a single system can have several different mechanisms for up-conversion. A rate equation model that would be applicable to all systems would be quite complex because of the number of levels required. An understanding of any one system would be difficult to comprehend for such a general analysis. Instead, we shall select a single system which illustrates all of the important general effects of other systems and yet remains simple enough to understand intuitively the important points.

The model we will use is shown in Fig. 4.7a. It is identical to the up-conversion mechanism shown in Fig. 4.7b for green fluorescence in Yb^{3+}, Er^{3+} doped systems. The rate equations needed to describe this model are

$$\dot{n}_e^d = \sigma^d \Phi n_g^d - c_{d2} n_e^d n_0 + c_{2d} n_g^d n_2 - c_{d4} n_e^d n_0 + c_{4d} n_g^d n_4 - \tau_d^{-1} n_e^d \qquad (4.27)$$

$$\dot{n}_1 = W_{21} n_2 - \tau_1^{-1} n_1 \qquad (4.28)$$

$$\dot{n}_2 = c_{d2} n_e^d n_0 - c_{2d} n_g^d n_2 - c_{d4} n_e^d n_2 + c_{4d} n_g^d n_4 - \tau_2^{-1} n_2 \qquad (4.29)$$

$$\dot{n}_3 = W_{43} n_4 - \tau_3^{-1} n_3 \qquad (4.30)$$

$$\dot{n}_4 = c_{d4} n_e^d n_2 - c_{4d} n_g^d n_4 - \tau_4^{-1} n_4. \qquad (4.31)$$

In writing these, we have made several approximations. Stimulated emission terms have been neglected, transitions from level 3 to level 2 are not considered to eliminate recycling which could then occur, and the relaxation of level 4 is assumed to occur only to level 3. The levels of the donor ion are denoted by a superscript d. A subscript g is used if the ground level is involved or e if the excited level is involved. Transitions between levels i and j of a single ion are denoted by W_{ij} and the lifetime of level i in the absence of ion-ion transfer is denoted τ_i. Transfer rates between the donor ion and the i-th level of the acceptor ion are denoted by c_{di}. The absorption cross-section of donor ions is σ^d and the incident pumping flux is Φ. The method to be used is similar to that of JOHNSON et al. [4.119] in their analysis of the $Yb^{3+}-Er^{3+}$ system. Their analysis divided the Yb^{3+} donor system into those ions that could transfer directly to the Er^{3+} acceptors and those more distant Yb^{3+} ions that must transfer their energy to the closer Yb^{3+} ions. This procedure is equivalent to the Perrin model of direct relaxation modified for exciton migration. It is a higher approximation than the model presented here but it is generally considered unnecessary for the concentrations normally considered.

If the steady state fluorescence intensity from level 3 is sought, the time derivatives of the level populations are set equal to zero and the equations can be solved simultaneously for n_3. From (4.30), we can relate n_3 to n_4

$$n_3 = \frac{W_{43}}{\tau_3^{-1}} n_4. \tag{4.32}$$

Using (4.31), n_4 can in turn be directly related to n_2, i.e.

$$n_4 = \frac{c_{d4} n_e^d n_2}{c_{4d} n_g^d + \tau_4^{-1}}. \tag{4.33}$$

One can see from the denominator of this equation that the transfer from the acceptor ion back to the donor ion causes an additional depopulation of level 4 and therefore a decrease in the fluorescence intensity from level 3. The value of n_2 can be obtained from (4.29). Since the transfer to and from level 4 is normally insignificant compared with the other processes affecting level 2, these terms in (4.29) can be neglected and a simple expression for n_2 is obtained, namely

$$n_2 = \frac{c_{d2} n_e^d n_0}{c_{2d} n_g^d + \tau_2^{-1}}. \tag{4.34}$$

Again, we see that the back transfer from level 2 will lower n_2 and therefore the output fluorescence. Combining (4.32–34) the fluorescence intensity from level 3 to the ground state can be written as

$$I = h\nu_{30} W_{30} n_3 = \frac{h\nu_{30} W_{30} W_{43} c_{d2} c_{d4} n_0 (n_e^d)^2}{\tau_3^{-1}(\tau_2^{-1} + c_{2d} n_g^d)(\tau_4^{-1} + c_{4d} n_g^d)}. \tag{4.35}$$

If we assume that the pumping intensity is sufficiently weak that the ground state is not depopulated, n_0 can be approximated by the total concentration of acceptor ions, N_a. A value for n_e^d can be determined from (4.27) and we will again assume that the transfer to and from level 4 can be neglected with respect to the other transfer rates. Then,

$$n_e^d = \frac{\sigma^d \Phi n_g^d}{\left(\tau_d^{-1} + \dfrac{\tau_2^{-1} c_{d2} n_0}{c_{2d} n_g^d + \tau_2^{-1}}\right)}. \tag{4.36}$$

The second term in the denominator describes the depopulation of the excited donor level by donor acceptor transfer. One can notice now from (4.35) that the maximum fluorescence is produced by a large transfer rate, $c_{d2} n_0$, and a large excited donor population. These two conditions are contradictory since (4.36) shows the maximum excited donor population is achieved when c_{d2} is sufficiently small that the second term in the denominator can be ignored. There is therefore, an optimum value for $c_{d2} n_0$. Combining (4.35) and (4.36) and neglecting depopulation of the acceptor and donor ground states,

$$I = h v_{30} \cdot \frac{W_{30}}{\tau_3^{-1}} \cdot \frac{W_{43} c_{d2} c_{d4} N_a N_d^2 (\sigma^d \Phi)^2}{(\tau_2^{-1} + c_{2d} N_d)(\tau_4^{-1} + c_{4d} N_d)\left(\tau_d^{-1} + \dfrac{\tau_2^{-1} c_{d2} N_a}{\tau_2^{-1} + c_{2d} N_d}\right)^2} \cdot$$

$$(4.37)$$

Several things should be noted at this point. The effects of back-transfer from a level always appear in the denominator in combination with the intrinsic depopulation rate of that level. The effect is to reduce the observed intensity by an amount that depends upon the ratio of the back-transfer rate to the intrinsic depopulation rate. If the inter-ion transfer rates are negligible compared with the intrinsic rates, the output fluorescence increases directly with the lifetime of the intermediate acceptor level 2 and excited donor level. In addition, the output intensity depends quadratically upon the donor concentration and linearly upon the acceptor concentration as one would expect from the law of mass action. However, if the back-transfer rates become larger than the intrinsic decay rates for some levels, the concentration dependences can be lowered or eliminated depending upon the levels. The advantage of a long intermediate acceptor level lifetime is also lost if the back-transfer rate from that level is high. The intensity depends quadratically upon the incident pump flux under the conditions stated above. Again, this is a consequence of the law of mass action. As stated earlier, there is an optimum value for the donor-acceptor transfer rate, $c_{d2} N_a$. The maximum fluorescence output is determined by differentiating (4.37) with respect to $c_{d2} N_a$. The optimum transfer rate that results is

$$c_{d2} N_a = \frac{\tau_d^{-1}}{\tau_2^{-1}} (\tau_2^{-1} + c_{2d} N_d). \qquad (4.38)$$

Physically, this condition arises because the rate of depopulating the excited donor level must be restricted to insure adequate excited donors for the second transfer step. This condition will be relaxed if excitation of the acceptor ions themselves can create a second excitation. The intensity produced by this optimum transfer rate is

$$I_{max} = h v_{30} \frac{W_{30}}{\tau_3^{-1}} \frac{W_{43} c_{d4} (\sigma^d)^2 \Phi^2 N_d^2}{4\tau_2^{-1} \tau_d^{-1} (\tau_4^{-1} + c_{4d} N_d)}. \qquad (4.39)$$

It is interesting to note there is no explicit dependence on the back transfer rate from level 2. As pointed out by JOHNSON et al. [4.119], this arises because the back transfer rate can be corrected for by making the forward transfer rate larger in accordance with (4.38).

The analysis above will only be applicable at pumping rates sufficiently low that the donor and acceptor ground states are not depopulated. At higher pumping rates, this approximation will be invalid. To understand the effects that would be expected, we will assume that levels 1 and 2 are metastable and their populations can build up to large values. Then,

$$n_0 = N_a - n_1 - n_2 - n_3 - n_4 \approx N_a - n_1 - n_2. \tag{4.40}$$

Combining (4.28), (4.34) and (4.40), one can find

$$n_0 \approx \frac{N_a}{1 + \dfrac{w_{21} c_{d2} n_e^d}{\tau_1^{-1}(c_{2d} n_g^d + \tau_2^{-1})} + \dfrac{c_{d2} n_e^d}{(c_{2d} n_g^d + \tau_2^{-1})}}. \tag{4.41}$$

The second and third terms in the denominator are responsible for the ground state depopulation and are appreciably larger than unity if depopulation is significant. The ground state depopulation will therefore be inversely proportional to the excited donor population. Combining the expression for n_0 in (4.41) with (4.35), one can see the output intensity will now be only directly proportional to the incident pumping rate. Since the excited donor population is directly proportional to the incident pumping rate, the effect of acceptor ground state depopulation has been to reduce the dependence of output fluorescence on pumping intensity from quadratic to linear. Experimentally, one could misinterpret a two-photon process as a single-photon process if ground state depopulation was occurring.

4.5 Experimental Work in Up-Conversion

4.5.1 Yb³⁺ Sensitized Systems

The most extensive work on up-conversion of infrared radiation (other than that connected with the IRQC) has been done in Yb³⁺ sensitized rare-earth systems. Two basic processes are responsible for up-conversion in these materials. The most efficient and therefore the most important scheme for up-conversion involves successive transfer from a Yb³⁺ donor system to the activator. The kinetics of this process was examined in the previous section. A second mechanism for up-conversion is the cooperative energy transfer. Two excited Yb³⁺ ions simultaneously transfer their energy to an activator ion exciting it to a higher state. It is thought that the cooperative transfer is not of importance whenever successive transfer can occur but does become important when there are no intermediate levels of the activator which can receive a single transfer of excitation. In this section, we will examine the important Yb^{3+} sensitized systems.

Er^{3+} Activated Systems

The greatest amount of experimental work has been performed on doubly-doped Er^{3+}, Yb^{3+} systems. The appropriate energy levels for this system are shown in Fig. 4.8a along with the levels used in the theoretical discussion of the previous section. Fluorescence is observed from either the $^4S_{3/2}$ manifold (green fluorescence at ~ 540 nm) or the $^4F_{9/2}$ manifold (red fluorescence at ~ 670 nm). The detailed mechanisms that populate each level after infrared excitation are different and will therefore be discussed separately.

The mechanism for excitation of the Er^{3+} green fluorescence is perhaps the best understood of any system studied. It was also the first example of up-conversion. AUZEL correctly proposed that the green fluorescence was excited by successive transfer from the Yb^{3+} donor system [4.12]. This mechanism is shown in Fig. 4.8 b. OVSYANKIN and FEOFILOV, however, proposed that up-converted fluorescence resulted from a cooperative transfer from two excited Yb^{3+} ions [4.14]. It was later shown that the cooperative transfer was too weak to satisfactorily account for the fluorescence [4.56]. This point will be examined in considerably more detail in the discussion of the $Yb^{3+}-Tm^{3+}$ system.

A rather complete study has been performed on $BaYF_5: Yb^{3+}, Er^{3+}$ and the important results will be outlined as a specific example of the Yb^{3+}, Er^{3+} system. Other host crystals have also been studied and the results are very similar [4.3, 12, 13, 120–131]. A key experiment in the study of up-conversion is measuring the fluorescence output dependence on the excitation intensity. The $^4S_{3/2}$ dependence is shown on the upper curve of Fig. 4.9 for this host [4.119]. The quadratic dependence observed is appropriate for the mechanism shown in Fig. 4.8b. Further confirmation of the proposed mechanism was provided by examining the transient response of the system to pulsed infrared excitation. The intrinsic lifetimes of the $^4S_{3/2}$ manifold in this material was 0.15 ms but the observed rise and fall of $^4S_{3/2}$ fluorescence was much longer when it was excited in the infrared. The time response expected for the proposed mechanism is quite complex but it can be closely approximated by the expression

$$\tau_D^{-1} = \tau_{^2F_{5/2}}^{-1} + \tau_{^4I_{11/2}}^{-1} + C_{2d} N_d$$

where τ_D is the observed time constant for decay, and the other time constants are the intrinsic decay rates. Note that the back transfer rate also enters the expression. Since the intrinsic decay rates can be measured in singly doped crystals, the back transfer rate can be determined. The back transfer rate was

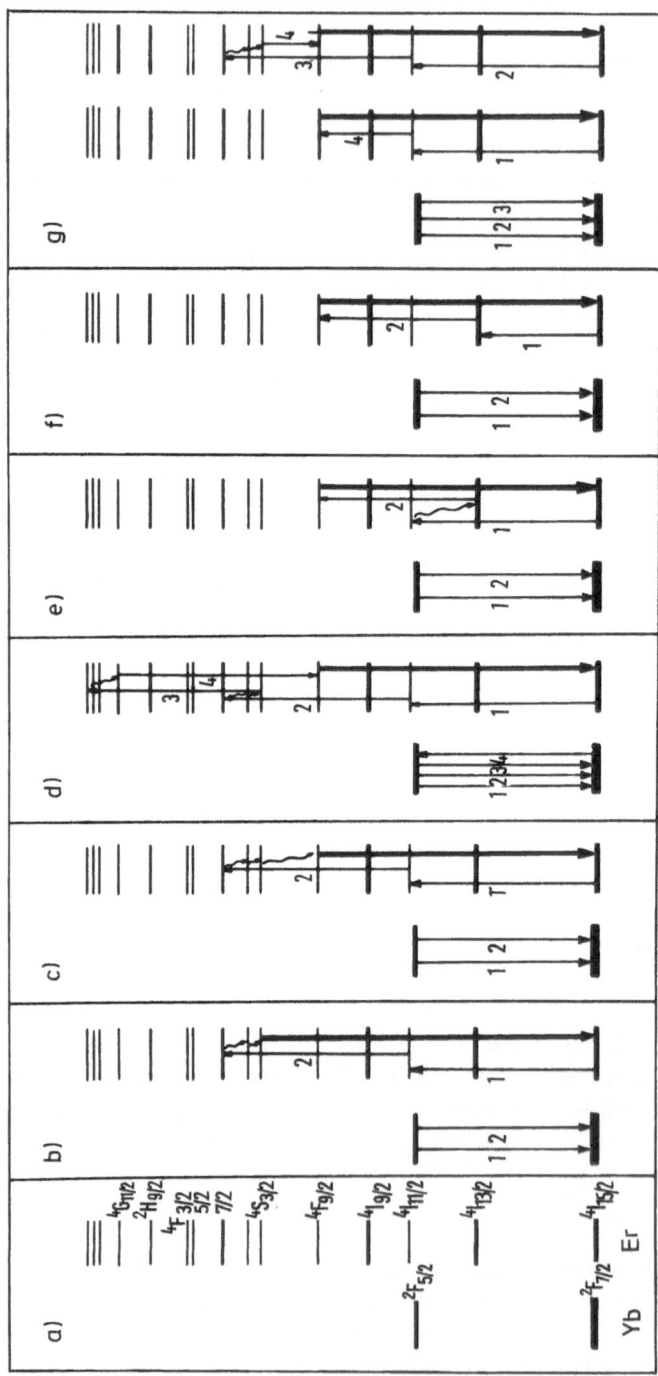

Fig. 4.8a–g. Different mechanism of up-conversion for Yb^{3+}, Er^{3+} systems. The double arrows represent transitions induced by the radiation field, the wavy arrows represent non-radiative transitions induced by the phonon field, and the single arrows represent transitions induced by the inter-ionic inter-action. The sequence of events in time is indicated by numbers to the side of each transition. The lowest numbers correspond to the earliest events

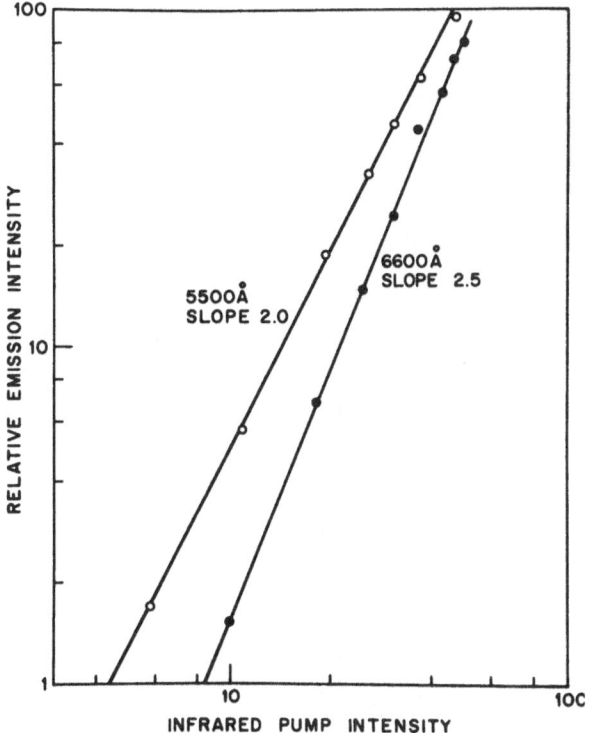

Fig. 4.9. Dependence of Er^{3+} $^4S_{3/2}$ fluorescence (upper curve) and $^4F_{9/2}$ fluorescence (lower curve) on the excitation intensity [4.119]

determined for several different concentrations and was found to vary linearly, implying that c_{2d} was a constant independent of concentration (see Fig. 4.10) [4.119]. This can be taken as justification for the rate equation model and supports the assumption that donor excitation diffusion is rapid. When similar analysis were performed for the rise of $^4S_{3/2}$ fluorescence, consistent values for $c_{2d} N_d$ were obtained. The forward transfer rate $c_{d2} N_a$ could be determined from a separate experiment. If the $^4I_{9/2}$ manifold of Er^{3+} is excited, internal relaxation occurs to the $^4I_{11/2}$ manifold. This level can either fluoresce directly or transfer its excitation to the Yb^{3+} $^2F_{5/2}$ manifold which can also fluoresce. The relative intensities of the $^4I_{11/2}$ and $^2F_{5/2}$ fluorescence depend upon the forward and reverse transfer rates as well as the intrinsic decay rates of the $^4I_{11/2}$ and $^2F_{5/2}$ manifolds. Since only the forward transfer rate is unknown at this point, it can be determined. For the most efficient phosphor $(BaYF_5: 5\% Er^{3+}, 45\% Yb^{3+})$, these measurements yielded the following values [4.119]

$$c_{2d}N_d = 417\,s^{-1} \qquad c_{d2} N_a = 42\,s^{-1}$$
$$\tau_2^{-1} = 100\,s^{-1} \qquad \tau_d^{-1} = 500\,s^{-1}.$$

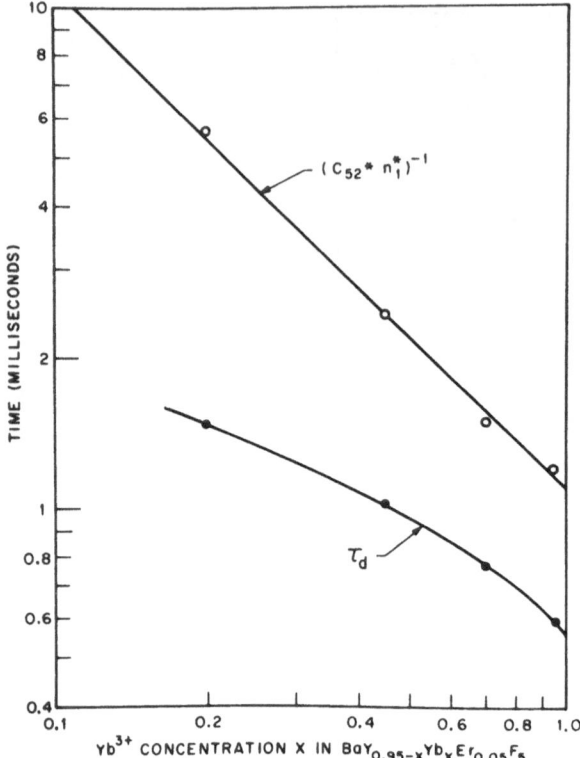

Fig. 4.10. The lower curve presents the time constant for $^4S_{3/2}$ fluorescence decay after infrared excitation. The intrinsic lifetime of the $^4S_{3/2}$ level is 0.15 ms, much shorter than that observed. The longer lifetime arises because the longer lived Yb^{3+} level. is feeding the $^4S_{3/2}$ level. The decay time varies with Yb^{3+} concentration in this figure because of increasing amounts of back-transfer from the Er^{3+}. The rate of back transfer $(C_{2d}N_d)^{-1}$ is shown in the upper curve of this figure [4.119]

The back transfer from the Er^{3+} $^4I_{11/2}$ manifold is larger than the intrinsic decay rate. As a consequence of the larger back transfer rate $c_{2d}N_d$, the intensity of up-converted fluorescence becomes linearly dependent upon the Yb^{3+} donor concentration instead of quadratic, see (4.37). In addition, the high back transfer rate affects the efficiency of up-conversion. Using the above values for the transfer rates in (4.38), one sees that

$$\frac{c_{d2}N_a}{\tau_d^{-1}/\tau_2^{-1}(\tau_2^{-1}+c_{2d}N_d)} \approx 0.016.$$

This ratio would be unity if the transfer rates were optimum. The reason the values are so far from the optimum value is that the back transfer rate is 10 times larger than the forward rate. Since the Yb^{3+} $^2F_{5/2}$ and the Er^{3+} $^4I_{11/2}$ manifolds are very close in energy (the $^4I_{11/2}$ may in fact be somewhat above the $^2F_{5/2}$ manifold), this is a reasonable result. To offset this poor forward transfer rate, one might expect that the acceptor concentration could simply be increased to increase $c_{d2} N_a$ to the optimum rate. However, Er^{3+} ion pair interactions become important as N_a is increased and lead to a rapid quenching of the $^4S_{3/2}$ fluorescence [4.119].

MITA has pointed out that the transfer rates required for such analysis can be obtained theoretically from the non-resonance energy transfer theories presented earlier in this chapter [4.132,133]. If the rate of resonance transfer and the effective phonon energies involved are known from experimental data in the host lattice to be used, any of the transfer rates should be obtainable. Several examples of such analyses have appeared in the literature. Although these theoretical calculations agree somewhat with experiment, the present state of knowledge about non-resonant energy transfer is not sufficiently complete to allow general application of the theory. This is particularly true for transfer processes that are out of resonance by much less than one phonon where no work has been done. It is not clear whether the same averaging over electronic states will occur to eliminate explicit dependences of transfer rate on the particular nature of the electronic state or whether the coupling mechanism itself will remain unchanged. Approaches such as these will become important however as the basic knowledge of energy transfer processes increases.

The donor concentration dependence of up-conversion efficiency has been studied by several authors [4.119, 123, 129, 130]. TAMATANI et al. [4.129] studied the green up-conversion in $Y_2O_2:Yb^{3+},Er^{3+}$ and found the fluorescence varied linearly with the Yb^{3+} concentration instead of quadratically, as might be expected. As explained earlier, the behavior results from the high back transfer rate from Er^{3+} to Yb^{3+} compared with the intrinsic Er^{3+} decay rate. Confirmation of this interpretation was provided by measuring the Yb^{3+} $^2F_{5/2}$ lifetime as a function of Yb^{3+} concentration. At concentrations that permit efficient transfer, the lifetime will actually be longer than the intrinsic lifetime because the back-transfer allows the Er^{3+} ion to act as a reservoir that feeds excitation back to the Yb^{3+}. At low donor concentrations where the back-transfer rate becomes small, the decay becomes intrinsic again and the concentration dependence of up-conversion efficiency becomes quadratic. One cannot go to very low concentrations though because donor migration does not occur rapidly enough to justify use of the rate equations. The concentration dependence would then depart from quadratic behavior again. These effects have been observed experimentally by HEWES and SARVER, as shown in Fig. 4.11 [4.123].

KINGSLEY et al. [4.125, 126] have examined the transient response of $LaF_3:Yb^{3+},Er^{3+}$ after pulse excitation of the Yb^{3+} transition. They observed a time dependence which changes as the excitation pulse width is increased. The observed behavior is shown in Fig. 4.12. The excitation pulse is applied at time $= 0$ and the arrow indicates when the pulse is removed. If the pulse is short, the up-converted fluorescence continues to increase after the excitation is

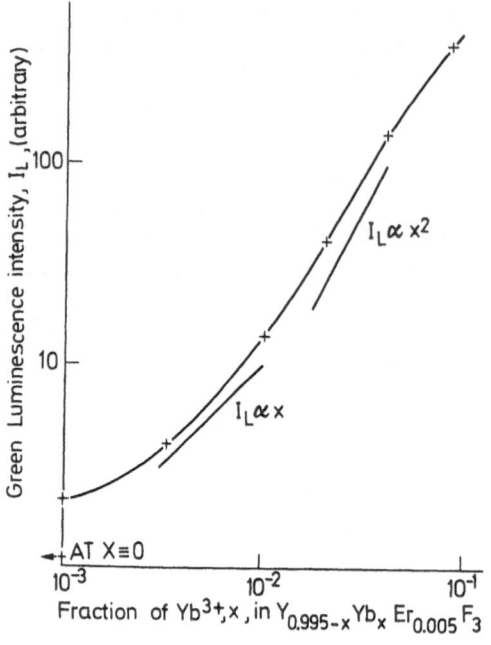

Fig. 4.11. Dependence of $^4S_{3/2}$ fluorescence intensity on the Yb^{3+} concentration [4.123]

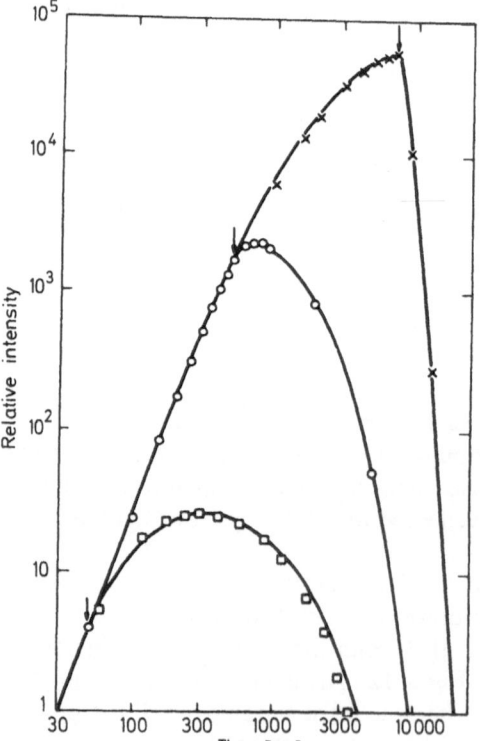

Fig. 4.12. Transient response of $^4S_{3/2}$ fluorescence after pulsed infrared excitation of varying pulse widths. The arrows indicate where the excitation was removed [4.125]

removed as the equilibrium population is sought by the system. However, since the system must also decay, the fluorescence will eventually begin to fall off. As the pulse width becomes much longer, the system is allowed to reach an equilibrium population. Then when the pulse is removed, decay must begin immediately. The temporal behavior has been explained very satisfactorily by assuming both the forward and backward transfer rates are much faster than the intrinsic decay rates. The figure is a graphic illustration of the differences expected between the decay after pulsed excitation and after a steady excitation has been removed.

The mechanism for up-conversion to the Er^{3+} $^4F_{9/2}$ manifold varies from material to material. A total of 7 different mechanisms have been suggested although it is likely that several are not of importance [4.3, 134]. The simplest mechanism suggested is the $^4F_{9/2}$ population is caused by simple non-radiative relaxation by multi-phonon emission from the $^4S_{3/2}$ manifold. It is shown schematically in Fig. 4.1c. Since the gap between the two states is large, this mechanism would only be operative in crystals with high energy phonons. It would require two transfers from excited Yb^{3+} donors and would therefore have a quadratic dependence upon pumping intensity. The $^4F_{9/2}$ and $^4S_{3/2}$ fluorescence would always have the same ratio of intensities regardless of pump intensity or excitation wavelength if only this mechanism is operative. This mechanism has been proposed to explain up-conversion in a number of crystals such as $LiYO_2, Y_2O_3, YPO_4$ [4.135], and also $YOCl$ or Y_3OCl_7 at low pump intensities [4.131].

The other mechanisms that have been proposed rely upon the existence of energy matches between levels that are not too different from the Yb^{3+} levels. (The one exception is a mechanism involving exciton annihilation in the Er^{3+} acceptor system.) An excited Yb^{3+} ion is capable of exciting an Er^{3+} from the $^4I_{13/2}$ to the $^4F_{9/2}$ or from the $^4S_{3/2}$ to the $^2G_{7/2}$. A non-resonant transfer is also allowed in some systems from the Er^{3+} ground state to the $^4I_{13/2}$ manifold. A ground state Yb^{3+} can accept an excitation from the Er^{3+} as it transfers excitation from the $^4G_{11/2}$ to the $^4F_{9/2}$ or in some lattices from the $^2H_{11/2}$ or $^4S_{3/2}$ to the $^4I_{13/2}$. The various mechanisms differ in which of these transfers are used and the method of populating the different intermediate levels.

An important mechanism involves population of the $^4S_{3/2}$ manifold by two transfers from the Yb^{3+}, as shown in Fig. 4.8b. A third Yb^{3+} transfer excites the Er^{3+} from the $^4S_{3/2}$ to the $^4G_{7/2}$ manifold which decays rapidly by multi-phonon relaxation to the $^2G_{11/2}$ manifold. This finally relaxes to the $^4F_{9/2}$ manifold by exciting a Yb^{3+} ion. The process is diagrammed in Fig. 4.8d. A mathematical treatment of the process is complex because the last transfer produces a Yb^{3+} excitation that can either be used in subsequent excitation processes or can be lost depending upon the efficiency of the Yb^{3+} donor system. If the excitation is lost, the mechanism will require three photons but if the excitation is used with 100% efficiency in a latter excitation, the mechanism requires only two photons. The intensity dependence observed could then vary anywhere from quadratic to cubic. The importance of this mechanism is unclear at this time. The mechanism has been proposed for the YOCl and Y_3OCl_7 systems when the pumping rates become large [4.124, 131] but other work in the same system contradicts the interpretation [4.136]. The reason for the discrepancy is not known.

If the $^4I_{11/2}$ manifold excited by an initial transfer relaxes by multi-phonon relaxation to the $^4I_{13/2}$ manifold, another excited Yb^{3+} ion can cause a transfer to the $^4F_{9/2}$ manifold. This mechanism is shown in Fig. 4.8e. It would be a two-photon process. The multi-phonon relaxation would not be expected to be efficient in many crystals because the gap involved is very large, larger in fact than the gap between the $^4S_{3/2}$ and $^4F_{9/2}$ manifolds. In crystals with strong phonon coupling and higher energy phonons, however, the relaxation can occur. A variation of this mechanism (shown in Fig. 4.8f) states that the $^4I_{13/2}$ manifold is populated directly by non-resonant transfer from the Yb^{3+} donor rather than by multi-phonon decay from the $^4I_{11/2}$ manifold. Evidence for the possible importance of this mechanism is shown in experiments by BROWN and SHAND [4.137] who showed that the Er^{3+} $^4I_{11/2}$ fluorescence is excited more efficiently by pumping the $^4I_{13/2}$ manifold than pumping the $^4I_{11/2}$ manifold directly. Their experiment demonstrated that a coupling exists between the two levels which is required for the proposed mechanism. These mechanisms have been suggested for the red emission in various scheelite [4,130] and fluoride hosts [4.127, 138] and phosphors such as $LiYO_2$, Y_2O_3 [4.135], YOCl, Y_3OCl_7 [4.136], and YPO_4 [4.135].

Another mechanism has been proposed that succeeds in populating the $^4I_{13/2}$ manifold without the large energy mismatch of the previous mechanisms [4.134]. After the excitation of the $^4S_{3/2}$ manifold is achieved by the normal mechanism, a transition occurs to the $^4I_{13/2}$ manifold exciting a nearby Yb^{3+} ion. (The transition could also begin from the $^2H_{11/2}$ manifold if it became thermally excited from the $^4S_{3/2}$ manifold.) This transfer requires absorption of lattice phonons and will not be important at low temperatures. Excitation of the $^4F_{9/2}$ manifold again results after donor transfer from a Yb^{3+} ion. The diagram is shown in Fig. 4.8g. This process can have a dependence between quadratic and cubic depending again upon the efficiency of the Yb^{3+} donor system. It has been suggested as a possible mechanism for fluoride lattices [4.134]. The transfer from the $^2H_{11/2}$ manifold to the $^4I_{13/2}$ and $^4I_{9/2}$ manifolds has been positively identified in YF_3:Er^{3+} while exciting in the ultra-violet [4.139, 140] but this is not sufficient for the mechanism to be operative. The $^2H_{11/2}$ or $^4S_{3/2}$ levels must first be populated. A variation on this mechanism has been proposed in which the de-excitation of the $^4S_{3/2}$ manifold to the $^4I_{13/2}$ manifold is caused by a neighboring Er^{3+} ion being excited to the $^4I_{9/2}$ manifold [4.3]. This mechanism could also have a quadratic or cubic intensity dependence depending upon whether the $^4I_{9/2}$ excitation was returned to the excited donor system.

A completely different mechanism has been proposed to explain the Y_2O_3:Yb^{3+},Er^{3+} phosphor but more work is required to establish its importance [4.141]. After the $^4F_{7/2}$ manifold is populated by two successive transfers from the Yb^{3+} ion, an exciton annihilation occurs in the Er^{3+} system between $^4F_{7/2}$ and $^4I_{11/2}$ excitations to form two ions in the $^4F_{9/2}$ manifold. This mechanism would exhibit a 1.5 power dependence on pump intensity. Such a dependence has not been observed experimentally.

A number of experimental techniques have been used to evaluate the importance of the different possible mechanisms proposed. The primary technique has been to measure the up-converted fluorescence intensity dependence upon pump intensity. Typically, powers between 2 and 2.5 have been obtained. This procedure is most efficient in eliminating schemes requiring two photons when dependences higher than quadratic are found but it is difficult to eliminate any schemes when a quadratic dependence is found. Another key experiment has been to measure the ratio of $^4S_{3/2}$ and $^4F_{9/2}$ fluorescence when various manifolds are excited. If the red fluorescence is excited more efficiently relative to the green fluorescence when pumping through the Yb^{3+} ion than when pumping above the $^4S_{3/2}$ manifold, mechanisms such as shown in Fig. 4.1c can be eliminated. Analysis of the temporal behavior after pulsed excitation can be a powerful method to distinguish between mechanisms. The temporal behavior will contain information on the intermediate excited states that are involved if the lifetimes of those states are longer or at least comparable to that of the final emitting level.

Kuroda et al. [4.136], performed a study of three Er^{3+}–Yb^{3+} phosphor systems to analyze the importance of all suggested mechanisms in three representative lattices and determine the lattice parameters that are of importance. YF_3 was selected as representative of hosts that produced fluorescence from the $^4S_{3/2}$ manifold efficiently and YOCl and Y_3OCl_7 were selected because the $^4F_{9/2}$ fluorescence was intense. They found that the $^4F_{9/2}$ fluorescence depended quadratically on the excitation intensity in all three lattices in contradiction to previous work [4.131]. The reason for the contradiction was not determined. Then all those mechanisms predicting dependences higher than quadratic were eliminated by assuming any excitation returned to the Yb^{3+} system by downward Er^{3+} transfers (see Fig. 4.8d or g, for example) would not be used efficiently. Multi-phonon decay from the $^4S_{3/2}$ manifold (Fig. 4.8c) was eliminated as a possible mechanism by the more efficient pumping of the $^4F_{9/2}$ when it was excited in the infrared. This left only the mechanisms shown in Fig. 4.8e and f in which the $^4I_{13/2}$ manifold is populated by either multi-phonon relaxation from the $^4I_{11/2}$ manifold or by direct non-resonant transfer from an excited Yb^{3+} donor. Time dependent work was not done to establish the $^4I_{13/2}$ manifold as an intermediate level. They found that since the lowest Yb^{3+} donor level is below the lowest Er^{3+} $^4I_{11/2}$ level in YF_3 and YOCl, energy transfer from Yb^{3+} donors is hindered at low temperatures. In YOCl, the phonon energies are sufficiently high that a non-resonant transfer to the lower $^4I_{13/2}$ manifold can still become important. In YF_3, non-resonant transfer can occur but with much less efficiency because of the lower phonon energies in this material. The interpretation is strengthened by the observation that direct pumping of the $^4I_{13/2}$ manifold produces a marked increase in $^4F_{9/2}$

fluorescence. In Y_3OCl_7, however, the Yb^{3+} donor level is higher than the Er^{3+} acceptor level and transfer to the $^4I_{11/2}$ state is efficient. The coupling between the electronic levels and the lattice is strong and the phonon energies are high enough in this material to cause multi-phonon relaxation from the $^4I_{11/2}$ state to the $^4I_{13/2}$ state. In all three lattices, the transition from the $^4I_{13/2}$ to $^4F_{9/2}$ manifold is caused by a second transfer from the Yb^{3+} donor. This study concluded that the efficiency of up-conversion to the $^4F_{9/2}$ manifold depends primarily upon strong coupling to the high energy phonons of the lattice and the relative positions of the $^4I_{11/2}$ Er^{3+} manifold and $^2F_{5/2}$ Yb^{3+} manifold. This study strongly suggests that the efficiency of red up-conversion depends upon whether the first transfer succeeds in populating the $^4I_{11/2}$ or $^4I_{13/2}$ manifolds. If the $^4I_{11/2}$ manifold is populated, red fluorescence will not be efficient if multi-phonon relaxation is not efficient. However, the work is not definitive and the reasons for discrepancies with previous work were not examined.

An intriguing application of the up-converted red fluorescence was shown by JOHNSON and GUGGENHEIM [4.2]. Using a single crystal of $BaY_2F_8:Yb^{3+}$, Er^{3+}, they succeeded in achieving lasing from the $^4F_{9/2}$ manifold while pumping the Yb^{3+} ion. This experiment demonstrated that lasing action could be sensitized with donor levels both below and above the lasing level.

Ho^{3+} Activated Systems

A doubly doped system containing Yb^{3+} and Ho^{3+} has also been used for up-conversion of infrared radiation [4, 14, 119, 121–123, 142–144]. Fluorescence is generally observed in the green from the $(^5S_2, ^5F_4)$ manifold. The excitation

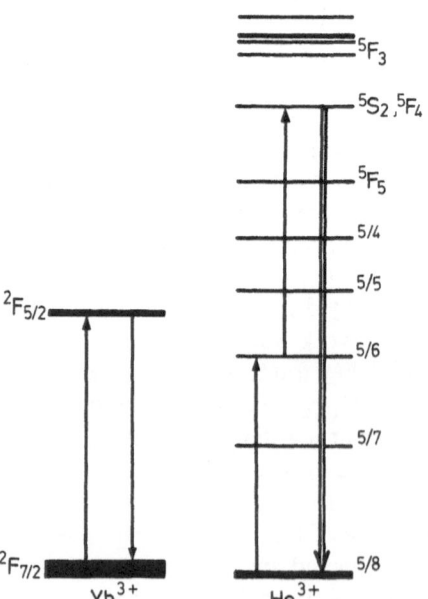

Fig. 4.13. Up-conversion mechanism in Yb^{3+}, Ho^{3+} systems

mechanism consists of two consequtive transfers from the Yb^{3+} ion. The first transfer results in excitation of the $^5I_6$ manifold and the second excitation excites it to the $(^5S_2, ^5F_4)$ manifold, as shown in Fig. 4.13. The first transfer is non-resonant requiring emission of $1100\ cm^{-1}$ to lattice phonons. The second transfer is nearly resonant. Both the transient response and intensity dependence of up-converted fluorescence are consistent with the mechanism shown.

Tm^{3+} Activated Systems

The up-conversion of infrared radiation in Yb^{3+}, Tm^{3+} systems was discovered independently by OVSYANKIN and FEOFILOV [4.14], and by AUZEL [4.13]. This was one of the first systems in which up-conversion had been found. AUZEL proposed the up-conversion was produced by successive non-resonant transfers from the Yb^{3+} donor system to a Tm^{3+} ion. OVSYANKIN and FEOFILOV, however, proposed that a cooperative transition involving two excited Yb^{3+} ions produced the observed fluorescence. This controversy stimulated a considerable amount of further work [4.119, 120, 145] and the behavior of the Yb^{3+}, Tm^{3+} is now reasonably well understood.

The states involved in the up-conversion are shown in Fig. 4.14. A Yb^{3+} excitation is first non-resonantly transferred to the $^3H_5$ manifold of Tm^{3+}.

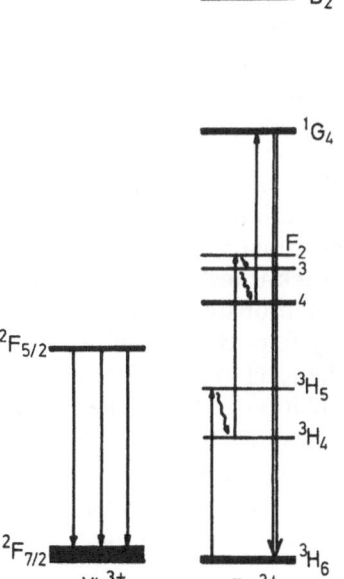

Fig. 4.14. Up-conversion mechanism in Yb^{3+}, Tm^{3+} systems

This transition requires the lattice to absorb $1650\ cm^{-1}$. This state relaxes non-radiatively by multi-phonon emission to the $^3H_4$ manifold from which a second transition to the $^3F_2$ manifold is caused by a second Yb^{3+} ion. This transfer releases 1000 cm^{-1} to the lattice phonons. The $^3F_2$ state again relaxes by multi-phonon relaxation to the $^3F_4$ state and a third non-resonant transfer excites it to the $^1G_4$ state with 1400 cm^{-1} of energy released. The $^1G_4$ can emit fluorescence at 475 nm. The initial experiments of AUZEL demonstrated the fluorescence has a cubic dependence on the pump intensity in agreement with the correct mechanism [4.13]. OVSYANKIN and FEOFILOV observed a quadratic dependence however and were therefore lead to the cooperative transfer model that required only two excited Yb^{3+} ions [4.14]. HEWES and SARVER performed a series of experiments that showed the difference arose from a difference in the excitation intensity used in the two experiments [4.123]. At low powers, the dependence is cubic but at higher powers, it departs from cubic and approaches a slope of ~ 2.27. They attributed this behavior to a ground state depletion caused by the population build-up of the Yb^{3+} excited level, and the Tm^{3+} $^3H_5$ and $^3H_4$ manifolds. JOHNSON et al. [4.119] disputed that interpretation and attributed the change in dependence to saturation of the $^3H_4$ manifold. The latter interpretation was strongly supported by OSTERMAYER et al. [4.145] who undertook a detailed study of $YF_3 : Yb^{3+}, Tm^{3+}$ and used the saturation of the $^3H_4$ manifold to determine one of the transfer rates. They first measured the lifetimes of all the intermediate levels involved in the up-conversion in singly doped crystals and as a function of concentration. This procedure allowed them to correct for ion pair relaxation processes within the Tm^{3+} or Yb^{3+} systems. They were able to determine the three transfer rates by another series of experiments. Since the lifetime of the Yb^{3+} $^2F_{5/2}$ level is dependent upon the transfer rate to the Tm^{3+} ions, a measurement of the lifetime as a function of Tm^{3+} concentration permits the first transfer rate to be determined. It was noted the procedure yielded a constant value for c_{d2} independent of the Tm^{3+} concentration again confirming the applicability of the rate equations. The rate constant for the second transfer was obtained by measuring the saturation of the $^3H_4$ level. This saturation effect occurs because the $^3H_4$ manifold is metastable and is sufficiently removed from other manifolds that back-transfer is not efficient. The population can build up to the point where the probability for the second transfer becomes comparable to the intrinsic decay rates. Since both the probability of the second transfer that depopulates the level and the first transfer that populates it depend linearly upon the population of the excited Yb^{3+} manifold, saturation will occur when the rate of transfer from the $^3H_4$ level is much greater than the intrinsic decay rate. The rate constant for the second transfer can therefore be extracted from measurements of $^3H_4$ emission as a function of pump intensity. Finally, the rate constant for the third transfer can be obtained by measuring the ratio of fluorescence from the $^3F_4$ manifold and the $^1G_4$ manifold. If ground state depopulation, back transfer, and recycling terms are ignored, this ratio is simply dependent upon the radiative rates from $^3F_4$ and $^1G_4$ and the total $^1G_4$ lifetime. With all the important parameters known, the authors calculated the concentration dependence of up-conversion efficiency. It was shown that the calculation is in very good agreement with experiment. The agreement is seen as justification of the proposed mechanism and the rate equation description for up-conversion.

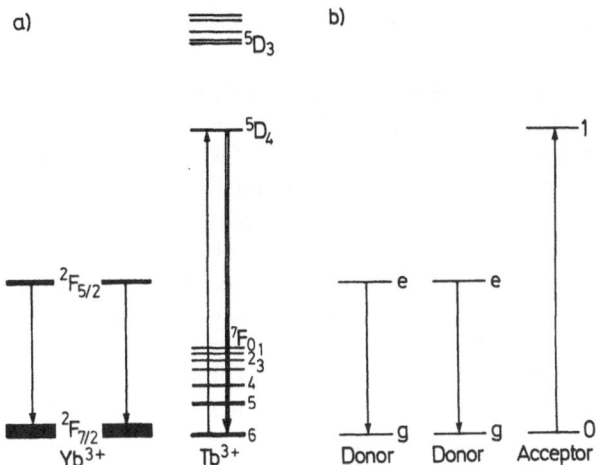

Fig. 4.15. (a) Cooperative up-conversion in Yb^{3+}, Tb^{3+} systems. (b) Theoretical model for cooperative up-conversion

Tb^{3+} Activated Systems—Cooperative Energy Transfer

The importance of cooperative vs. successive transfers from excited Yb^{3+} donor ions has been discussed theoretically by MIYAKAWA and DEXTER [4.56], and later by KUSHIDA [4.114–116]. The predictions of both authors differed by three orders of magnitude but even with this disagreement, both conclude that cooperative transitions are less efficient than successive transfers and therefore would not be an important mechanism in systems which allow successive transfers. MIYAKAWA und DEXTER pointed out that the successive transfer mechanism could not be operative for Yb^{3+}–Tb^{3+} systems because the gap between the fluorescent $^5D_4$ level and the next lower level was much too large to be spanned by a single Yb^{3+} transfer [4.56]. It could be spanned, however, by a cooperative transition of two excited Yb^{3+} ions. Working independently, both LIVANOVA et al. [4.25], and OSTERMAYER and VAN UITERT [4.26] positively identified cooperative transitions in Yb^{3+}–Tb^{3+} systems. The levels involved are shown in Fig. 4.15 a.

The cooperative transfer can be described by a rate equation model with the same assumptions of rapid diffusion. The levels to be used in the model are shown in Fig. 4.15 b. The rate equations become

$$\dot{n}_e^d = -\tau_d^{-1} n_e^d - c_{d1}(n_e^d)^2 n_0 + c_{1d}(n_g^d)^2 n_1 + \sigma^d \Phi n_g^d \tag{4.42}$$

$$\dot{n}_1 = -\tau_1^{-1} n_1 + c_{d1}(n_e^d)^2 n_0 - c_{1d}(n_g^d)^2 n_1 . \tag{4.43}$$

Neglecting stimulated emission, ground state depopulations, and treating the transfer terms as small perturbations upon the intrinsic processes, one finds for the steady state value

$$n_1 = \tau_1 \tau_d c_{d1} (\sigma^d \Phi)^2 N_a N_d^2 \tag{4.44}$$

where the same notation is used as in Section 4.4. Both LIVANOVA et al., and OSTERMAYER and VAN UITERT observed the $^5D_4$ fluorescence depended quadratically on the pump intensity as predicted by (4.44) [4.26]. They also measured both the $^5D_4$ and $^2F_{5/2}$ lifetimes as a function of the Tb^{3+} concentration and showed that (4.44) was satisfied over the entire concentration range used. LIVANOVA et al. [4.25] measured the time dependence of the cooperatively excited $^5D_4$ fluorescence and showed that it could be explained using the known lifetimes of the $^5D_4$ and $^2F_{5/2}$ states and a cooperative transition. These studies show conclusively the existence of three body processes in rare-earth crystals. OSTERMAYER and VAN UITERT [4.26] estimated that such processes are ~130 times weaker in the Tb–Yb system than the efficient Er–Yb system. Although this appears relatively weak, one must bear in mind that the Er^{3+} system involves almost resonant processes. When only non-resonant decays are allowed for pair processes but resonant schemes are possible for three-body processes, the three-body process might indeed be important. As explained in Subsect. 4.3.2, multi-body interactions may be important in previously performed fluorescence quenching studies. Further work is required to understand the importance of such processes.

4.5.2 Singly Doped Systems

Up-conversion in singly doped crystals differs from the Yb^{3+} sensitized crystals described in the last subsection because the levels that act as donors are not always populated directly by the pumping radiation but must sometimes be populated indirectly by non-radiative relaxation processes. Since the same ion is acting as donor and acceptor and since the ion possesses a more complex level structure than the Yb^{3+} donor, ion pair interactions become increasingly important at higher concentrations and it is usually not advantageous to go to the high donor concentrations that were typical of Yb^{3+} sensitized systems. However, when one works at relatively low concentrations, exciton migration becomes less efficient and more spatial inhomogeneities can exist. Ion pairs that are close together produce up-conversion more readily than those more distant. More importantly, the pairs that are near each other will have different lifetimes when self-quenching occurs than those pairs that are more distant. Such behavior makes the systems difficult to study because the temporal behavior of fluorescence generated by direct pumping will not necessarily be related to the temporal behavior generated by up-conversion since the ensemble average of ions involved in the two process is different. This effect could become quite severe at low concentrations. Studies in fact have been performed which demonstrate the inactivity of most ions in a crystal in producing up-conversion [4.11a]. Three examples of these singly doped crystals will be given.

Pr^{3+} Activated Systems

Up-conversion has been studied in singly doped $LaF_3:Pr^{3+}$ crystals and $LaCl_3:Pr^{3+}$ crystals by ZALUCHA et al. [4.146]. Although both systems have similar behavior, the $LaCl_3:Pr^{3+}$ is more complex and will be the one considered here. The up-conversion in this system was discovered as a result of the interference it caused in obtaining a practical infrared quantum counter [4.5]. An example of this is shown in Fig. 4.16. This spectrum was obtained

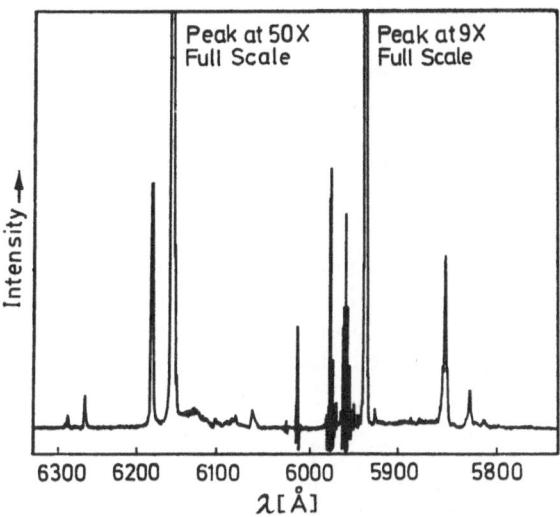

Fig. 4.16. Double resonance excitation spectrum of Pr^{3+} $^3P_0$ fluorescence under broad-band infrared excitation and varying dye laser excitation wavelengths. The fluorescence is detected synchronously with the infrared excitation and fluorescence that is generated without the infrared source appears as noise such as observed around 6000 Å in this figure [4.5]

by exciting the infrared Pr^{3+} levels with a chopped infrared excitation source and detecting the green up-converted fluorescence from the $^3P_0$ state synchronously with the infrared source. The up-conversion was produced by scanning a dye laser over the wavelength region that could excite a Pr^{3+} ion from a lower infrared level to the upper levels. (The scheme is sketched in Fig. 4.17a.) Each peak represents a resonance between a lower infrared level and an upper state. At certain frequencies, however, a large noise component appears which obscures any possible IRQC signal. The noise occurs because the strong laser pump is directly exciting unchopped $^3P_0$ fluorescence at these

a)   b)   c)

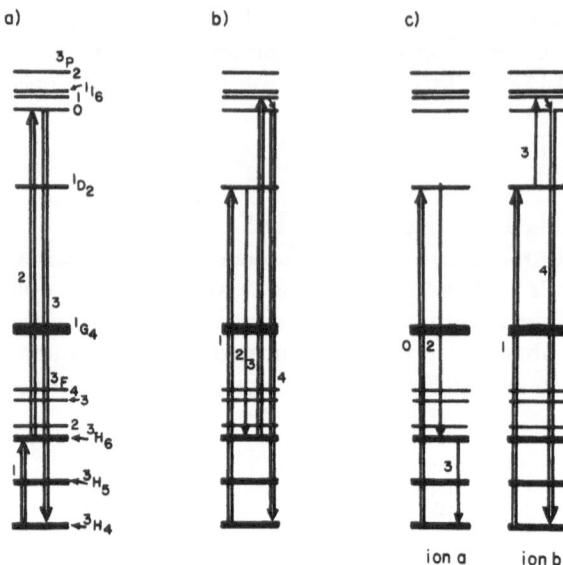

ion a   ion b

Fig. 4.17 a–c. Mechanism of up-conversion in LaCl$_3$: Pr^{3+}. Double arrows represent transitions induced by the radiation field and single arrows represent transitions induced by inter-ionic coupling. The temporal sequence of transitions is indicated by arrows beside each transition

frequencies. If the laser excited fluorescence is monitored directly without the infrared pump present, spectra such as shown in Fig. 4.18 are obtained. Each line represents an excitation wavelength where up-converted green fluorescence ($\sim$490 nm) from the Pr^{3+} $^3P_0$ state is obtained. Both the wavelength and po-

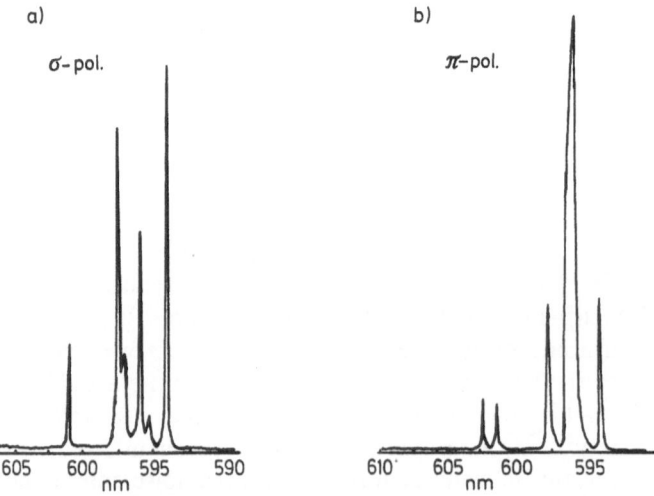

Fig. 4.18a and b. Excitation spectrum of the $^3P_0$ up-converted fluorescence in LaCl$_3$: Pr^{3+}. The polarizations indicate the orientation of the c-axis with the polarization of the excitation source

larization of each line can be identified with a known transition in $LaCl_3: Pr^{3+}$. These are tabulated in Table 4.4. Two mechanisms have been determined for the observed up-conversion [4.147]. These are sketched in Fig. 4.17b and c. The first mechanism begins with an absorption of a photon to the Pr^{3+} $^1D_2$

Table 4.4.

Wavelength [nm]	Polarization	Transition
602.4	π	$^3H_4(3) \rightarrow {}^1D_2(0)$
601.3	σ	$^3H_4(2) \rightarrow {}^1D_2(0)$
597.7	σ	$^3H_4(2) \rightarrow {}^1D_2(2)$
597.1	σ	$^3H_4(3) \rightarrow {}^1D_2(1)$
596.3	π	$^3H_6(2) \rightarrow {}^3P_1(1)$
595.9	π	$^3H_4(2) \rightarrow {}^1D_2(1)$
595.2	σ	$^3H_6(2) \rightarrow {}^3P_1(0)$
594.0	σ	$^3H_6(3) \rightarrow {}^3P_1(1)$

manifold. The absorption need not correspond with a pure electronic transition but can involve lattice phonons. This level decays to the lower levels eventually populating the $^3H_6$ manifold. A second photon then excites the ion from the $^3H_6$ state to the $^3P_1$ state which can either fluoresce or decay by multiphonon emission to the $^3P_0$ state which can also fluoresce. The entire process involves a single ion and is not really different from an IRQC scheme. The second mechanism differs from the first because the excitation to the $^3P_1$ is caused by a two-ion pair interaction between an ion in the $^1D_2$ and one in a lower state such as the $^3H_6$. (There are actually a number of lower states that can resonantly cause excitation to upper levels.) This mechanism is identical to exciton annihilation seen in organic systems. Both mechanisms have a quadratic dependence of up-converted fluorescence on pump power as is observed experimentally for all the transitions.

ZALUCHA et al. [4.147] examined the time dependence of several of these transitions to determine the mechanism for up-conversion. If the excitation mechanism that populates the $^3P_0$ requires photon absorption, the $^3P_0$ state will decay with its intrinsic time dependence. If the excitation mechanism is annihilation of two lower energy excitations, the decay will be a function of the two lower level lifetimes and the $^3P_0$ lifetime. Their analysis demonstrated that the excitation at 595.9 nm produced up-conversion by absorption of two photons while excitation at 602.4 nm produced up-conversion by annihilation of two lower energy excitations.

The detailed fitting of the observed time dependences in this system is complex. If one assumes the 602.4 nm excitation is caused by annihilation of $^3H_6$ and $^1D_2$ excitations, the time dependence for the up-converted fluorescence can be predicted from the rate equation model. The lifetimes of the $^3H_6$, $^1D_2$ and $^3P_0$ levels can be obtained by measuring their time dependence after direct excitation. They were determined to be 14, 73.2, and 17.3 μs, respectively, for a 1% $LaCl_3$: Pr^{3+} crystal. The time dependence obtained is shown in Fig. 4.19 along with the experimental values [4.147]. It can be seen that

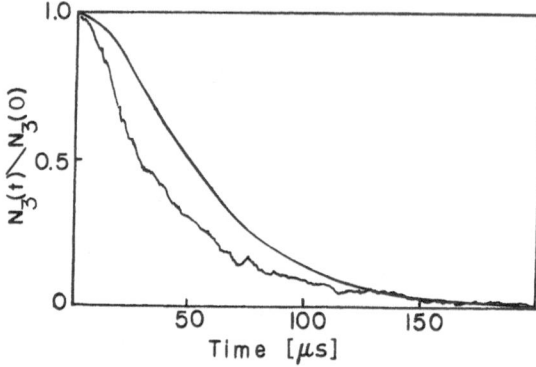

Fig. 4.19. Decay of $^3P_0$ up-converted fluorescence after excitation at 602.4 nm along with the theoretical decay expected from measurement of intrinsic level lifetimes [4.147]

the behavior after ~60 μs is that predicted but before that time, it is different. The experimental curve decays much more rapidly than would be expected for the theoretical prediction of the rate equation model. This is exactly the behavior expected if the excitation migration was restricted. As seen in the previous section, the diffusion constant for migration will not be large at 1% concentrations. The rapid decay during the initial time period is caused by direct annihilation processes between nearby excited Pr^{3+} ions. This interpretation is supported by the non-exponential decay that is observed from the $^3H_6$ and $^1D_2$ manifolds when they are directly excited showing that the limit of rapid diffusion has not been reached.

Er^{3+} Activated Systems

Singly doped Er^{3+} crystals have been studied in different crystalline hosts by a number of groups [4.140, 148–154]. Up-converted fluorescence can be generated from the $^4F_{9/2}$ manifold and the $^4S_{3/2}$ when either the $^4I_{11/2}$ or $^4I_{13/2}$ manifolds are excited. JOHNSON et al. [4.119] have observed a series of successive transfers from Er^{3+} ions excited to the $^4I_{13/2}$ manifold and to the $^4I_{9/2}$, $^4F_{9/2}$, $^4F_{7/2}$, and $^2H_{9/2}$ manifolds, as shown in Fig. 4.20. Van DER ZIEL et al. [4.140] excited the $^4I_{11/2}$ manifold and observed fluorescence from the $^4S_{3/2}$ and $^4F_{9/2}$ manifolds. The up-conversion mechanism for the $^4S_{3/2}$ emission was very similar to that of the Yb^{3+} sensitized systems. Two $^4I_{11/2}$ excitations annihilate to cause population of the $^4F_{7/2}$ or $^2H_{11/2}$ manifolds which decay nonradiatively to the $^4S_{3/2}$ manifold. Studies of the time dependence of $^4S_{3/2}$ fluorescence again show disagreements between the predicted behavior using the directly measured level lifetimes and the temporal behavior actually observed. The disagreement persisted until a concentration of about 10% where agreement is reached. The mechanism for populating the $^4F_{9/2}$ manifold

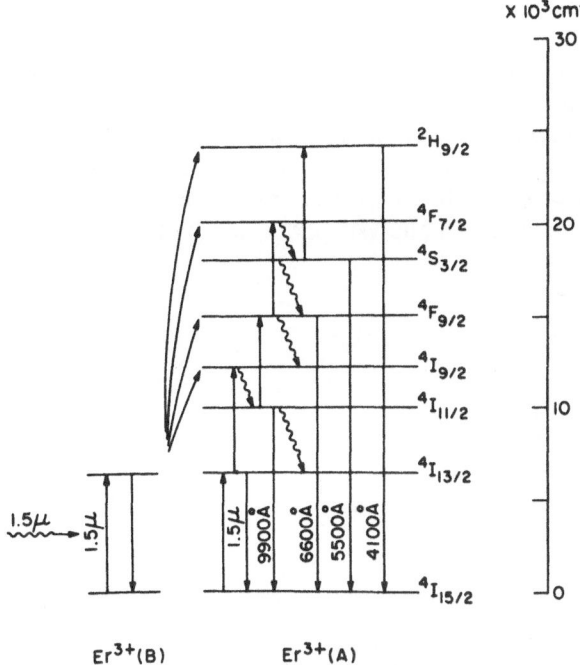

Fig. 4.20. Successive transfers observed in $BaYF_5:Er^{3+}$ when the $^4I_{13/2}$ manifold is excited [4.119]

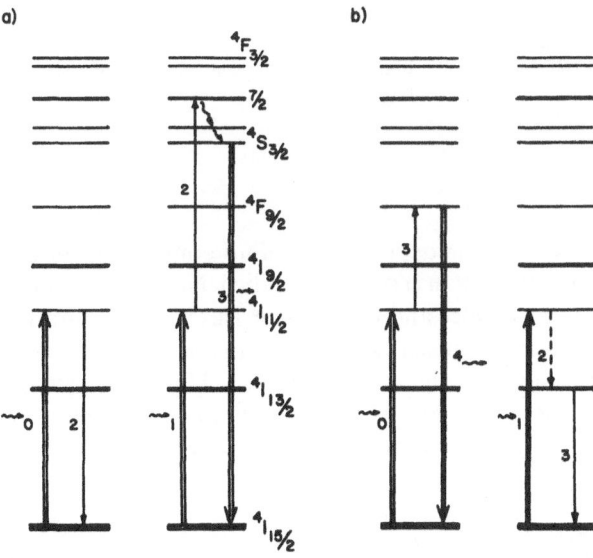

Fig. 4.21a and b. Mechanism of up-conversion in Er^{3+} crystals

was again not clear. Since the ratio of $^4S_{3/2}$ to $^4F_{9/2}$ fluorescence was smaller when fluorescence was excited through the $^4I_{11/2}$ manifold than through the $^4F_{7/2}$ manifold, multi-phonon decay from the $^4S_{3/2}$ level was not considered to be an important mechanism. It was suggested that the $^4I_{13/2}$ manifold could be populated either by multi-phonon decay from the $^4I_{11/2}$ manifold or an ion pair decay from the $^2H_{11/2}$ and this excitation could combine with another $^4I_{11/2}$ excitation to produce the $^4F_{9/2}$ excitation. These schemes are shown in Fig. 4.21. They agree with the analysis performed by KURODA et al. [4.136].

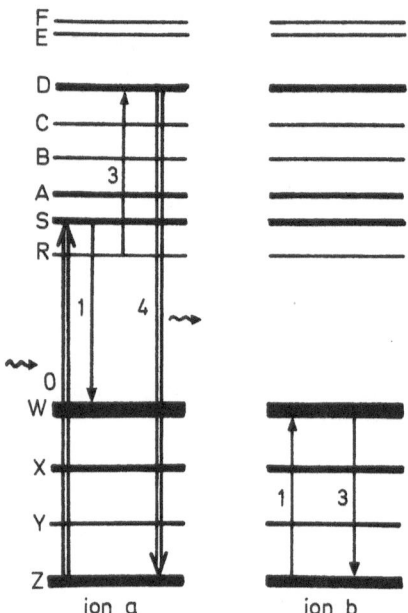

Fig. 4.22. Mechanism of up-conversion in NdCl$_3$

NdCl$_3$

In high concentration crystals, the time dependences of up-conversion should closely obey a rate equation model. PARTLOW has studied NdCl$_3$ and found excellent agreement between the model and experiment [4.155]. The fluorescence from the D level was excited by pulsed excitation of the A, S, and R levels. The mechanism that was proposed is shown in Fig. 4.22. After the S and R levels are excited, the S will decay by ion pair processes to the W level. An excitation of the W and R level can combine to excite the D level. The time dependence for this process is

$$n_D = \frac{A}{\tau_D^{-1} - \tau^{-1}} \left[\exp\left(-\frac{t}{\tau}\right) - \exp\left(-\frac{t}{\tau_D}\right) \right] \tag{4.45}$$

where

$$\tau^{-1} = \tau_R^{-1} + \tau_W^{-1} \qquad\qquad (4.46)$$

and τ_R, τ_W and τ_D are the intrinsic level lifetimes, and A is a constant. The intrinsic decay from the R, W, and D levels was exponential with time constants of 161, 1985, and $<3\,\mu s$, respectively. These values then predict an exponential decay with a time constant of 149 μs. This was compared with a measured value of 139 μs. The values agreed within experimental error and demonstrate the appropriateness of the rate equation model and the proposed mechanism.

PARTLOW pointed out that the procedure just described is not very sensitive to the W level lifetime since it is much longer than the R level [4.155]. To prove the W level was involved in the mechanism, a double resonance experiment was performed in which a second pulsed source excited the R manifold after a variable delay from the first excitation. The D fluorescence that was produced after the second excitation decayed exponentially with increasing delay times and had a time constant of 2003 μs. Since this is the same value as the W state lifetime, the proposed mechanism is further confirmed. This technique of double resonance has not been used in other systems to establish proposed mechanism but it is potentially a very powerful technique, as PARTLOW showed. Its use could eliminate many of the uncertainties in studying a system and establish mechanisms with greater certainty than has been done in the past.

4.5.3 Energy Transfer Between Ions in Isolated Sites

All of the systems described to this point involve energy transfer between ions that can have different radial separations. A completely different approach has recently been developed that allows the study of direct energy transfer between two ions in well-defined positions relative to each other. In crystals such as $CaF_2:Er^{3+}$, the rare-earth dopant is charge compensated by a fluoride interstitial ion [4.156]. The dipolar interactions between two rare-earth interstitial pairs can cause clustering of pairs to occur [4.106]. Such a process has been definitively identified in $CaF_2:Er^{3+}$ [4.105, 153]. The charge compensated systems have been difficult to study spectroscopically because a large number of different crystallographic sites are available in such crystals and the composite spectra of all sites is very complex. It has recently been shown that this problem can be overcome by selectively exciting a specific site with a narrow-band tunable dye laser [4.153, 157]. Energy transfer has not been observed between crystallographic sites and the fluorescence spectra resulting from this technique belongs to one specific site. In particular, single sites associated with rare-earth interstitial clustering can be selectively excited and studied. The technique is illustrated in Fig. 4.23. The spectra shown in Fig. 4.23a is the excitation spectrum of the $^4F_{5/2}$ manifold of Er^{3+} that results after monitoring fluorescence from all possible sites with a broad bandwidth monochromator. If a specific fluorescence line of a specific site is monitored however, spectra such, as shown in Fig. 4.23 b–f are obtained. Both spectra in Fig. 4.23 b and c arise from single pair sites while that shown in

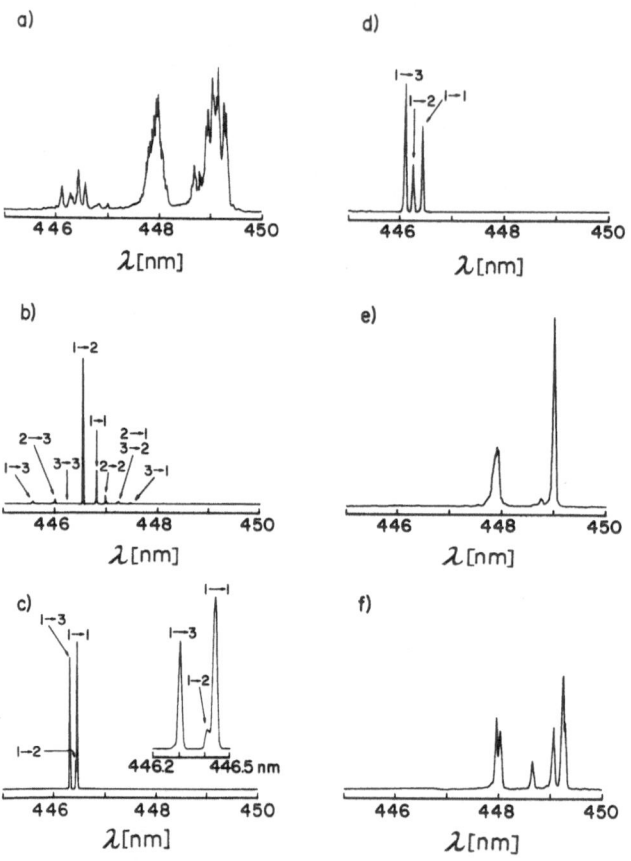

Fig. 4.23 a–f. $^4I_{15/2} \to {}^4F_{5/2}$ ($Z \to H$) excitation spectrum of CaF_2: Er^{3+} at 10 K. Crystal field transition assignments are indicated by numbers, e. g., $1\to2$ indicates the $Z_1 \to H_2$ transition of a given site [4.153]. (a) All sites present in the 0.2% CaF_2: Er^{3+} crystal are contributing to the spectrum in this figure because the fluorescence of the $^4S_{3/2} \to {}^4I_{15/2}$ transition was monitored with a very wide bandwidth instrument. (b) A site excitation spectrum obtained under identical conditions to the "a" figure but while monitoring the $E_1 \to Y_1$ transition of the A site at 835.4 nm. (c) B site excitation spectrum obtained by monitoring the $E_1 \to Y_1$ transition of the B site at 835.8 nm. (d) C site excitation spectrum obtained by monitoring the $E_1 \to Y_1$ transition of the C site at 836.1 nm. e) $D(1a)$ site excitation spectrum obtained by monitoring a $D(1a)$ fluorescence transition at 843.1 nm. This spectrum is representative of one of sixteen different sites that have very similar spectra and together make up the broad-band shown in the "a" figure. (f) $D(2a)$ site excitation spectrum obtained by monitoring a $D(2a)$ fluorescence transition at 654.4 nm. This spectrum is representative of one of five different sites that have very similar spectra and together make up the broad-band shown in the "a" figure

Fig. 4.23 d–f arise from pair clustering. The spectra in Fig. 4.23 e,f are actually just representative spectra of two general classifications of cluster sites that overlap to form the broad bands in Fig. 4.23a. There are a total of sixteen sites that have been identified and have similar spectral splittings to that shown in Fig. 4.23 e. Four sites have been identified that have similar lines to those shown in Fig. 4.23 f.

The cluster sites are particularly interesting because the ions within the site are very close to each other and demonstrate efficient energy transfer. Since one is dealing with a specific site and not an ensemble average over many ion pairs with different radial separations, one can look at direct energy transfer processes without the need to perform theoretical modeling of the experimental data to abstract information [4.158]. In addition, the averaging effects of the ensemble of ion pair separations are not present to obscure some of the details of energy transfer mechanisms. A particularly illustrative example of this was observed in analyzing the excitation mechanism of $^4F_{9/2}$ fluorescence in $CaF_2 : Er^{3+}$ when the $^4F_{5/2}$ manifold was excited. The temporal dependence of the $^4F_{9/2}$ fluorescence is shown in Fig. 4.24 [4.153]. One might explain the two

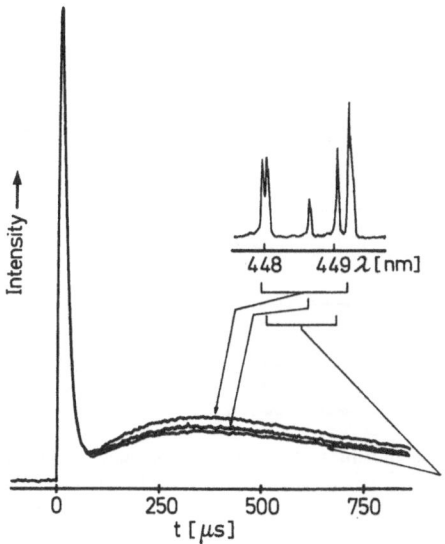

Fig. 4.24. Time dependence of $D(2a)$ site fluorescence [4.157]. Traces represent signal-averaged fluorescence decay curves for the five $^4I_{15/2} \rightarrow {}^4F_{5/2}$ $(Z \rightarrow H)$ excitation lines of the $D(2a)$ site (see insert). For height of the first peak in the time dependence was normalized for each trace. All the traces and the excitation spectrum (insert) were obtained by monitoring the fluorescence line at 654.4 nm. Zero time was synchronous with the laser pulse. The figure shows that the two outer lines in the excitation spectrum are associated with one of the two ions that make up the dimerized $D(2a)$ sites, that the next two inner lines are associated with the other ion of the dimer, and that the middle line is a composite of both [4.153]

peaks observed in this figure by two excitation mechanisms. The first would be a direct excitation of the $^4F_{9/2}$ manifold by an ion pair relaxation in which the $^4F_{5/2}$ manifold excitation decays into $^4F_{9/2}$ and $^4I_{13/2}$ excitations. The second would arise from cascading of the $^4F_{5/2}$ excitation downward by multi-phonon relaxation eventually exciting the $^4F_{9/2}$ manifold. A detailed analysis of the temporal behavior reveals that the mechanism is considerably more complex.

The direct population of the $^4F_{9/2}$ manifold does result from an ion pair decay from the $^4F_{5/2}$ manifold. The lifetime of the $^4F_{9/2}$ manifold is too long however to explain the rapid decay time of the first peak in Fig. 4.24. The rapid decay results from an efficient recombination of the neighboring $^4F_{9/2}$ and $^4I_{13/2}$ excitations causing up-conversion to the $^4F_{7/2}$ manifold. This process has been confirmed from the temporal behavior of $^4F_{7/2}$ fluorescence. The $^4F_{7/2}$ excitation decays rapidly by multi-phonon relaxation to the $^4S_{3/2}$ manifold. A fraction of the $^4S_{3/2}$ excitation can relax directly to the $^4F_{9/2}$ manifold but a large fraction relaxes by ion pair decay from the $^4S_{3/2}$ manifold to the $^4I_{11/2}$ and $^4I_{13/2}$ manifolds of the two ions. These excitations in turn recombine efficiently to excite the $^4F_{9/2}$ manifold. The latter steps have been confirmed by fitting the temporal behavior of $^4I_{11/2}$ fluorescence. It is significant that the $^4I_{11/2}$ and $^4I_{13/2}$ excitations recombine efficiently over a time scale of 200–300 μs because it means that exciton diffusion away from the cluster site is not very efficient on that time scale. The various steps in the mechanism have been sketched in Fig. 4.25. The theoretical prediction of this mechanism matches

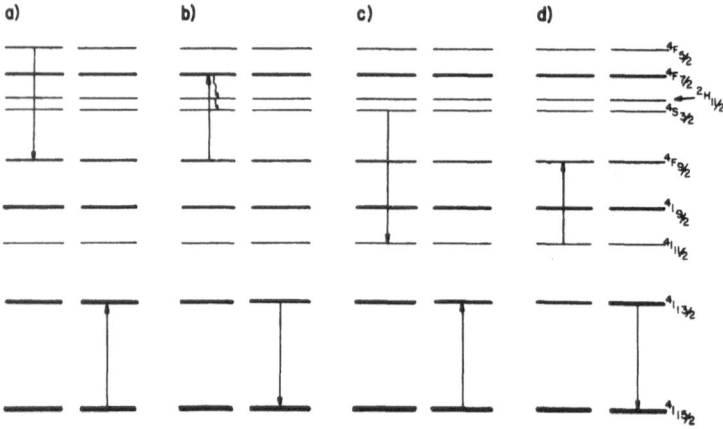

Fig. 4.25a–d. Excitation mechanism responsible for the fluorescence time dependence seen in Fig. 4.24 [4.158, 159]. The initial excitation of the Er^{3+} $^4F_{5/2}$ manifold decays by an ion pair relaxation exciting the $^4F_{9/2}$ fluorescence monitored in Fig. 4.24. This process produces the first peak in the time dependence. The $^4F_{9/2}$ excitation is then up-converted to the $^4F_{7/2}$ state resulting in the rapid decay of the first peak in Fig. 4.24. The $^4F_{7/2}$ excitation relaxes by multi-phonon decay to the $^4S_{3/2}$ manifold. A large fraction of the $^4S_{3/2}$ excitation decays by an ion-pair relaxation to the $^4I_{11/2}$ manifold and is then up-converted to the $^4F_{9/2}$ manifold. This results in the second peak in the time dependence. A small fraction of $^4S_{3/2}$ excitation can also relax directly to the $^4F_{9/2}$ manifold by multi-phonon decay

the observed $^4F_{9/2}$ fluorescence time dependence shown in Fig. 4.24 exactly over the entire time span [4.159]. The example demonstrates the details of excitation mechanism that can be obtained in these systems and the unique ability to look directly at energy transfer between ions in a single defined environment.

4.6 Concluding Remarks

The basic processes that are important in obtaining up-conversion of energy in rare-earth systems are known but their description requires more detailed knowledge. In those crystals where concentrations are sufficiently high that rapid exciton diffusion makes all ions equivalent, the up-conversion mechanisms can be described by a rate-equation model if intermediate levels and their lifetimes are known and if the rate constants for the energy transfer steps can be measured. The model can then be used to optimize the system. Although it is possible to describe systems at the present time, it would be desirable to predict how a new system will perform theoretically without the need for experimental measurements. The major difficulty with this approach at the present time is the basic models for energy transfer are not well established experimentally. Theoretical models have been proposed to describe the effect of non-resonance, exciton migration, the coupling constants of electronic states to the lattice, lattice phonon modes, and the individual nature of electronic states on the rate of energy transfer and there has been experimental work that confirms much of the theory. The theory is not able to predict rate constants without the use of parameters that must be determined by experiment. A very limited number of experiments have been done to provide this information but there is need for much more work before theoretical analysis of new systems can be attempted. MITA has demonstrated the basic principles required for a theoretical analysis when sufficient information is obtained [4.132, 133].

At low concentrations where the inequivalence of ion sites is seen in the non-exponential decay of levels, the problem is more complex and therefore difficult. To describe these systems, the ion coupling mechanism becomes important because it determines the ions that participate in the up-conversion. A simple rate-equation model cannot be used except as a rough approximation and one must instead perform an ensemble average over all configurations of ions. This will require description of ion pair decays and multi-ion decay processes as a function of ion separation and orientation which is not feasible at the present time. In fact, the possible occurrence of multi-ion decays has only been recently demonstrated and more work is required to evaluate their importance.

Experimentally, many studies have not identified but only suggested mechanisms for up-conversion. Since a system cannot be described until the mechanism is known, its determination is an important contribution. Many experimenters have not used the transient behavior of their system to determine mechanisms although measurement of the temporal build-up and decay can give direct information on the intermediate levels involved in the mechanism if the lifetimes of the intermediate levels are known. Additionally, double resonance experiments such as PARTLOW performed [4.155] can provide further information on intermediate levels that are not definitively determined by the transient behavior. A modification of this technique could be equally important. If a second pulsed excitation was delayed from the first by a variable amount and was tuned to a transition from an intermediate state to a fluorescent state much higher than those participating in the up-conversion, the temporal behavior of the intermediate state could be related directly to the up-converted fluorescence by monitoring the upper level fluorescence as a function of the delay between the excitations. Such a technique could permit each proposed intermediate level be examined individually to see if its population is consistent with the mechanism under consideration.

References

4.1   G. H. DIEKE: *Spectra and Energy Levels of Rare Earth Ions in Crystals* (Wiley-Interscience, New York 1968)
4.2   L. F. JOHNSON, H. J. GUGGENHEIM: Appl. Phys. Lett. **19**, 44 (1971)
4.3   F. E. AUZEL: Proc. IEEE **61**, 758 (1973)
4.4   L. ESTEROWITZ, J. NOONAN, J. BAHLER: Appl. Phys. Lett. **10**, 126 (1967)
4.5   J. C. WRIGHT, D. J. ZALUCHA, H. V. LAUER, D. E. COX, F. K. FONG: J. Appl. Phys. **44**, 781 (1973)
4.6   N. A. TOLSTOI, A. P. ABRAMOV: Opt. Spectrosc. **20**, 273 (1966)
4.7   N. A. TOLSTOI, A. P. ABRAMOV: Opt. Spectrosc. **22**, 272 (1967)
4.8   N. BLOEMBERGEN: Phys. Rev. Lett. **2**, 84 (1959)
4.9   W. B. GANDRUD, H. W. MOOS: J. Appl. Phys. **39**, 4777 (1969)
4.10   J. C. WRIGHT, F. K. FONG, M. M. MILLER: J. Appl. Phys. **42**, 3806 (1971)
4.11   M. R. BROWN, W. A. SHAND: *Adv. Quant. Electron.*, ed. by D. W. GOODWIN (Academic Press, New York 1970)
4.12   M. F. AUZEL: Compt. Rend. **262**, 1016 (1966)
4.13   M. F. AUZEL: Compt. Rend. **263 B**, 819 (1966)
4.14   V. V. OVSYANKIN, P. P. FEOFILOV: Sov. Phys. JETP Lett. **3**, 317 (1966)
4.15   V. V. OVSYANKIN, P. P. FEOFILOV: Sov. Phys. JETP Lett. **3**, 322 (1966)
4.16   S. SINGH, J. E. GEUSIC: In *Optical Properties of Ions in Crystals*, ed. by H. W. CROSSWHITE and H. W. MOOS (Wiley-Interscience, New York 1967)
4.17   V. ERN, P. AVAKIAN, R. E. MERRIFELD: Phys. Rev. **148**, 862 (1966)
4.18   R. G. KEPLER, J. C. CARIS, P. AVAKIAN, E. ABRAMSON: Phys. Rev. Lett. **10**, 400 (1963)
4.19   N. A. TOLSTOI, A. P. ABRAMOV: Sov. Phys. – Solid State **9**, 255 (1967)
4.20   D. L. DEXTER: Phys. Rev. **108**, 630 (1957)
4.21   P. J. BOTDEN, Th.: Philips Res. Rep. **7**, 197 (1952)
4.22   D. L. DEXTER, J. H. SCHULMAN: J. Chem. Phys. **22**, 1063 (1954)
4.23   B. M. ANTIPENKO, A. V. DMITRUK, V. S. ZUBKOVA, G. O. KARAPETYAN, A. A. MAK: In *Luminescence of Crystals, Molecules, and Solutions*, ed. by F. WILLIAMS (Plenum Press, New York 1973)

4.24 B. M. Antipenko, A. V. Dmitryuk, G. O. Karapetyan, V. S. Zubkova, V. I. Kosyakov, A. A. Mak, N. V. Mikhailova: Opt. Spektrosk. 35, 540 (1973) [English Translation in Opt. Spectrosc. 35, 315 (1973)]

4.25 L. D. Livanova, I. G. Saitkulov, A. L. Stolov: Sov. Phys. – Solid State 11, 750 (1969)

4.26 F. W. Ostermayer, L. G. Van Uitert: Phys. Rev. B1, 4208 (1970)

4.27 R. S. Meltzer: Phys. Rev. B2, 2398 (1970)

4.28 R. S. Meltzer, H. W. Moos: Phys. Rev. 6, 264 (1972)

4.29 V. A. Arkhangelskaya, P. P. Feofilov: Opt. Spectrosc. 28, 657 (1970)

4.30 M. D. Sturge: J. Chem. Phys. 51, 1254 (1969)

4.31 M. J. Taylor: Phys. Rev. Lett. 23, 405 (1969)

4.32 J. P. van der Ziel, L. G. Van Uitert: Phys. Rev. B8, 1835 (1973)

4.33 F. Varsanyi, G. H. Dicke: Phys. Rev. Lett. 7, 442 (1961)

4.34 P. P. Feofilov, A. K. Trofimov: Opt. Spectrosc. 27, 291 (1969)

4.35 P. P. Feofilov: Opt. Spectrosc. 31, 462 (1971)

4.36 J. Ferguson, H. J. Guggenheim, Y. Tanable: J. Chem. Phys. 45, 1134 (1966)

4.37 E. Nakazawa, S. Shinoya: Phys. Rev. Lett. 25, 1710 (1970)

4.38 J. P. van der Ziel, L. G. Van Uitert: Phys. Rev. B8, 1889 (1973)

4.39 J. P. van der Ziel: J. Lumin. 1/2, 807 (1970)

4.40 J. P. van der Ziel, L. G. Van Uitert: Solid State Commun. 7, 819 (1969)

4.41 J. P. van der Ziel, L. G. Van Uitert: Phys. Rev. Lett. 21, 1334 (1968)

4.42 J. P. van der Ziel, L. G. Van Uitert: Phys. Rev. 186, 332 (1969)

4.43 J. P. van der Ziel, L. G. Van Uitert: Phys. Rev. 180, 343 (1969)

4.44 M. Altarelli, D. L. Dexter: Opt. Commun. 2, 36 (1970)

4.45 V. I. Bilak, G. M. Zuerev, G. O. Karapetyan, A. M. Onishanko: Sov. Phys. JETP Lett. 14, 199 (1971)

4.46 A. L. Schawlow, D. L. Wood, A. M. Cloyston: Phys. Rev. Lett. 3, 271 (1959)

4.47 J. P. van der Ziel: Phys. Rev. B4, 2888 (1971)

4.48 J. C. Wright, H. W. Moos, J. H. Colwell, B. W. Mangum, D. D. Thornton: Phys. Rev. B3, 843 (1971)

4.49 G. G. P. Van Gorkam: Phys. Rev. B8, 1827 (1973)

4.50 D. L. Dexter, Th. Förster, R. S. Knox: Phys. Status Solidi 34, K 159 (1969)

4.51 D. L. Dexter: Phys. Rev. 126, 1962 (1962)

4.52 V. V. Ovsyankin, P. P. Feofilov: In Luminescence of Crystals, Molecules, and Solutions, ed. by F. Williams (Plenum Press. New York 1973)

4.53 J. D. Dow: Phys. Rev. 174, 962 (1968)

4.54 F. K. Fong, D. J. Diestler: J. Chem. Phys. 56, 2875 (1972)

4.55 J. C. W. Grant: Phys. Rev. B4, 648 (1971)

4.56 T. Miyakawa. D. L. Dexter: Phys. Rev. B1, 70 (1970)

4.57 D. L. Dexter: J. Chem. Phys. 21, 836 (1953)

4.58 Th. Förster: Naturwissenschaften 33, 166 (1946)

4.59 Th. Förster: Ann. Phys. 2, 55 (1948)

4.60 R. M. Birgeneau, M. T. Hutchings, R. N. Rogers: Phys. Rev. 175, 1116 (1968)

4.61 P. M. Levy: Phys. Rev. Lett. 20, 1366 (1968)

4.62 P. M. Levy: Phys. Rev. 177, 509 (1969)

4.63 J. M. Baker, R. J. Birgeneau, M. T. Hutchings, J. D. Riley: Phys. Rev. Lett. 21, 620 (1968)

4.64 R. J. Birgeneau: Appl. Phys. Lett. 13, 193 (1968)

4.65 R. M. Birgeneau: J. Chem. Phys. 50, 4282 (1969)

4.66 G. F. Imbusch: Phys. Rev. 153, 326 (1967)

4.67 Th. Förster: Disc. Faraday Soc. 27, 7 (1959)

4.68 Th. Förster: In Modern Quantum Chemistry P. 3: Action of Light and Organic Crystals, ed. by Oktay Sinanoglu (Academic Press, New York 1965)

4.69 D. L. Dexter: In Luminescence of Crystals, Molecules, and Solutions, ed. by F. Williams (Plenum Press, New York 1973)

4.70 T. Miyakawa, D. L. Dexter: Phys. Rev. B1, 2961 (1970)

4.71   R. ORBACH: In *Optical Properties of Ions in Crystals*, ed. by H. W. MOOS and H. W. CROSS-WHITE (Wiley-Interscience, New York 1967)

4.72   N. YAMADA, S. SHINOYA, T. KUSHIDA: J. Phys. Soc. Japan **32**, 1577 (1972)

4.73   K. B. EISENTHAL, S. SIEGEL: J. Chem. Phys. **41**, 652 (1964)

4.74   M. INOKUTI, F. HIRAYAMA: J. Chem. Phys. **43**, 1978 (1965)

4.75   L. G. VAN UITERT: *Luminescence of Inorganic Solids*, ed by GOLDBERG (Academic Press, New York 1966)

4.76   L. G. VAN UITERT: J. Electrochem. Soc. **114**, 1048 (1967)

4.77   L. G. VAN UITERT, L. F. JOHNSON: J. Chem. Phys. **44**, 3514 (1966)

4.78   L. G. VAN UITERT, E. F. DEARBORN, H. M. MARCOS: Appl. Phys. Lett. **9**, 255 (1966)

4.79   L. G. VAN UITERT, E. F. DEARBORN, J. J. RUBIN: J. Chem. Phys. **47**, 3653 (1967)

4.80   L. G. VAN UITERT, E. F. DEARBORN, J. J. RUBIN: J. Chem. Phys. **47**, 547 (1967)

4.81   L. G. VAN UITERT, E. F. DEARBORN, J. J. RUBIN: J. Chem. Phys. **47**, 1595 (1967)

4.82   L. G. VAN UITERT, E. F. DEARBORN, J. J. RUBIN: J. Chem. Phys. **46**, 420 (1967)

4.83   L. G. VAN UITERT, E. F. DEARBORN, J. J. RUBIN: J. Chem. Phys. **45**, 1578 (1966)

4.84   C. HSU, R. E. POWELL: J. Luminescence **10**, 273 (1975)

4.85   R. C. POWELL, R. G. KEPPER: Phys. Rev. Lett. **22**, 636 (1969), Correction in Phys. Rev. Lett. **22**, 1232 (1969)

4.86   R. C. POWELL: Phys. Rev. **B4**, 628 (1971)

4.87   M. YOKOTA, O. TANIMOTO: J. Phys. Soc. Japan **22**, 779 (1967)

4.88   J. P. VAN DER ZIEL, L. KOPF, L. G. VAN UITERT: Phys. Rev. **B6**, 615 (1972)

4.89   J. D. AXE, P. F. WELLER: J. Chem. Phys. **40**, 3066 (1964)

4.90   J. HEBER, U. KOBLER: In *Luminescence of Crystals, Molecules, and Solutions*, ed. by F. WILLIAMS (Plenum Press, New York 1973)

4.91   L. F. JOHNSON, L. G. VAN UITERT, J. J. RUBIN, R. A. THOMAS: Phys. Rev. **133**, A 494 (1964)

4.92   L. F. JOHNSON, J. E. GEUSIC, L. G. VAN UITERT: Appl. Phys. Lett. **8**, 200 (1966)

4.93   A. D. PEARSON, G. E. PETERSON, W. R. NORTHOVER: J. Appl. Phys. **37**, 729 (1966)

4.94   A. D. PEARSON, G. E. PETERSON: Appl. Phys. Lett. **5**, 222 (1964)

4.95   A. D. PEARSON, G. E. PETERSON: Appl. Phys. Lett. **8**, 210 (1966)

4.96   G. E. PETERSON, P. M. BRIDENBAUGH: J. Opt. Soc. Am. **54**, 644 (1964)

4.97   G. E. PETERSON, A. D. PEARSON, P. M. BRIDENBAUGH: J. Appl. Phys. **36**, 1962 (1965)

4.98   J. F. PORTER, Jr., H. W. MOOS: Phys. Rev. **152**, 300 (1966)

4.99   E. J. SHARP, J. E. MILLER, M. J. WEBER: J. Appl. Phys. **44**, 4098 (1973)

4.100 M. J. WEBER: Phys. Rev. **B4**, 3153 (1971)

4.101 M. J. WEBER, E. J. SHARP, J. E. MILLER: J. Phys. Chem. Solids **32**, 2275 (1971)

4.102 M. J. WEBER: J. Appl. Phys. **44**, 4058 (1973)

4.103 M. J. WEBER, M. BASS, G. A. DEMARS: J. Appl. Phys. **42**, 301 (1971)

4.104 M. J. WEBER: J. Chem. Phys. **44**, 3205 (1973)

4.105 J. B. FENN, J. C. WRIGHT, F. K. FONG: J. Chem. Phys. **59**, 5591 (1973)

4.106 S. L. NABERHUIS, F. K. FONG: J. Chem. Phys. **56**, 1174 (1972)

4.107 W. B. GANDRUD, H. W. MOOS: J. Chem. Phys. **49**, 2170 (1968)

4.108 N. KRASUTSKY, H. W. MOOS: Phys. Rev. **B8**, 1010 (1973)

4.109 R. J. BENNETT: J. Chem. Phys. **41**, 3037 (1964)

4.110 E. NAKAZAWA, S. SHINOYA: J. Chem. Phys. **47**, 3211 (1967)

4.111 H. NISHIMURA, M. TANAKA, M. TOMURA: J. Phys. Soc. Japan **28**, 128 (1970)

4.112 R. K. WATTS, H. J. RICHTER: Phys. Rev. **B6**, 1584 (1972)

4.113 M. J. WEBER: Phys. Rev. **B4**, 2932 (1971)

4.114 T. KUSHIDA: J. Phys. Soc. Japan **34**, 1318 (1973)

4.115 T. KUSHIDA: J. Phys. Soc. Japan **34**, 1327 (1973)

4.116 T. KUSHIDA: J. Phys. Soc. Japan **34**, 1334 (1973)

4.117 J. C. BOURCET, F. K. FONG: J. Chem. Phys. **60**, 34 (1974)

4.118 P. AVAKIAN, V. ERN, R. E. MERRIFIELD, A. SUNA: Phys. Rev. **165**, 974 (1968)

4.119 L. F. JOHNSON, H. J. GUGGENHEIM, T. C. RICH, F. W. OSTERMAYER: J. Appl. Phys. **43**, 1125 (1972)

4.120 M. F. Auzel, O. Deutschbein: Z. Naturforsch. **249**, 1562 (1969)
4.121 H. J. Guggenheim, L. F. Johnson: Appl. Phys. Lett. **15**, 51 (1969)
4.122 R. A. Hewes: J. Lumin. **1/2**, 778 (1970)
4.123 R. A. Hewes, J. F. Sarver: Phys. Rev. **182**, 427 (1969)
4.124 L. F. Johnson, J. E. Geusic, H. J. Guggenheim, T. Kushida, S. Singh, L, G. Van Uitert: Appl. Phys. Lett. **15**, 48 (1969)
4.125 J. D. Kingsley, G. E. Fenner, S. V. Galginaitis: Appl. Phys. Lett. **15**, 115 (1969)
4.126 J. D. Kingsley: J. Appl. Phys. **41**, 175 (1970)
4.127 N. M. P. Low, A. L. Major: J. Lumin. **4**, 357 (1971)
4.128 Y. Mita, E. Nagasawa, K. Shiroki, Y. Ohno, T. Matsubara: Appl. Phys. Lett. **23**, 173 (1973)
4.129 M. Tamatani, K. Yokota, T. Nishimura: J. Phys. Soc. Japan **29**, 1099 (1970)
4.130 J. P. van der Ziel, L. G. Van Uitert, W. H. Grodkiewicz: J. Appl. Phys. **10**, 3308 (1970)
4.131 L. G. Van Uitert, S. Singh, H. J. Levinstein, L. F. Johnson, W. H. Grodkiewicz, J. E. Geusic: Appl. Phys. Lett. **15**, 53 (1969)
4.132 Y. Mita: J. Appl. Phys. **43**, 1772 (1972)
4.133 Y. Mita, F. Nagasawa: Jap. J. Appl. Phys. **12**, 540 (1973)
4.134 J. L. Sommerdijk, A. Bril: *Luminescence of Crystals, Molecules and Solutions*, ed. by F. Williams (Plenum, New York 1973)
4.135 J. L. Sommerdijk, W. L. Wanmaker, J. G. Verriet: J. Lumin. **4**, 404 (1971)
4.136 H. Kuroda, S. Shinoya, T. Kushida: J. Phys. Soc. Japan **33**, 125 (1972)
4.137 M. R. Brown, W. A. Shand: Phys. Lett. **18**, 95 (1965)
4.138 J. L. Sommerdijk: J. Lumin. **4**, 441 (1971)
4.139 M. R. Brown, H. Thomas, J. S. S. Whiting: J. Chem. Phys. **50**, 881 (1969)
4.140 J. P. van der Ziel, F. W. Ostermayer, Jr., L. G. Van Uitert: Phys. Rev. **B2**, 4432 (1970)
4.141 J. P. Wittke, I. Ladany, P. N. Yocum: J. Appl. Phys. **43**, 595 (1972)
4.142 V. V. Ovsyankin, P. P. Feofilov: Opt. Spectrosc. **31**, 510 (1971)
4.143 C. M. Verber, D. R. Grieser, W. H. Jones, Jr.: J. Appl. Phys. **42**, 2767 (1971)
4.144 R. K. Watts: J. Chem. Phys. **53**, 3552 (1970)
4.145 F. W. Ostermayer, J. P. van der Ziel, H. M. Marcos, L. G. Van Uitert, F. F. Geusic: Phys. Rev. **B3**, 2698 (1971)
4.146 D. J. Zalucha, J. C. Wright, F. K. Fong: J. Chem. Phys. **59**, 997 (1973)
4.147 D. J. Zalucha, J. A. Sell, F. K. Fong: J. Chem. Phys. **60**, 1660 (1974)
4.148 N. E. Byer, T. C. Ensign, W. M. Mularie, S. W. Stokowski: J. Appl. Phys. **44**, 1733 (1973)
4.149 E. Chicklis, L. Esterowitz: Phys. Rev. Lett. **21**, 1149 (1968)
4.150 P. P. Feofilov, V. V. Ovsyankin: Appl. Opt. **6**, 1828 (1967)
4.151 V. V. Ovsyankin, P. P. Feofilov: J. Appl. Spectr. (USSR) **7**, 340 (1967)
4.152 V. V. Ovsyankin: Opt. Spectrosc. **28**, 112 (1970)
4.153 D. R. Tallant, J. C. Wright: J. Chem. Phys. **63**, 2075 (1975)
4.154 R. J. Woodward, J. M. Williams, M. R. Brown: Phys. Lett. **22**, 435 (1966)
4.155 W. D. Partlow: Phys. Rev. Lett. **21**, 90 (1968)
4.156 F. K. Fong: In *Progress in Solid State Chemistry, Vol. 3* (Pergamon Press, New York 1966)
4.157 D. R. Tallant, J. C. Wright: Proc. 11th Rare Earth Research Conference (1974) p. 788
4.158 D. R. Tallant, M. M. Miller, J. C. Wright: Presented at 169th Am. Chem. Soc. Nat. Meet. Philadelphia, Pa. (1975)
4.159 D. R. Tallant, M. M. Miller, J. C. Wright: J. Chem. Phys. **65**, 510 (1976).

5. Exciton Percolation in Molecular Alloys and Aggregates

R. KOPELMAN

With 18 Figures

Exciton percolation—the migration of excitons in disordered materials—is based on the availability of a disordered, exciton conducting, quasi-lattice A. Simultaneously there exists a disordered, exciton insulating, quasi-lattice B. The condition of exciton percolation, i. e. efficient migration of the excitation, requires an effectively connected A quasi-lattice. The simplest case of exciton percolation is that of a binary mixed crystal with random substitutional disorder and where the B quasi-lattice is energetically inaccessible to the A excitons.

Exciton percolation is operationally tied in with the method of exciton sensing. We have formulated, theoretically, the exciton percolation problem for a number of situations, with emphasis on that where the sensors are distributed throughout the material (i. e. impurity quenching, fusion, photosynthetic reactive centers). We have solutions of the percolation problem for a wide range of topologies and physical conditions. Special emphasis has been placed on the aspect of exciton incoherence (Förster-Dexter type) vs coherence (coherence time and lengths). Cluster and surface aspects have been handled, too.

Experimentally, we have investigated randomly mixed, multi-component molecular crystals, consisting of naphthalene and some of its isotopic and chemical derivatives. The spectroscopic investigations focused on the first singlet and triplet excitons, at temperatures from 1.8 to 30 K. The all-important variable has been the concentration of the components (the two major ones as well as the minor components serving as sensors). Quantitatively, the results compare well with the theory (with no adjustable parameters, only independently known data). Furthermore, the singlet exciton coherence time turns out to be close to the lower limit provided by the spectral linewidth. Also, the triplet exciton migration is shown to be based on superexchange (tunneling).

In the primary process of photosynthesis, the transfer of excitation from the "antenna" pigment to the reactive centers is of major importance, especially in the green plants. In our quasi-random model, most of the light-absorbing pigments are distributed in a substitutionally nearly-random fashion. The light absorbed by the higher energy pigments (B) quickly cascades down to the nearby lowest energy pigment (A), and then this A excitation migrates in the A quasi-lattice, until it reaches a reactive center. The same migration occurs for light absorbed directly by A. Based on reasonable assumptions, one gets an upper limit for the ratio of the major pigments

$$B/A < 1 \pm 0.5.$$

The minor component, i. e. the reactive center, should be at a concentration of about 10^{-2} or less. According to our model there is an important analogy between the green plant photosynthetic system and our laboratory mixed crystals. In both cases the exciton is quickly confined to the A quasi-lattice, starting thereby the migration that should result in its trapping by a sensor or reactive center.

5.1 General Theory*

5.1.1 Exciton Migration

The dynamics of exciton transfer [5.1, 2] in the tight-binding ("Frenkel exciton") limit [5.3] is a case of energy conduction in which neither charge nor mass is transferred. It also differs significantly from the heat flow process, involving phonons, in condensed phases. First, from a thermodynamic aspect, in contrast to phonons, exciton energy can be considered as a form of free energy or non-thermal energy, which can be utilized, in principle, without the restrictions of the Second Law. Second, again in contrast to both phonons and electrons, Frenkel excitons can usually be considered to be confined momentarily to a single molecular site [5.4]. Thus the excitons in an ideal solid solution (i. e. $C_{10}H_8/C_{10}D_8$) are capable of behaving as if they were in a heterogeneous conglomerate composed of conducting and insulating materials. (On the molecular level, any solution is a heterogeneous conglomerate.) The ideal solution thus becomes a most finely printed random assembly of microcircuits, i. e. one with molecular dimensions (<1 nm).

In standard problems involving conduction, one is used to considering a gradient of some potential across some bulk of material, thus defining a direction of flow in laboratory coordinates. This is not the only possible way of defining conductivity, and in the case of excitons probably also not the most practical one. For instance, in the primary process of photosynthesis [5.5–7], exciton migration is somewhat similar to that of a liquid in a large sink, flowing down a multitude of drains. We call this exciton migration.

Operationally, one measures the exciton flow (migration) with the help of microsensors, fixed or moving, distributed (often at random) throughout the bulk of the alloy (or pure substance). These "sensors" may be a defect site, (physical or chemical) a doped-in impurity site, another exciton, a domain boundary, etc. We discuss here a randomly distributed sensor with a given

---

* Based on work with J. HOSHEN (to be published).

concentration (equilibrium or steady state). The exciton flow rate (transport), monitored by these sensors, is therefore proportional to the number of distinct lattice sites visited by the exciton, within its lifetime. We notice that only conducting sites (including "sensors") are visited, and not insulating sites (except in the case of tunneling or by thermal promotion, i. e. modification of the exciton into a more energetic one). We also notice that at the sensor the particular exciton may become annihilated, with a probability as large as unity, while at other conducting sites this probability is usually quite low. (Annihilation of an exciton means its modification into a lower—or higher—energy exciton, a photon, a multi-phonon, etc.)

5.1.2 The Algorithm

Let the sensor concentration (mole fraction) be C_S. We also need to define C'_S, which is the effective sensor concentration,

$$C'_S \equiv C_S \gamma, \tag{5.1}$$

where γ is the sensor's registering (trapping) efficiency. We define the conducting site concentration C_g, where the guest (g) site may be an ordinary conducting site (t) or a sensor (S),

$$C_g = C_t + C_S. \tag{5.2}$$

Usually, the sensor sites are a small minority

$$C_t \gg C_S. \tag{5.3}$$

The concentration (mole fraction) of the insulating (ballast or host) material (h) is obviously

$$C_h = 1 - C_g. \tag{5.4}$$

With a total number N of lattice sites, the number of guest sites is

$$G = N C_g \tag{5.5}$$

and that of the sensors is

$$Z = N C_S. \tag{5.6}$$

The number of ways of distributing Z sensors among G guest sites is

$$C_Z^G = G(G-1)(G-2)\dots(G-Z+1)/Z!. \tag{5.7}$$

The number of ways of distributing Z sensors among $(G-m)$ guest sites is

$$C_Z^{G-m} = (G-m)(G-m-1)\dots(G-m-Z+1)/Z!. \tag{5.8}$$

We are obviously focusing our attention on a particular set of m guest sites. From the above two equations it is obvious that the probability F_m' of *not* having any of the Z sensors included in a set of m sites is

$$F_m' = C_Z^{G-m}/C_Z^G = (1-m/G)[1-m/(G-1)]\dots[1-m/(G-Z+1)]. \tag{5.9}$$

If the fraction of sensors is very small (if not, one can use Stirling's approximation) one gets

$$F_m' = (1-m/G)^Z \quad \text{iff} \quad Z \ll G. \tag{5.10}$$

Finally, the probability F_m for having at least one sensor included in the set of m sites is

$$F_m = 1 - F_m', \tag{5.11}$$

and

$$F_m = 1 - (1-m/G)^Z \quad \text{iff} \quad Z \ll G. \tag{5.12}$$

Note that if the above set is small compared to the ratio of $C_g/C_S = G/Z$ one gets, with the help of the binomial expansion

$$F_m = Zm/G \quad \text{iff} \quad 1 \le m \ll G/Z. \tag{5.13}$$

One practical aspect of the above results pertains to the probability of having one of the Z sensors included in a cluster of m sites, irrespective of the way one defines such a cluster (see, i. e., [5.8]). Furthermore, having an exciton confined to an m-cluster, its probability $\bar{F}_m$ of registering at any sensor is

$$0 \le \bar{F}_m \le F_m. \tag{5.14}$$

In the limit of efficient exciton transport, as well as high registration efficiency (the supertransfer case) one gets

$$\bar{F}_m = F_m \quad \text{iff} \quad \text{supertransfer}. \tag{5.15}$$

In the limit of high registration efficiency $(\gamma \to 1)$, (5.15) is always justified for very small clusters $(m \to 1)$. We notice that the exciton transfer efficiency depends both on the exciton transport efficiency within the conducting cluster and the sensor registration efficiency. The transport efficiency, in its turn, is a function of both the propagation rate and the excitation lifetime. In very small clusters ($m \rightarrow 1$) the propagation is essentially instantaneous. Provided that the registration time is also much shorter than the excitation lifetime, (5.15) will hold.

The probability of confining an exciton onto an m-cluster, provided it cannot leak out, is the probability of having it enter the cluster. Assuming that the guest-exciton creation is simply proportional to the number of guest molecules, the probability of any guest exciton being created in the given m-cluster is m/G. If the total number of m-clusters ("m-frequency") is i_m then the probability P_m of creating the exciton in *any* m-cluster is

$$P_m = i_m m/G . \tag{5.16}$$

Note that

$$\sum_m P_m = G^{-1} \sum_m i_m m = 1 . \tag{5.17}$$

Finally, the probability B_m of any guest exciton being created and registered inside *any* m-cluster is

$$B_m = P_m \bar{F}_m , \tag{5.18}$$

or

$$B_m = \bar{F}_m i_m m/G . \tag{5.19}$$

The total probability of a guest exciton registering on a sensor is therefore

$$P = \sum_m B_m = \sum_m P_m \bar{F}_m , \tag{5.20}$$

or

$$P = \sum_m \bar{F}_m i_m m/G . \tag{5.21}$$

We note that the cluster frequency i_m and the cluster registration probability $\bar{F}_m$ have to be solved as a function of C_g, for the given topology of exciton interactions (and the other factors relevant to exciton transfer and therefore affecting $\bar{F}_m$).

Limit of Low Guest Concentration. Well below the site percolation concentration C_c^S, $i_m \rightarrow 0$ for large m and so (5.15) will hold for the guest "miniclusters", provided that the transfer efficiency is large enough so as to result in a transfer time small compared with the exciton lifetime. We can then rewrite (5.20) and (5.21) as

$$P = \sum_m F_m P_m = \sum_m F_m i_m m/G \quad \text{iff miniclusters,} \quad \gamma \gg 0 . \tag{5.22}$$

Under these conditions, (5.13) is also likely to hold, giving thus

$$P = Z \sum_m i_m (m/G)^2 \quad \text{iff} \quad C_g \ll C_c^S, \quad m \ll G/Z, \quad \gamma \gg 0, \quad (5.23)$$

or, alternatively,

$$B_m = Z i_m (m/G)^2 = (i_m m^2/G)(Z/G) \quad \text{iff} \quad C_g \ll C_c^S, \ m \ll G/Z, \ \gamma \gg 0. \quad (5.24)$$

Limit of High Guest Concentration. Sufficiently above the site percolation concentration C_c^S for the given topology, i. e. $C_g > \bar{\bar{C}}_g \equiv C_c^S + \delta$ ($\delta \approx 0.05$), most of the guest sites belong to the "infinite cluster" (maxicluster) [5.9]. Simultaneously, *all* the other clusters are very small (miniclusters). We can thus substitute in (5.20), for all the miniclusters, i. e. $m \neq m'$, where m' refers to the size of the maxicluster, (5.22), giving

$$P = P_{m'} \bar{F}_{m'} + \sum_{m \neq m'} P_m F_m \quad \text{iff} \quad \gamma \gg 0, \quad C_g > C_g. \quad (5.25)$$

Noticing [5.9] that $i_{m'} = 1$, and using the definition [5.10]

$$\bar{P}_\infty \equiv m'/G, \quad (5.26)$$

one gets, with the aid of (5.16)

$$P = \bar{F}_{m'} \bar{P}_\infty + \sum_{m \neq m'}' F_m i_m m/G \quad \text{iff} \quad C_g > \bar{\bar{C}}_g \gg C_S \quad \text{and} \quad \gamma \gg 0 \quad (5.27)$$

or, utilizing (5.13) for the miniclusters,

$$P = \bar{F}_{m'} \bar{P}_\infty + (Z/G^2) \sum_{m \neq m'} i_m m^2 \quad \text{iff} \quad C_g > \bar{\bar{C}}_g \gg C_S \quad \text{and} \quad \gamma \gg 0. \quad (5.28)$$

Remembering (5.24), the second term in (5.28) is small, giving

$$P = \bar{F}_{m'} \bar{P}_\infty \quad \text{iff} \quad C_g \gg C_c \gg C_S \quad \text{and} \quad \gamma \gg 0. \quad (5.29)$$

However, at the limit where

$$\bar{P}_\infty \to 1 \quad \text{iff} \quad C_g \to 1, \quad (5.30)$$

(5.27) gives

$$P = \bar{F}_{m'} \quad \text{iff} \quad C_S \ll C_c^S \ll C_g \to 1. \quad (5.31)$$

Below we shall deal with the evaluation of $\bar{F}_{m'}$. Once this quantity is available one has solved not only (5.31), but also (5.29), as $\bar{P}_\infty$ is often available in the literature [5.9, 10] or can easily be evaluated. This also leads to the evaluation of (5.25), (5.27) and (5.28), provided that the cluster frequency distribution (the set i_m) is available. The latter, as mentioned above, is only a function of topology (for a given C_g).

Intermediate Guest Concentration. This is essentially the C_g region close to the percolation concentration

$$C_g \approx C_c^s. \tag{5.32}$$

Well below the percolation concentration, (5.22) is a good approximation to the exact expression (5.20, 21). For intermediate low $C_{g'}$, (5.23) and (5.24) are not recommended. One should substitute (5.12), rather than (5.13), into (5.22). Thus (5.22) is replaced by

$$P \approx \sum_m F_m P_m = \sum_m [1-(1-m/G)^Z] i_m m/G \quad \text{iff} \quad C_s \ll C_g < \bar{C}_g, \quad \gamma \gg 0, \tag{5.33}$$

which should now be considered as a good approximation to (5.21), having now defined $\bar{C}_g \equiv C_c^s - \delta$ ($\delta \approx 0.05$).

An expression symmetric to (5.33) can be used above the percolation concentration, based on (5.27), but now including lower concentrations, compared to (5.28)

$$P = \bar{F}_{m'} \bar{P}_\infty + \sum_{m \neq m'} [1-(1-m/G)^Z] i_m m/G \quad \text{iff} \quad C_g > \bar{\bar{C}}_g \gg C_s, \gamma \gg 0 \tag{5.34}$$

which should again be considered as only an approximation to (5.21). Here again m' designates the maxicluster.

We can both remove that discontinuity at $\bar{C}_g \leq C_g \leq \bar{\bar{C}}_g$ and improve on the validity of (5.33) and (5.34) by realizing that there is a range of m for which (5.15) does not hold, only (5.14). We now designate this whole range as m'' including, above percolation, the maxicluster. We now get

$$P = \sum_{m \neq m''}'' P_m \bar{F}_m + \sum_{m''} P_{m''} \bar{F}_{m''}, \tag{5.35}$$

where $\sum''$ is a summation excluding the whole set m''. This expression can now be written as

$$P = \sum_{m \neq m''}'' P_m F_m + \sum_{m'' \neq m'}' P_{m''} \bar{F}_{m''} + \bar{P}_\infty \bar{F}_{m'} \quad \text{iff} \quad \gamma \gg 0. \tag{5.36}$$

where m' again designates the maxicluster, if it exists, i.e. if $C_g > C_c^s$. Again, further manipulation gives

$$P = \sum_{m \neq m''}'' [1-(1-m/G)^Z] i_m m/G + \sum_{m'' \neq m'}' \bar{F}_{m''} i_{m''} m''/G + \bar{P}_\infty \bar{F}_{m'}$$
$$\text{iff} \quad G \gg Z \quad \text{and} \quad \gamma \gg 0. \tag{5.37}$$

The evaluation of $\bar{F}_{m''}$ will be described together with that of $\bar{F}_{m'}$. In opportune cases both the set m'' and the values of $i_{m''}$ will be small. We note that the breakdown of (5.15) for the set m'' depends not only on C_g and the topology, but also on C_S and on the intimate details of the exciton transfer mechanism and parameters.

Evaluation of $\bar{F}_m$. The probability $\bar{F}_m$ of an exciton (confined to a given m-cluster) registering on any sensor obviously depends on many factors: C_g, C_S, γ, the interaction topology, the size of m, the shape of the m-cluster, the exciton interactions, lifetime and transport mode (coherence length, scattering behaviour), etc. One way of deriving $\bar{F}_m$ is by computer simulation, specifying all the above parameters and conditions, and averaging over crystal configurations (random lattice), cluster shapes, point of origin, sensor distribution, mode of propagation, lifetime, etc. Such "games" can be conducted, but a significant saving in effort and money is achieved with the help of the following algorithm, which gives $\bar{F}_m$ as an analytical expression of the sensor concentration C_S, with only one quantity in this expression to be derived from computer simulations that do NOT specify sensor concentrations and registration efficiencies.

Our algorithm is based on the very general derivation of (5.9) to (5.12). For simplicity we assume first a perfect sensor registration efficiency ($\gamma \to 1$). For arbitrary exciton transport efficiency (5.14) holds. We define n_m to be the average number of distinct sites visited within its lifetime by an exciton confined to a given m-cluster with the condition of $\gamma = 0$ (or $Z = 0 = C_S$). Obviously, one has

$$n_m \leq m . \tag{5.38}$$

The probability F_n of this set of n_m sites including at least one sensor is (analogously to (5.12) and its derivation)

$$F_n = 1 - (1 - n_m/G)^Z \quad \text{iff} \quad Z \ll G , \tag{5.39}$$

where the restriction can also be written as

$$Z/G \ll 1 . \tag{5.40}$$

Now, if we let $\gamma \to 1$ this means that the above exciton probability to register on a sensor is

$$\bar{F}_m = F_n \quad \text{iff} \quad \gamma \to 1 . \tag{5.41}$$

We note that the above equation holds irrespective of (5.39) and (5.40). However, utilizing these equations one gets our algorithm

$$\bar{F}_m = 1 - (1 - n_m/G)^Z \quad \text{iff} \quad Z \ll G, \quad \gamma \to 1 . \tag{5.42}$$

We note that n_m has to be derived for the given C_g, interaction topology, m'-cluster (size and shape), the exciton interaction parameters, its lifetime and its mode of propagation and scattering. In addition n_m also has to be averaged over cluster shape and composition as well as the exciton origin, propagation and lifetime.

For the case of $\gamma \neq 1$, two methods of correction suggest themselves. One is to substitute Z in (5.42) with

$$Z' \equiv Z\gamma = N C_S'.\tag{5.43}$$

A better alternative, especially for low γ, is to substitute $\bar{n}_m$ for n_m in the maxi-cluster, where $\bar{n}_m$ is the average number of distinct sites visited at least γ^{-1} times by the confined exciton (with the condition $Z=0$ and with γ^{-1} rounded off to an integer).

Some Limits. Rewriting (5.39), utilizing (5.5) and (5.6),

$$F_n = 1 - (1 - n_m/N C_g)^{N C_S}\tag{5.44}$$

or

$$F_n = 1 - \left(1 - \frac{n_m}{N C_g}\right)^{\frac{N C_g}{n_m}\frac{C_S}{C_g}n_m}.\tag{5.45}$$

For very large N, i.e. a large crystal,

$$F_n = 1 - \exp(-C_S n_m/C_g) \quad \text{iff} \quad N C_g/n_m \to \infty.\tag{5.46}$$

Furthermore, if

$$C_g/C_S = G/Z \gg n_m,\tag{5.47}$$

then

$$F_n = C_S n_m/C_g = Z n_m/G \quad \text{iff} \quad C_S n_m/C_g \ll 1 \quad \text{and iff} \quad N C_g/n_m \to \infty.\tag{5.48}$$

We notice that by the first conditional relation of (5.48) we get a restriction

$$Z \ll G,\tag{5.49}$$

which was crucial to our original derivation of (5.10). This consistency check indicates that (5.46) and (5.48) can be utilized under the appropriate conditions. We notice the close analogy between (5.48) and (5.13). However, while we utilized (5.13) mainly for the case of small m-clusters, (5.48) ist most appropriate for use, via (5.41), for the case of large (and very large) clusters with

inefficient exciton transport, the latter resulting from small exciton interactions and/or short exciton lifetimes and/or an unfavourable cluster topology.
We notice that (5.46) holds if the dimension of the sample is large compared
to the sampling radius of the exciton, while (5.48) holds if the average distance
to a sensor is also large compared to this exciton radius. The above state-
ments have to be modified appropriately for $\gamma \neq 1$, either according to the
spirit of (5.43) or the alternative proposed in the discussion following it.   ·

In the limit of efficient transfer (supertransfer case), we can rewrite (5.38) as

$$n_m = m \quad \text{iff} \quad \text{supertransfer,} \tag{5.38a}$$

for *all* m. This implies a combination of long exciton lifetime and/or large
exciton interactions and/or efficient trapping. In this case (5.15) is valid. Actually, if one has a large enough sensor concentration (but not too large)
and $\gamma \to 1$, one always gets (5.15) to be valid, i. e. supertransfer, if $n_m \gg G/Z$.
This gives, similar to (5.33), but for *all* C_g [using (5.12, 15, 20)].

$$P = \sum_m F_m P_m = \sum_m P_m [1 - (1 - m/G)^Z] \quad \text{iff} \quad Z \ll G \quad \text{and supertransfer,}$$
$$\tag{5.33a}$$

which reduces to [see (5.16 and 26)]

$$P = F_{m'} \bar{P}_\infty + \sum_{m \neq m'}{}' [1 - (1 - m/G)^Z] i_m m/G \quad \text{iff} \quad C_g \gg C_s \tag{5.34a}$$
$$\text{and supertransfer.}$$

We notice that, using (5.12),

$$F_{m'} = 1 - (1 - \bar{P}_\infty)^Z \quad \text{iff} \quad Z \ll G, \text{ and supertransfer,} \tag{5.12a}$$

and even

$$F_{m'} = 1 \quad \text{iff} \quad 1 \ll Z \ll G, \quad C_g > \bar{\bar{C}}_g > C_c^S \quad \text{and supertransfer.} \tag{5.12b}$$

Thus (5.34a) turns into

$$P = \bar{P}_\infty + \sum_{m \neq m'}{}' [1 - (1 - m/G)^Z] i_m m/G \quad \text{iff} \quad 1 \ll Z \ll G, \quad C_g > \bar{\bar{C}}_g > C_c^S$$
$$\text{and supertransfer.} \tag{5.34b}$$

We thus get, similar to (5.29),

$$P = \bar{P}_\infty \quad \text{iff} \quad C_g \gg C_c^S \gg C_s \quad \text{and supertransfer,} \tag{5.29a}$$

giving, in the limit of (5.30), the trivial answer

$$P = 1 \quad \text{iff} \quad C_s \ll C_c^S \ll C_g \to 1 \quad \text{and supertransfer.} \tag{5.31a}$$

We can also rewrite (5.34 b) for regions of small miniclusters, i. e. for both the ranges $C_g \ll C_c^S$ and $C_g \gg C_c^S$, utilizing (5.13), similar to its use in (5.23) and (5.28),

$$P = \bar{P}_\infty + (Z/G^2) \sum_{m \neq m'}{}' i_m m^2 \quad \text{iff} \quad 1 \ll Z \ll G, \quad \text{supertransfer and}$$
$$\text{iff} \quad C_g \ll C_c^S \quad \text{or} \quad C_g \gg C_c^S. \qquad (5.34\text{c})$$

Note that the region covered by (5.34 b) but not by (5.34 c) is that of C_g close to C_c^S (whether smaller or larger). Eq. (5.34 c) is still more rigorous than (5.29 a) and can also be used below the percolation concentration (where $\bar{P}_\infty = 0$), as an approximation to (5.33).

5.1.3 Discussion

The basic philosophy given in the above theoretical derivation is to make the problem soluble with minimum utilization of computer simulation. That means avoiding unusually structured averaging procedures. An example of such an unusual averaging procedure is one where nine out of ten times the result will be close to unity while one out of ten times it will be close to zero, on the average. Unawareness of such a pitfall (an extremely non-Gaussian distribution) might lead to computer simulation experiments giving 15 times in a row a value close to unity. The student stops right there and derives a result like 0.9999 while the correct answer may be 0.900. In our particular situation, above the percolation concentration, the exciton is most likely to be trapped in a maxicluster (in a simulation) and only very rarely in a mini-cluster. However, that does not mean that the minicluster contribution can be avoided. If, alternatively, we rely on an exhaustive computer simulation, we can supplement all our algorithms with a very simple one. For instance, one of our most useful algorithm forms, for guest concentrations well above percolation and small sensor concentrations, can be derived by combining (5.28) in conjunction with (5.42), thus giving (for a specific topology, etc.)

$$P(C_g) = \bar{P}_\infty [1 - (1 - n_{m'}/G)]^Z + (Z/G^2) \sum_{m \neq m'}{}' i_m m^2 \quad \text{iff} \quad C_g > C_c \gg C_S,$$
$$\gamma \to 1. \qquad (5.50)$$

Here, for a given topology, it is trivial to compute, by simulation, $\bar{P}_\infty$ (or take it from the literature) and the set i_m. With an appropriate model, $n_{m'}$ can also be computed, by simulation. None of these computations are likely to give wide statistical fluctuations (if computed wisely...). On the other hand, we could replace this algorithm (5.50), or any other algorithm, with the following "simple" algorithm (for all C_g)

$$P(C_g) = 1 - (1 - n/G)^Z \quad \text{iff} \quad C_S \ll C_g. \qquad (5.51)$$

Now n is the whole-lattice average number of distinct guest sites visited by a guest exciton within its lifetime (for $\gamma = 0$ and any arbitrary exciton transfer model). Obviously one has to average again over all lattice configurations, exciton origins, exciton propagation paths and exciton lifetimes. The first two averaging procedures are especially tricky, as here an exciton originating on a maxicluster will give one kind of n ($\leq m'$), while one originating on a mini-cluster will give a very different n ($= m$). The discrepancy in n can easily be a factor of 10^6. Indeed, algorithm (5.51) is only a little improvement over having no algorithm at all and simply letting the computer simulation game derive $P(C_g)$ for every possible sensor concentration ($C_S = Z/N$) and distribution (among maxicluster and minicluster sites). The above discussion may serve as an example of a paradigm underlying the utilization of computer simulation in the physical sciences.

The classification of the sensor sites as "guests" also requires some clarification. One situation is a sensor comprised simply of an excited guest site (guest + exciton). Another case will be a defective guest site (such as an X-trap, domain boundary or other crystal imperfection). A third, very important case is that of a third component—an impurity (chemical or isotopic) supertrap. One is tempted to proportionally "distribute" such an impurity among guest (component A) and host (B) domains. This creates a problem of cluster boundaries, as the supertrap will be effective not only if found embedded in a cluster but also on its domain boundary. While it is well known in nucleation theory, as well as in most other situations, that the ratio of surface-to-volume decreases with size and tends to zero with infinite size, this is not true in our situation of maxiclusters. Actually one can easily convince himself that when $\bar{P}_\infty \to 1$ (say about $C_g \geq 0.7$), the volume of the "surface layer" of the maxicluster in an infinite crystal tends toward infinity, or more precisely, the concentration of surface layer sites C_l becomes

$$C_l \to C_h = 1 - C_g,\tag{5.52}$$

and the number of surface layer sites

$$L = C_l N \to N - G = H,\tag{5.53}$$

where H is the number of host sites. If now H includes $C_h Z$ sensor sites, then practically all of them are on the surface. By contrasting our two methods of handling the sensor sites, it becomes obvious that the number of sensor sites Z totally embedded in the host (i. e. in no contact with maxicluster) is

$$\bar{Z} = (1 - \bar{P}_\infty)Z,\tag{5.54}$$

i. e. an insignificant fraction of sensors for our case of $\bar{P}_\infty \to 1$. Eq. (5.54) is simply derived by remembering our original definition (5.2) and noticing that the total number $\bar{G}$ of guest sites in miniclusters is

$$\bar{G} = (1 - \bar{P}_\infty)G, \tag{5.55}$$

and that

$$\bar{Z}/Z (= \bar{G}/G) \to 0 \quad \text{iff} \quad \bar{P}_\infty \to 1. \tag{5.56}$$

From this discussion it is clear that when $C_g > C_c^S$ nearly all of the sensors activity is due to exciton flow through the guest lattice and very little is due to migration through the host.

Another important consideration is the connectivity of clusters. Most of the mathematical and physical literature [5.9, 10] defines clusters with a connectivity of nearest neighbors only or nearest and next nearest neighbors. In a real physical situation there is no such thing as an interaction cut off completely after nearest or next nearest neighbors, no matter how steeply it falls off with distance. There is, therefore, always a chance for a very long range exciton "hop". If the exciton lifetime were infinite, one would have to take into account every such possible hop, resulting in a definition of cluster that encompasses every possible guest molecule. This would result in one single cluster (maxicluster) even for a very small guest concentration ($C_g \ll 1$). The end result is an apparently absurd situation where

$$C_c^S \to 0 \quad \text{iff} \quad \text{lifetime} \to \infty, \tag{5.57}$$

a mathematical case which is as trivial as the opposite case of a linear lattice with nearest neighbor connectivity only where

$$C_c^S = 1 \quad \text{iff} \quad \text{n.n. 1-dim.} \tag{5.58}$$

Thus it is the finiteness of the exciton lifetime which makes percolation theory non-trivial. Actually, as we shall mention later, there are two more physical processes, in addition to long-range interactions, namely the classical "thermal" climbing over the host barrier and the quantum mechanical tunneling under it (see Sect. 5.4). Each one of these leads essentially to the result of (5.57), for an infinite exciton lifetime.

5.1.4 Graphical Illustrations

Here we illustrate both the trivial and some of the less trivial points via graphical representations. Figures 5.1a and b give a computer simulation of a random binary 60×60 square lattice, emphasizing the largest cluster. It is obvious that the percolation concentration C_c^S is about $C_g = 0.58$, even though C_c^S is not well defined here, due to the small size of the lattice and the absence of

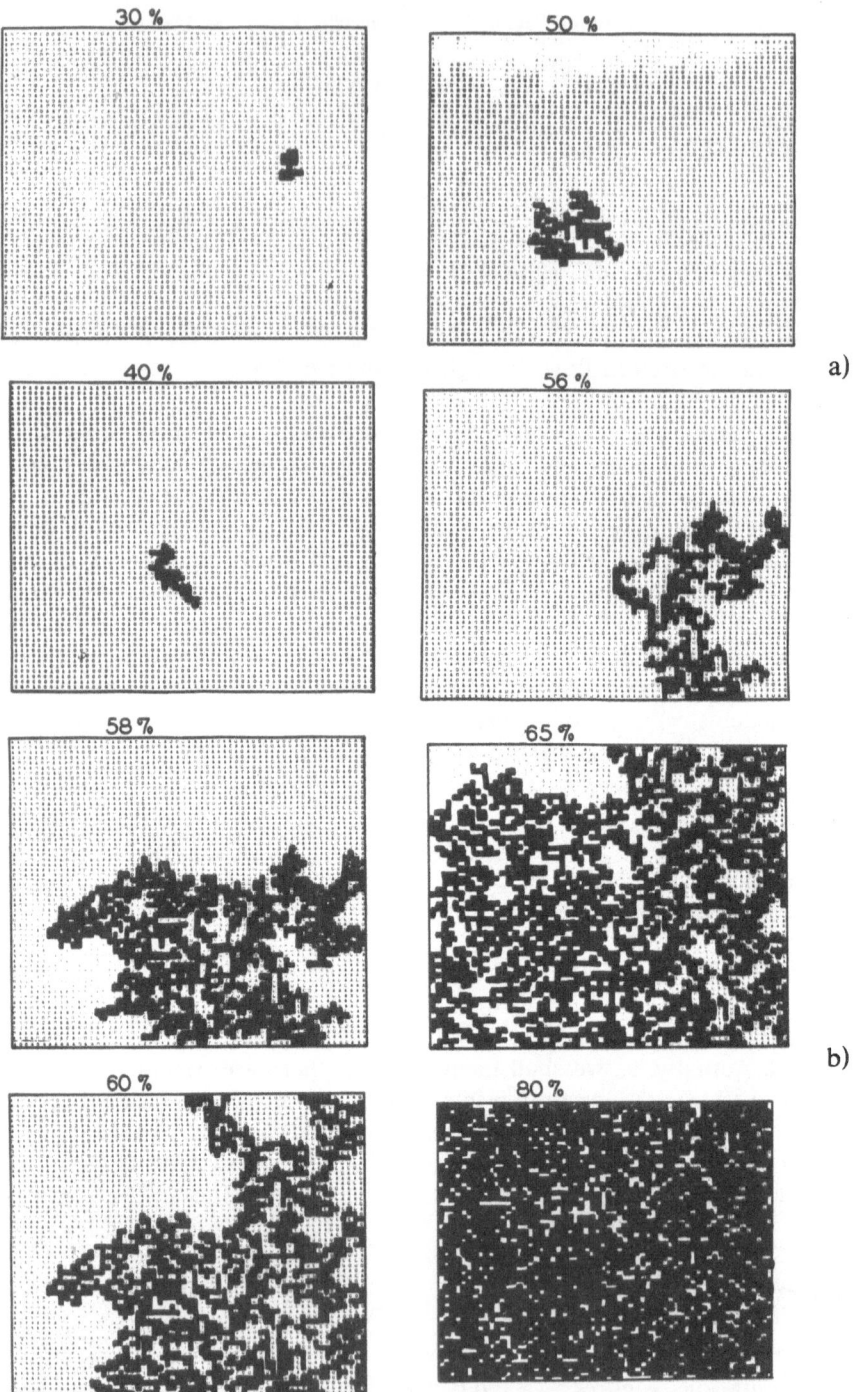

Fig. 5.1 a and b. The largest guest cluster as a function of guest concentration (in % mole for a 60 × 60 square lattice). A computer simulation for illustration purposes
a) Below the percolation concentration
b) About and above the percolation concentration

boundary conditions and, most importantly, an averaging procedure. Moreover, the same set of random numbers was used for all values of C_g, as can be guessed from the persistent two locations of the largest cluster.

Figure 5.2 is a "naive" computer simulation of an exciton searching for a supertrap in a small random lattice, utilizing a "random walk" involving nearest neighbor only jumps in a square lattice. Neither Fig. 5.1 nor Fig. 5.2 have been used for anything except pictorial illustrations.

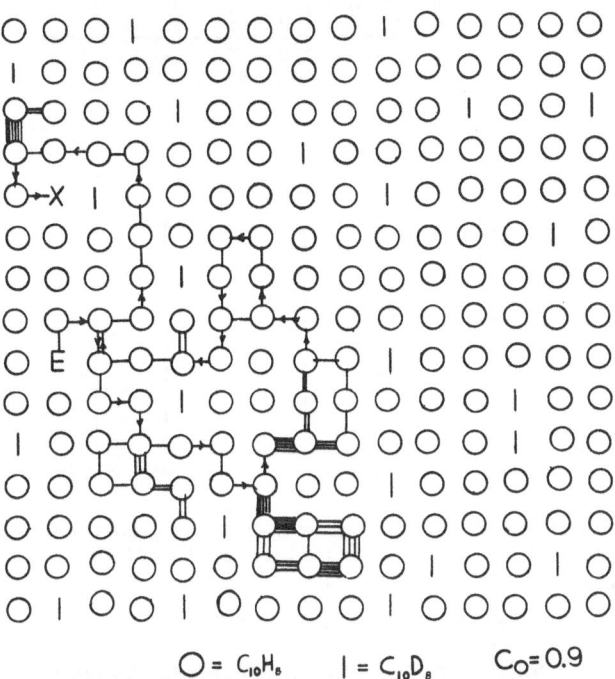

$$O = C_{10}H_8 \qquad | = C_{10}D_8 \qquad C_0 = 0.9$$

Fig. 5.2. Exciton in search of a supertrap, via random walk in a random ternary lattice. A computer simulation for illustration purposes. Notice that in this small (15×15) random square lattice (nearest neighbor only interactions), with reflective boundary conditions, the guest concentration, $C_g = 0.90 + 0.004$, includes both $C_{10}H_8$ ($C_0 = 0.90$ mole fraction) and the supertrap ($C_S = 1/225 = 0.004$). No steps are spent on the $C_{10}D_8$ host (barrier). The exciton started on the site E and "arrived" at the supertrap X (after about 80 steps)

Figure 5.3 shows the weighted cluster frequency distribution in the form of a histogram for the Square (1, 2) lattice, i. e. a 200×200 square lattice including both nearest and next-nearest neighbor interactions. Again these histograms are for illustration purposes only. P_m is the probability of a guest molecule belonging to a cluster of size m, compare (5.16–17). Note that the probability definition of P_m is normalized to unity, i. e.

$$P_m = i_m m / \sum_m i_m m. \qquad (5.59)$$

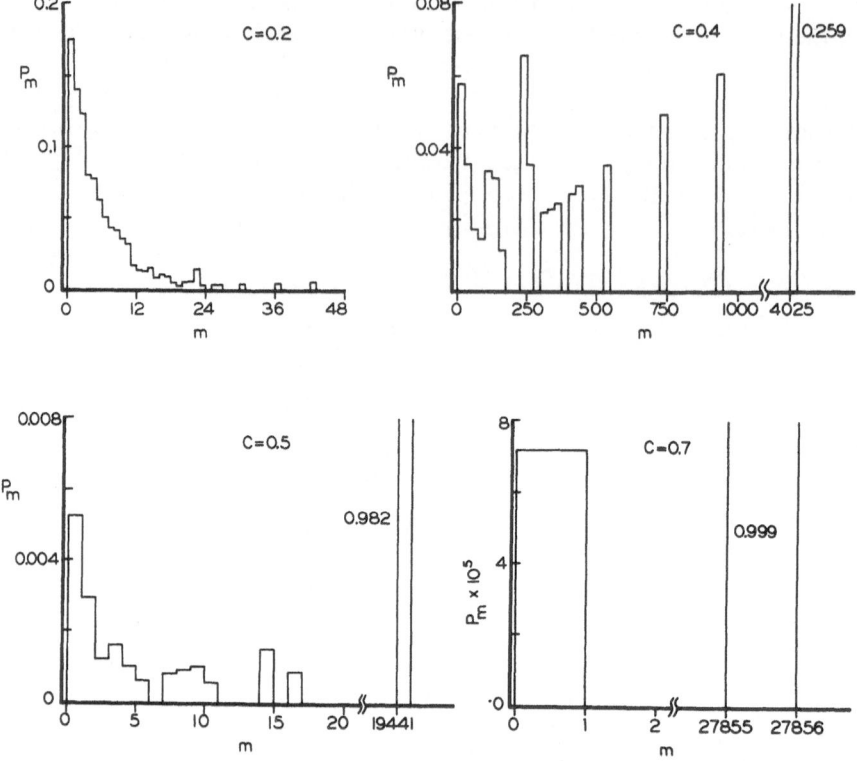

Fig. 5.3. Weighted cluster frequency distribution for the Square (1, 2) lattice (square lattice with nearest and next nearest neighbors): The probability P_m for a weighted cluster frequency as a function of m, the cluster size. Note that $P_m = i_m m / \sum_m i_m m$. The size of this lattice was 200×200.

Note that the last distribution contains just 2 monomers and one maxicluster

We see that for both $C_g \ll C_c^S$ and $C_g \gg C_c^S$, the majority of clusters are small ones (i. e. monomers). However, this characterization is not true for the fluctuation region, just below C_c^S.

Our graphical method of actually obtaining the percolation concentration C_c^S for various lattices is based on a computation of the average cluster size I_{AV} and a related quantity $I_{AV}^{\backslash}$, the reduced average cluster size. Referring to (5.23), (5.28) and (5.34 c), we now use the following definitions of the average cluster size,

$$I_{AV} \equiv G^{-1} \sum_m i_m m^2 = \sum_m P_m m , \tag{5.60}$$

and the reduced average cluster size

$$I_{AV}^{\backslash} \equiv G^{-1} \sum_{m \neq m'}^{\backslash} i_m m^2 = I_{AV} - P_{m'} m' \tag{5.61}$$

where m' always designates the largest cluster.

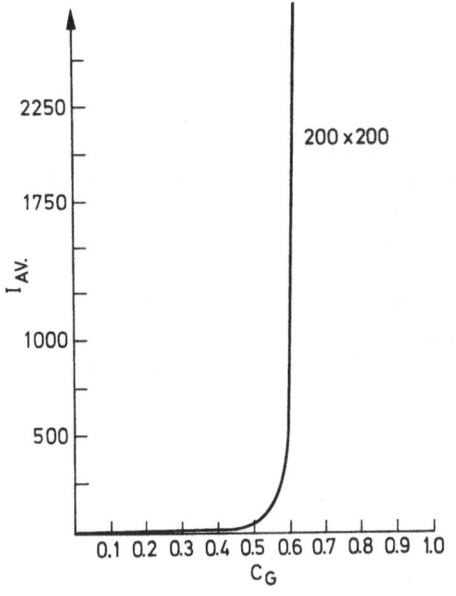

Fig. 5.4. Average cluster size I_{AV} vs guest concentration C_g in a 200×200 square lattice, see (5.60). Notice that the sharp rise of the curve gives the percolation concentration C_c^s (roughly)

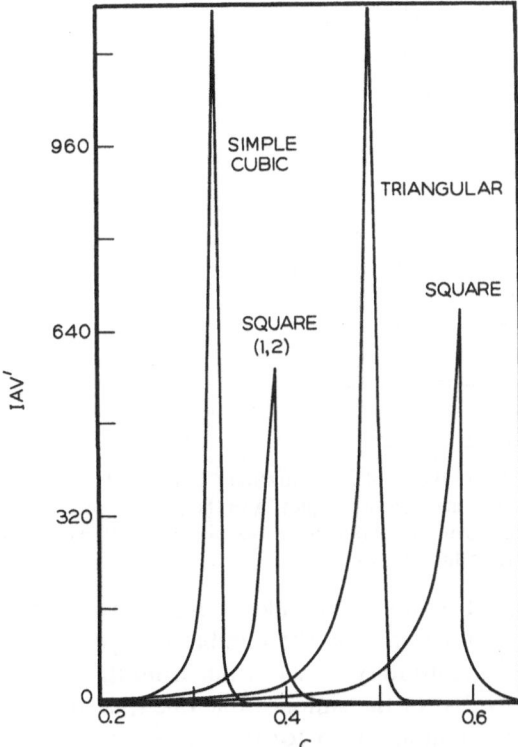

Fig. 5.5. The reduced average cluster size I'_{AV} (5.61), vs guest concentration C_g; for square, triangular, Square (1, 2) and simple cubic lattices. The sharp maxima give the respectice critical percolation concentrations C_c^s. The three two-dimensional topologies refer to a lattice of 200×200 sites, while the simple cubic to one of $100 \times 100 \times 30$ sites (see Fig. 5.6). Note that the absolute heights of the peaks are determined by statistical fluctuations

Figure 5.4 shows the average cluster size I_{AV} vs the guest concentration C_g for a 200×200 square lattice. We can see how $I_{AV} \to \infty$, asymptotically with the percolation concentration C_c^S, which is about 0.6. This presentation does not lend itself to a precise determination of C_c^S and a better method is shown below (Fig. 5.5), giving $C_c^S \approx 0.59$ for the square lattice.

Figure 5.5 shows I_{AV}' vs the guest concentration C_g for a number of lattice topologies. This function is the best we have found for bringing out the site percolation concentration, C_g^S, for these finite lattices ($100 \times 100 \times 30$ for the simple cubic and 200×200 for the others).

An interesting problem, related to surface excitons and thin crystals, is shown next (Fig. 5.6). A collection of parallel square lattice layers will even-

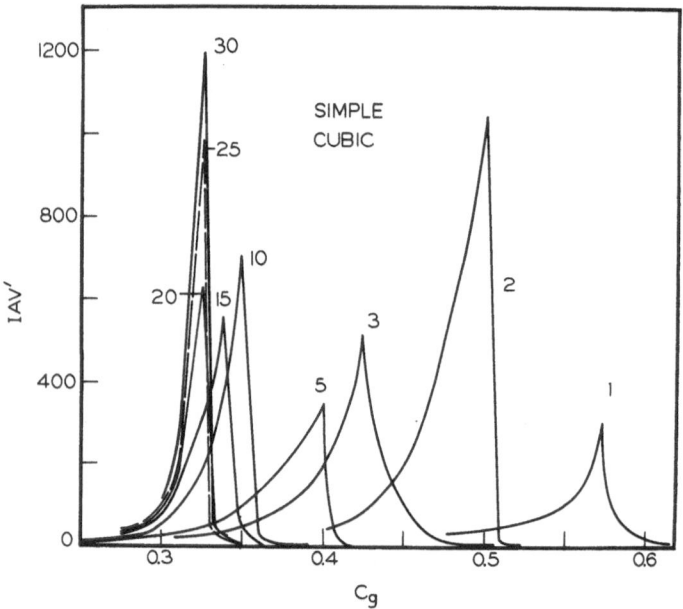

Fig. 5.6. The reduced average cluster size I_{AV}' vs the guest concentration C_g, for a 3-dimensional, simple cubic lattice of $100 \times 100 \times n$ sites. The maxima in the plots give the percolation concentration (C_c^S) for various values of the third dimension parameter, n (see also Fig. 5.7). Note that the peak heights are determined by statistical fluctuations

tually create a simple cubic lattice. Therefore it should be instructive to follow the percolation behaviour as a function of n, the number of layers, from $n = 2$ (or 1) to $n = 35$. We can see a continuity of behaviour from that of the square lattice (compare Fig. 5.5), to that of the simple cubic one (the limiting value being $C_c^S \approx 0.32$). It is most encouraging to see the above given series approach the correct limiting value. We note that the width of each curve is a function of the number of lattice sites (in the sample). As this size goes up with the increase in crystal thickness, the width decreases. The height, on the other hand, is mainly due to statistical fluctuations. There is an interesting similarity between the I'_{AV} curves and the heat capacity measurements for phase transitions, especially higher order ones.

Figure 5.7 summarizes the results from Fig. 5.6 and shows that 20 layers of a square lattice are practically all that is necessary to mimic a simple cubic behaviour for a $200 \times 200 \times n$ lattice. Note that between $n=20$ and $n=35$ the C_c^S concentration is practically constant.

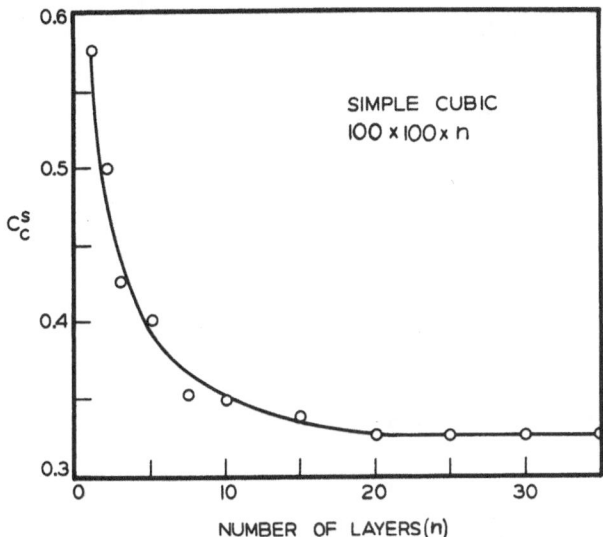

Fig. 5.7. The percolation concentration C_c^S vs crystal thickness, i. e. the number of 2-dimensional layers, n, in a simple cubic lattice (see Fig. 5.6). It appears that 20 to 25 such layers of 100×100 sites each are sufficient to approximate C_c^S for $n \to 100$, i. e. a perfect cube

Figure 5.8 shows the all important maxicluster guest occupation probability $\bar{P}_\infty$ curves as a function of guest concentration C_g. This is the probability that any guest site is part of the "infinite" maxicluster—the size of the "infinite" maxicluster being 2000, 2500, 5000 and 10,000 sites, or larger, respectively.

Figure 5.9 gives the supertransfer limit for the probability P of exciton transfer and registration at the sensor, with sensor concentrations of C_S in the range 8×10^{-3} to 2×10^{-4}, as a function of C_g. P is calculated from (5.34 b) and its equivalent expression

$$P = \sum_m [1-(1-m/G)^Z] i_m m/G \quad \text{iff} \quad 1 \ll Z \ll G, \text{ supertransfer}$$
$$\text{and } C_g < C_c^S \qquad (5.34\,\text{d})$$

or an equivalent equation using the Stirling formula, instead of (5.10), for high C_S.

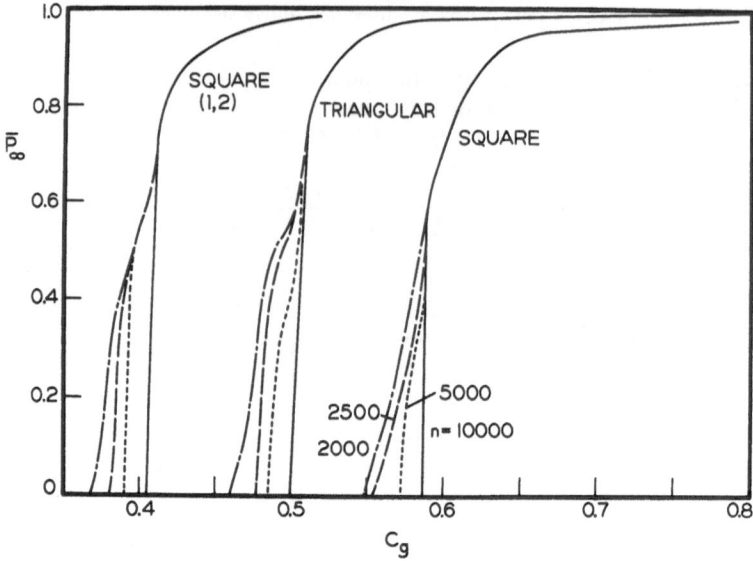

Fig. 5.8. The maxicluster occupation probability $\bar{P}_\infty$ vs guest concentration C_g for various topologies and maxicluster sizes. Notice that Square $(1, 2)$ is the square lattice with nearest and next-nearest neighbor interactions. The maxicluster sizes are 10,000 (full line), 5000 (dotted), 2500 (dashed) and 2000 (dash-dotted) sites. All lattices have 150×150 sites. The square lattice has a coordination number of 4, the triangular one of 6 and the Square $(1, 2)$ of 8

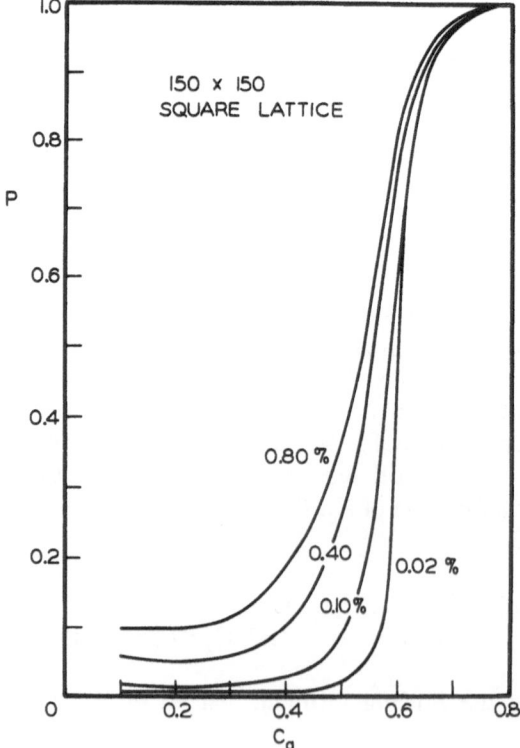

Fig. 5.9. Exciton supertransfer probability P vs guest concentration C_g (mole fraction), with sensor concentration C_S as a parameter (all concentrations in mole percent). These 150×150 square lattice curves are calculated from (5.34b) for $C_g > C_c^S$ and from (5.34d) at $C_g < C_c^S$. However, for high C_S the Stirling formula is used instead of (5.10)

Figure 5.10 is a computer simulation of the exciton transfer in the following way: The computer generates the previously described 3-component random lattice and the criterion used is essentially that of the previous figure. The question we are asking is whether or not the guest cluster is linked with a sensor. In the limit of supertransfer the end result should be equivalent to that of Fig. 5.9. The discrepancies arising will be discussed at a later date.

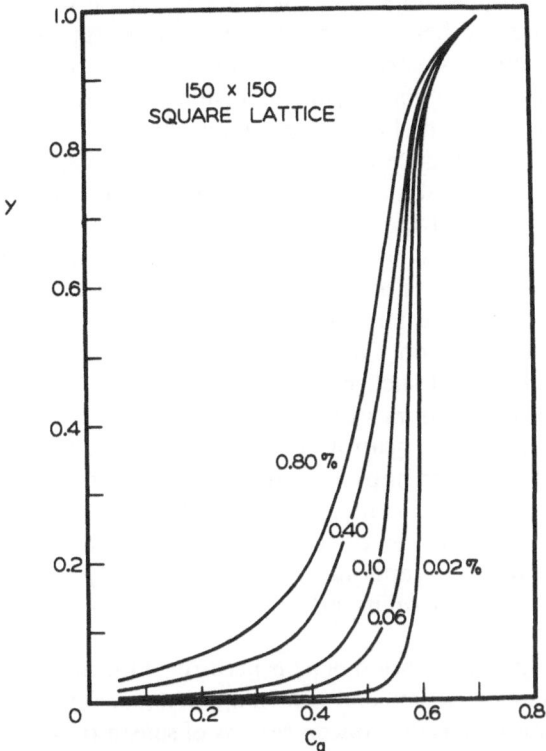

Fig. 5.10. A computer simulation of the exciton supertransfer vs guest concentration C_g (mole fraction) with sensor concentration C_S (mole percent) as a parameter. The 150×150 square lattice is generated by the computer, with the help of a random number generation subroutine, as a *ternary* system, including C_S mole percent of supertrap (sensor). Y is the fraction of total guest which is connected via a succession of nearest-neighbor guest interactions to a sensor

5.2 Incoherent and Semicoherent Transport

5.2.1 Hopping Model—Random Walk

Most mathematical works have considered random walks limited to nearest neighbors [5.11]. Here one uses a stochastic (markovian) energy transfer process. The exciton hops at random with no memory. This is an incoherent mode of transport [5.4, 6]. In addition to the topology of the lattice the only "physical" parameter is the number of steps (equal to the exciton lifetime divided by the hopping time). Fairly recently an attempt has been made to generalize the random walk theory from pure crystals to mixed crystals, using a coherent potential approximation Green's function approach [5.12]. However, as is the situation with energy states [5.3, 8], such averaged potential techniques cannot account for the clustering effects, which are even more important in energy transfer considerations than they are for static energy states and spectra. We therefore use here a computer simulation approach to calculate n_m, the average number of sites visited by the exciton confined to a given cluster of m guest molecules, for a number of steps (t). The result depends, of course, both on the size and shape of the guest cluster.

As pointed out in Section 5.1, in practice we only have to calculate $n_{m'}$, the number of distinct sites visited by an exciton in the maxicluster. This is justified by the fact that we usually avoid the fluctuation region, i. e. the concentration region just at and below percolation, and rely on the number of steps (t) to be large enough so as to assure at least one visit (better yet, more than γ^{-1} visits) to each site of the miniclusters, whether C_g is above or significantly below the percolation concentration. Obviously we consider the probability of exciton confinement to any of such miniclusters, (5.16).

We have further generalized the random walk approach to include direct hops to next nearest neighbors, next-next nearest neighbors, etc. The random walk problem with non-nearest neighbor hops has only recently been solved analytically and this only for one-dimensional pure lattices [5.13]. Our computer simulation program enables us to do calculations for two and three-dimensional, substitutionally mixed (at random) lattices. We simply assign a different hopping probability (and concomittantly a different hopping time) for each kind of hop.

As explained in Section 5.1, the computation of n_m is performed for the condition of $\gamma = 0$ (vanishing trapping efficiency) and is then applied (usually with the mere help of a minicalculator) to specific concentrations of sensor (C_S) and trapping efficiencies ($\gamma \leq 1$). One does have to choose boundary conditions (in this context we always used cyclic ones), and to be careful that the number of steps (t) is not significantly larger than the size of the lattice. The binary lattices are generated by the computer (the substitutional configuration being chosen at random). The computer also controls the randomness of the random walk process. Both selections are done with the help of standard random number subroutines.

5.2.2 Semicoherent Model—Extended Random Walk

In the limit of the coherent model (band model) [5.4, 14] the exciton propagates from site to site with perfect directional memory. In our semicoherent model the exciton preserves perfect directional memory over a mean-free-path (coherence length) l (measured in lattice spacings), with a standard deviation d defining the actual Gaussian distribution of free-paths. At the end of such a free-path the exciton gets strongly scattered, i.e. suffers a complete loss of memory. In practice, it is advantageous to use a mean free time (coherence time), measured in steps (time steps), with both a mean value and a standard deviation. In addition, a complete loss of memory (strong scattering) is produced by a host site interrupting the projected free path and is effective at the guest site preceding the interruption.

Due to the cyclic boundary conditions usually employed, no interruptions are associated with the lattice boundaries. Also, we still consider a free path of l' lattice spacings to consume l' times the number of random lattice steps consumed along the same direction with $l' = 1$. We notice that our semi-coherent model includes as a special case $(l = 1, d = 0)$ the incoherent hopping model. The limit $(l \to \infty, d = 0)$ is one possible definition for complete coherence. The limit $(l = l, d \to \infty)$ is also of interest.

We notice that analytical calculations of extended random walks with exponentially distributed steps have been performed for pure, one-dimensional lattices [5.13]. In spite of these limitations they have resulted in very significantly different results, compared to the traditional results [5.11]. Other recent work [5.15] on semicoherent excitons in linear chains has demonstrated that partial coherence often results in a less efficient energy transfer, compared to an incoherent process. We would like to emphasize that for the case including non-nearest neighbor interactions one should actually define a coherence time rather than a coherence length (mean free path). The coherence time is assumed to be independent of the direction of movement (in the neat crystal) while the coherence length may depend on it (even if defined in lattice units). Specific examples illustrating this point are given below.

5.2.3 Experimental Criteria and Parameters

We propose to use the above model to fit experimental data. The experimental variation of both C_g and C_S should provide ample data points to test such a model with two (l, d) or even only one parameter $(l, d = 1)$. As will be pointed out below, the total number of steps (t), as well as the relative transport probabilities (nearest neighbor vs next nearest neighbor, etc.) are available, in opportune cases, from independent experimental determinations. The same holds for the trapping efficiency γ, or at least it can be assumed to be constant over a series of both guest and sensor concentrations.

More important, however, than the fitting of parameters, is the test of the basic model. We notice here that the molecular alloy has a number of advantages, as a test case for energy conduction models, compared to the traditional "perfect" crystal. Any imperfections (defects, impurities) might drastically affect the energy transfer properties of a "neat" crystal. However, our typical alloy has already incorporated in it, purposely, a very efficient sensor (quencher, supertrap) to compete with the quenching of natural defects and impurities. Moreover, for every quenching impurity or defect, the typical neat crystal has many "non-quenching" impurities and defects that are efficient scatterers. Our alloy has already a large but known proportion of scatterers ("antitraps"), purposely incorporated in the form of host sites. The "natural" impurity and defect scatterings thus become insignificant. This is probably still true even when the "host" concentration is reduced to 1% ($C_h = 0.01$). *By having an abundance of scatterers compared to quenchers we have thus effectively decoupled the two processes.*

We further notice than an ideal alloy contains no localized or pseudo-localized phonons [5.16] but only amalgamated phonons [5.17] (see below). Thus the phonon-caused scattering, which is a prime contributor to non-coherence, should be nearly identical in the ideal alloy and in a neat crystal containing only the energy conducting component (guest). The alloy components that differ from the host only by isotopic substitution, or by "minor" chemical substitution, meet the requirement of phonon amalgamation [5.17]. As to how a "minor" chemical substitution is defined, we notice that a methyl substitution on a naphthalene is an example in case. In general, we here define a chemical substitution as being "minor" if the amalgamated phonon limit is preserved. We thus argue that the phonon contribution to the coherence time (mean free time) should be independent of guest (and sensor) concentrations. Therefore the phonon contributed scattering (incoherence) can be decoupled from the defect and impurity scattering. This can be achieved at a fixed temperature, by a variation of alloy concentrations.

Our choice of lattice topology and parameters has been guided by the naphthalene first singlet exciton [5.3]. An exciton exchange parameter of 18 cm^{-1} has been used for the nearest-neighbor pairwise interaction [5.18]. This value includes a superexchange correction [5.19], which is justified for high C_g, while for low C_g the uncorrected value [5.20] of about 16 cm^{-1} may be more justified. This pairwise interaction by itself gives a square lattice topology (with nearest neighbor connectivity, i. e. coordination number 4). Adding one next-nearest neighbor interaction (in-plane) gives a quasi-triangular lattice (coordination number 6). This interaction is the naphthalene crystal a (or b) axis pairwise interaction [Ref. 5.3, Fig. 5.7], which is about [5.18, 20] 4 cm^{-1}, corrected for superexchange, but about 8 cm^{-1} for the isotopic dilute guest [5.3, 20] (low C_g). The next-next nearest in-plane interaction is about 2 cm^{-1} in the pure crystal ("corrected for superexchange") but about [5.20] 3 to 5 cm^{-1} in the isotopic dilute guest. We have therefore chosen, somewhat arbitrarily, a ratio of in-plane pairwise interactions of $10:5:2$. This combination of pairwise interactions gives a quasi-Square $(1, 2)$ lattice with nearest and next-nearest interactions (coordination number 8). Note that here we only discuss the in-plane interactions.

The lifetime of the first excited naphthalene singlet is about [5.21] 10^{-7} s. We believe that the scattering time (coherence time) in the neat naphthalene crystal at 2 K is at the very least as long as the jump-time derived from the nearest-neighbor pairwise interaction, and most probably four or more times longer. This latter conclusion is based on the linewidth of the 0–0 singlet emission of the neat crystal at 4 K, which is about [5.22] 4 cm^{-1} or less (FWHM), compared to the pairwise interaction of 16 to 18 cm^{-1}. Thus, even for the next-next nearest neighbor direct jump we feel that the scattering time is not significantly shorter than the jump time. [Actually [5.22a], the absorption linewidth is ≤ 2 cm^{-1}.]

Using the relationship [5.4] $T_j = (4M)^{-1}$, where T_j is the jump time and M the absolute pairwise interaction in Hertz (1 cm$^{-1} = 3 \times 10^{10}$ Hertz), we get, for $M = 17$ cm^{-1}, $T_j = 4.9 \times 10^{-13}$ s. Using an exciton lifetime [5.21] τ of 10^{-7} s one gets the total number of "steps",

$$t = \tau / T_j \tag{5.62}$$

to be $t = 2 \times 10^5$. Thus we have an average of 200,000 hops and the same number of "steps" (we usually computed our results for the range of 40,000 to 700,000 steps). We note that a hop across one next nearest lattice spacing takes up 2 "steps" and along the next-next one 5 "steps". If our coherence time is 10 "steps", then along the nearest neighbor direction $l = 10$, along the next nearest one $l = 5$ and along the next-next nearest one $l = 2$.

5.2.4 Graphical Illustrations

We illustrate some of our above points with sample calculations. Fig. 5.11a shows the probability P [$= B_{m'}$ of (5.18)] of an exciton registering in the maxi-cluster, see (5.29), as a function of guest concentration (C_g) and sensor concentration (C_S), assuming maximum trapping efficiency ($\gamma = 1$). The topology is that of a nearest-neighbor square lattice (coordination number 4). The number of steps (t) is 200,000, and the coherence length is $l = 1$ ($d = 0$, no standard deviations), which also means a coherence time of one step, i.e. random walk. The limiting line (highest P) in the figure gives $\bar{P}_\infty$ for this topology (5.29a), which is the curve approached at a sensor concentration (mole fraction) of $C_S \geq 10^{-3}$. Obviously, the registration probability $\bar{F}_m$, and thus the total probability P, decreases with decreasing C_S. The limiting $\bar{P}_\infty$ curve is the same as in Fig. 5.8, with $C_c^S = 0.59$. The $\bar{F}_m$ values were calculated with the help of the algorithm given by (5.42). The number $n_m(C_g)$ of distinct sites visited was computed for random lattices of $500 \times 500 = 250,000$ sites, with cyclic boundary conditions, as a function of C_g, with the above given random walk assumption and parameters.

Figure 5.11b is closely related to Fig. 5.11a. Here the number of steps (t) has been varied from 160,000 to 640,000, giving $P(C_g)$ for a fixed parameter $C_S = 2 \times 10^{-5}$.

Figure 5.12a is analogous to Fig. 5.11a, with the following changes. A next-nearest in-plane interaction has been added (along *one* diagonal of the

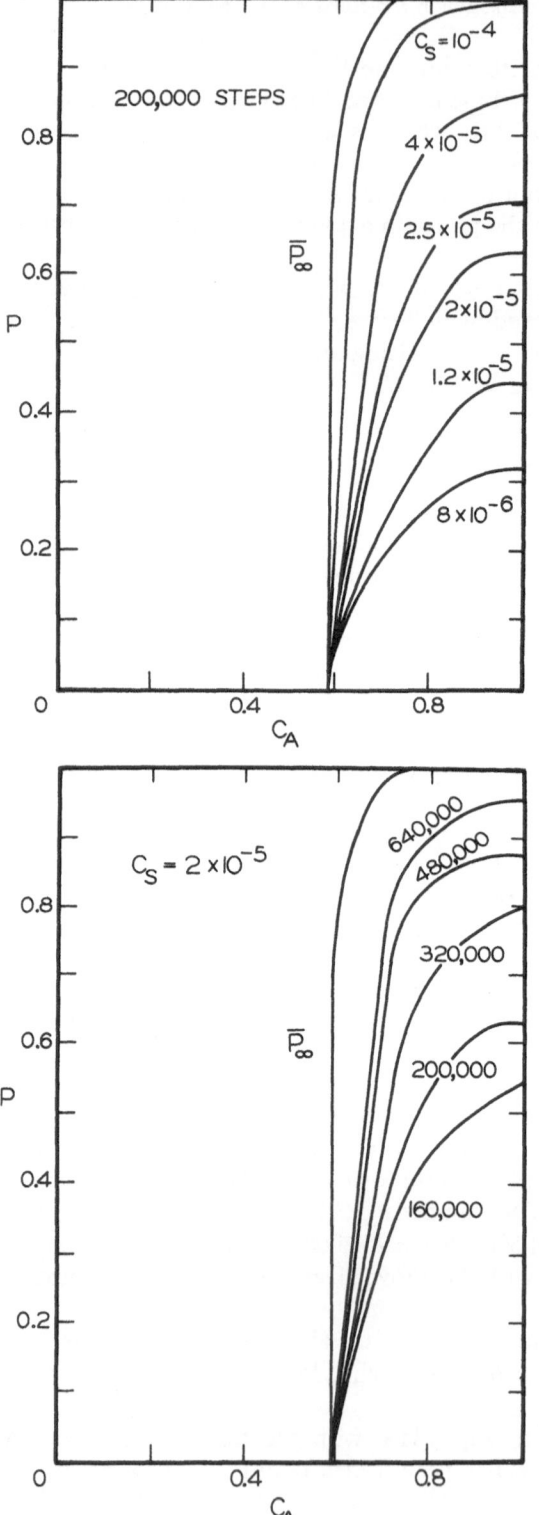

Fig. 5.11a. Maxicluster registration probability $P [= B_{m'}$ of (5.18)] vs guest concentration C_g (mole fraction) with sensor concentration C_S (mole fraction) as variable parameter and random walk ($l=1$, $d=0$) on a 500×500 square lattice ($t=200,000$). The supertrapping efficiency is $\gamma=1$. The $\bar{P}_\infty$ limiting curve is the result for $10^{-3} < C_S \ll C_g$ and is taken from Fig. 5.8. Cyclic boundary conditions and the algorithm of (5.42) were used (see text)

Fig. 5.11b. Same as a) but with fixed sensor concentration C_S, and a variable number of random walk steps t (160,000 to 640,000). $\bar{P}_\infty$ gives the result for $t \to \infty$

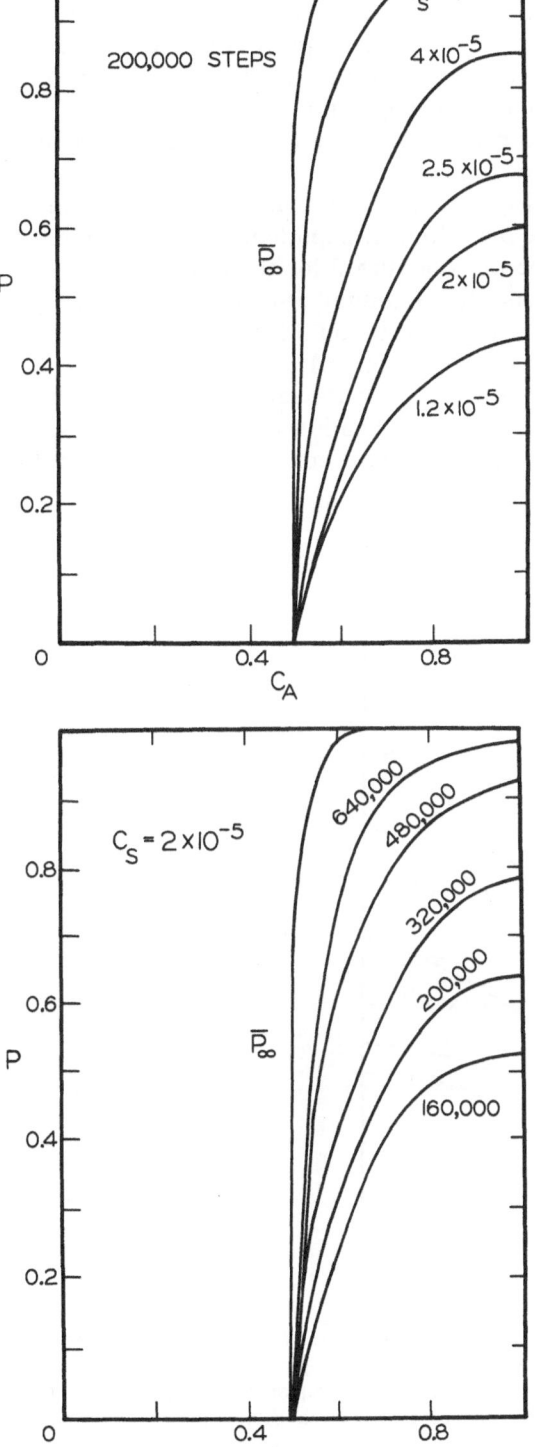

Fig. 5.12a. Maxicluster registration probability P (cf. Fig. 5.11a) for a 500×500 quasi-triangular lattice (the ratio of nearest-neighbor-to-next-nearest-neighbor interaction is 2:1) vs guest concentration C_g, with sensor concentration C_S (mole fraction) as variable parameter and the number of random walk steps, $t = 200,000$, as fixed parameter. The $\bar{P}_\infty$ limiting curve is taken from Fig. 5.8 (triangular lattice). See caption of Fig. 5.11 for additional information

Fig. 5.12b. Same as a) but with a fixed sensor concentration C_S and a variable number of random walk steps t (160,000 to 640,000). $\bar{P}_\infty$ gives the result for $t \to \infty$

square). This is equivalent to adding the a-axis translational interactions to the $1/2 (a+b)$ interchange ones in naphthalene [5.3]. The ratio of nearest-neighbor to next-nearest neighbor interactions is 2:1. This is our "coordination number 6" topology case ("quasi-triangular").

Figure 5.12b is analogous to Fig. 5.11b, except for the same change introduced in Fig. 5.12a compared to Fig. 5.11a.

Figure 5.13a is similar to Fig. 5.12a, except that now another pairwise interaction has been added (the b-axis one in naphthalene [5.3]), reduced by another factor of 2.5, giving the ratio of 10:5:2 for the three in-plane interactions. This is therefore the "coordination-number-8" topology case.

Figure 5.13b is analogous to Figs. 5.11b and 5.12b, except for the same changes that differentiate Fig. 5.13a from Figs. 5.11a and 5.12a.

Comparing Figs. 5.11a to 5.13a, we notice the rather obvious conclusion that the probability P increases with coordination number in the lower C_g range, close to where the percolation concentration C_c^s goes down from 0.59

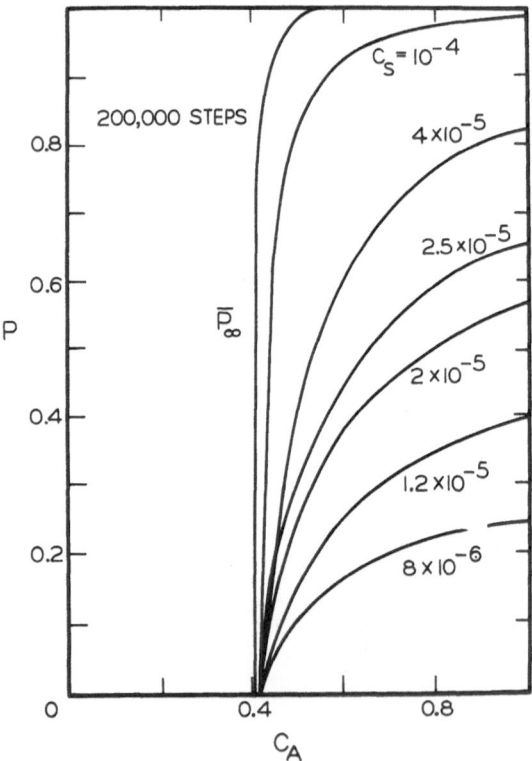

Fig. 5.13a. Maxicluster registration probability P (cf. Fig. 5.11a) for a 500×500 quasi-Square (1, 2) lattice. The ratio of inplane nearest neighbor interactions [$1/2 (a+b)$ in naphthalene]: next-nearest neighbor interactions (a in naphthalene): next-next nearest neighbor interactions (b in naphthalene) is 10:5:2. The $\bar{P}_\infty$ limiting curve is taken from Fig. 5.8 (Square (1, 2) lattice). Guest concentration C_g, sensor concentration C_s and all other conditions are the same as in Figs. 5.11a and 5.12a

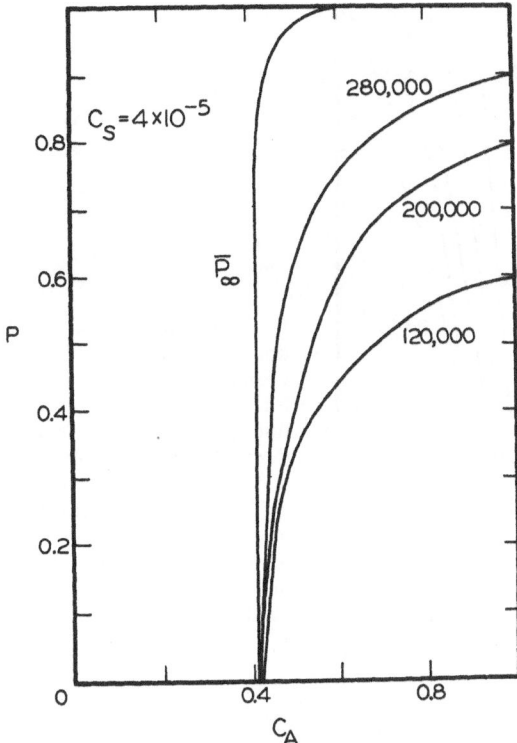

Fig. 5.13b. Same as a) but with a fixed sensor concentration C_S, and a variable number of random walk steps t (120,000 to 280,000). $\bar{P}_\infty$ gives the result for $t \to \infty$

to 0.50 to 0.41. There is no such tendency at high concentrations $(C_g \geq 0.8)$, and, actually, an opposing trend might be indicated.

Figure 5.14a shows a semicoherent calculation of P. The parameters are the same as for Fig. 5.11a (square lattice), except that $l=10$, $d=3$. Notice that while the limiting ("supertransfer") curve $(\bar{P}_\infty)$ is still the same as in Fig. 5.11a (random walk) this is not so for the other curves, with lower C_S, where the exciton registration is transport limited.

Figure 5.14b is related to Fig. 5.14a in the same way that Fig. 5.11b is related to Fig. 5.11a, except that now the number of steps (t) ranges only from 80,000 to 320,000 (for the same $C_S = 2 \times 10^{-5}$).

The striking behaviour shown in the semicoherent transfer case (Figs. 5.14a and b) is the difference in the P (but not $\bar{P}_\infty$) curves from the random walk case (Figs. 5.11a and b). While semicoherence makes the energy transfer probability higher at high guest concentrations $(C_g > 0.8)$, as expected intuitively for pure crystals $(C_g = 1)$, the opposite happens for intermediate guest concentrations $(C_c^S \leq C_g < 0.7)$. The same qualitative trend has been documented for a "delta function" coherence length $(l=10, d=0)$. It appears that with few (or none)

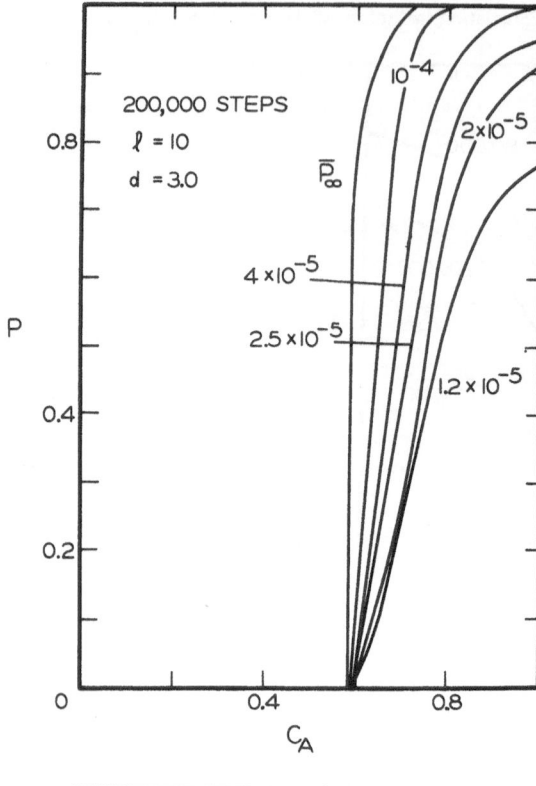

Fig. 5.14a. Maxicluster registration probability P (cf. Fig. 5.11a) vs guest concentration C_g (mole fraction) with sensor concentration C_s as variable parameter (1.2×10^{-5} to 10^{-4} mole fraction) and semicoherent walk ($l = 10$, $d = 3$) on a 500×500 square lattice ($t = 200,000$). The $\bar{P}_\infty$ limiting curve is the same as in Fig. 5.11a (but not the other curves). All other conditions are the same as for Fig. 5.11a

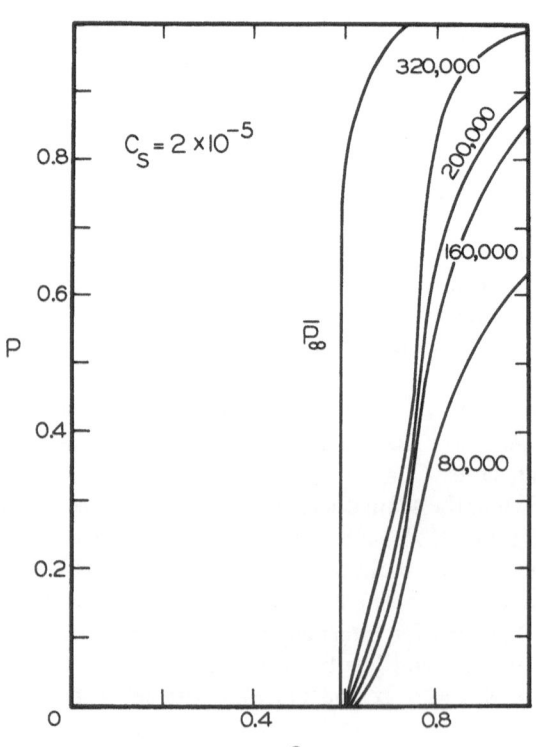

Fig. 5.14b. Same as a) but with a fixed sensor concentration C_s and a variable number of semicoherent steps t (80,000 to 320,000). $\bar{P}_\infty$ gives the result for $t \to \infty$

scattering sites the semicoherent motion is more efficient, i.e. the exciton visits more distinct sites n_m (with fewer repeat visits). However, with a multitude of scattering sites the semicoherent exciton has more difficulties finding and penetrating the "narrow and winding passages" of the maxicluster (compare Fig. 5.1).

5.3 Supertransfer Experiments

5.3.1 Introduction

We have seen in the discussion leading to (5.29a) how one can test the percolation concept without getting involved in the difficult problem of coherence vs incoherence discussed above. This limit of supertransfer is independent of the nature of exciton transport and registration (i.e. trapping), as can be seen both from the extremely simple equation (5.29a) as well as from the more rigorous equation (5.34b). We note, however, that it is a mathematical, not physical approximation, that leads from (5.34b) to (5.29a), based on the fact that $m/G \ll 1$ for the miniclusters of size m surviving above percolation. The physical implication of this mathematical approximation is that there is a very small probability of a minicluster containing a sensor as $C_S \ll m^{-1}$. On the other hand, it is guaranteed that an exciton created on the maxicluster will register, i.e. find a sensor within its lifetime, thus ending prematurely its role of an energy conducting, guest exciton. The latter "guarantee" is based on having a high enough sensor concentration C_S, so as to ensure an appreciable number of sensors in the maxicluster. There is a trade-off between the "mathematical need" for a small Z, see (5.10), and a large enough $C_S = Z/N$, where the total number of lattice sites N is constant. However, with an efficient enough exciton transfer, based on efficient exciton transport (long lifetime, fast jump time, etc.) and efficient sensor registration (trapping), there may be a range in C_S covering several orders of magnitude where supertransfer is a valid limit. In this limit the exciton transfer depends only on the topology of the exciton interactions. There is an analogy between excitonic supertransfer and electronic superconductivity.

5.3.2 The Naphthalene Singlet Exciton Experiment

We have found out in Section 5.2 that the naphthalene lowest singlet exciton can make about 2×10^5 lattice jumps within its lifetime, i.e. visit about 2×10^5, not necessarily distinct, guest sites. This result is based on nearest neighbor interactions. Our discussion in Section 5.2 has probably exaggerated the actual importance of non-nearest neighbor interactions, i.e. the non-nearest jump probabilities, as these interactions should be corrected for superexchange (see Sect. 5.2) and thereby reduced in magnitude. (There are additional reasons for reducing the probability of non-nearest jumps, both at high and low temperature.)

From the above-mentioned average number of lattice sites visited ($\approx 2 \times 10^5$) it is obvious that with about $C_s \approx 10^{-3}$ and a good trapping efficiency ($\gamma \rightarrow 1$), the maxicluster exciton is ensured of finding the sensor and registering. Thus (5.12b) is satisfied and this leads to (5.34b) and (5.29a), i.e. supertransfer. Therefore, if one knows the topology of exciton interactions (determining $\bar{P}_\infty$ as well as the second term in (5.34b)), one can check the applicability of the exciton percolation concept. On the other hand, accepting the percolation concept and assuming supertransfer one can work back from the experimental results (see Fig. 5.16) and get a good idea of the interaction topology.

In the specific experiment described here we used as a sensor the efficient (and notorious) naphthalene quencher, betamethylnaphthalene, at saturation concentrations. Our two-component guest-host system was $C_{10}H_8/C_{10}D_8$, with a $C_{10}H_8$ concentration range from 0.01 to 1.0. As mentioned, $C_s \approx 10^{-3}$. Our difficulties in eliminating the large variation in the "saturation" C_s was quite frustrating and both the crystal growing and analytical procedures are reported elsewhere [5.20, 23, 24]. We notice that not only does the alloy $C_{10}H_8/C_{10}D_8$ meet the phonon amalgamation requirement [5.17], but so does [5.25] the alloy betamethylnaphthalene/naphthalene. The trapping efficiency γ of the betamethylnaphthalene "supertrap" (in $C_{10}H_8$ or $C_{10}D_8$) is high and very close [5.23, 25] to that of $C_{10}H_8$ in $C_{10}D_8$. We thus indeed expect a case of supertransfer.

The exciton conduction in this experiment is monitored by the betamethyl-naphthalene "excitometer". The photon emission (fluorescence) from this supertrap is mostly due to exciton migration representing "energy conduction". There is, however, a residual "no-conduction emission", analogous to the "dark current" in a phototube, originating from fluorescence due to direct photon absorption by the supertrap, regardless of the transfer process. We recapitulate here the proposed transfer process: 1) absorption by the naphthalene sites and creation of naphthalene excitons; 2) exciton transfer (migration) throughout the lattice (maxicluster); 3) exciton supertrapping; 4) supertrap fluorescence.

For $C_g = 1$ all the fluorescence, within the experimental sensitivity, originates from the supertrap (at 2 K). Therefore, we simply take I_S/I_{total} as a measure of relative exciton transfer (conduction), meaning relative to a sample with $C_g = 1$. Thus at $C_g = 1$ we have $I_S/I_{total} = 1$. On the other hand, at $C_g = 0.1$ there is effectively no exciton conduction. (Practically all excitons are trapped in $C_{10}H_8$ miniclusters with no BMN, resulting in naphthalene fluorescence.) The major exception is the no-conduction supertrap fluorescence (the dark current analog) mentioned above, which should be down roughly by $Z/G = C_s/C_g \gtrsim \approx 10^{-2}$. The exact value should fluctuate about an order of magnitude, as does the supertrap concentration in the nominally saturated crystals. We monitor [5.23] a naphthalene vibronic fluorescence and compare it with the supertrap 0–0 fluorescence (see Fig. 5.15). The absolute intensity (molar extinction) ratio of $C_{10}H_8 (510 \, cm^{-1})/BMN (0–0)$ is about 0.5. The relative

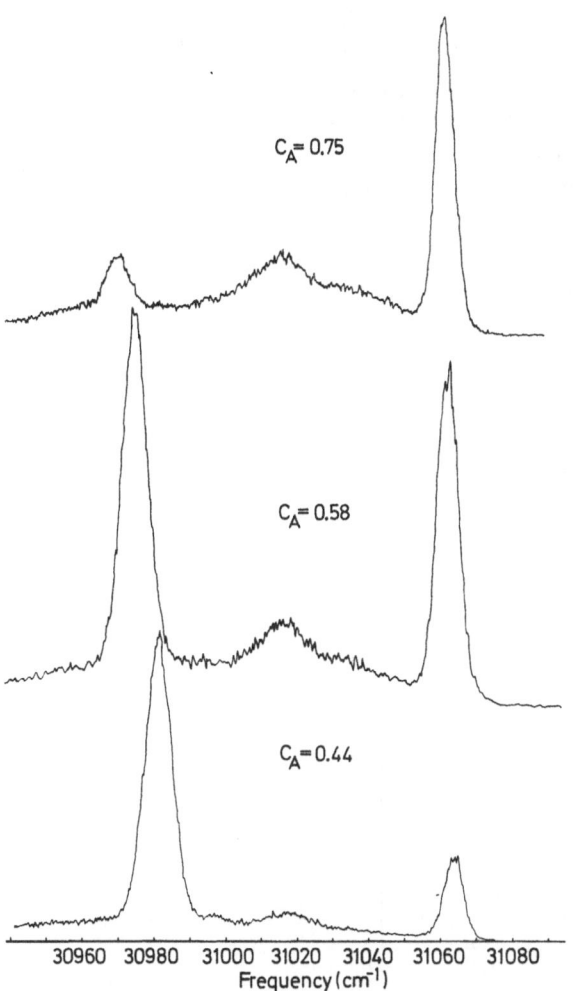

$C_A = 0.75$

$C_A = 0.58$

$C_A = 0.44$

30960 30980   31000 31020   31040 31060   31080
Frequency (cm^{-1})

Fig. 5.15. Trap and supertrap fluorescence spectra of the ternary system: $C_{10}H_8$–$C_{10}D_8$–BMN. Since the concentration of BMN is held constant at about 10^{-3} mole fraction throughout, the guest concentration C_A is essentially the $C_{10}H_8$ concentration. The most striking observation is that as the $C_{10}H_8$ concentration increases, the $C_{10}H_8$ (vibronic, $510 \, cm^{-1}$) fluorescence ($30{,}970$–$30{,}980 \, cm^{-1}$) decreases drastically, with the BMN emission (0–0 band at $31{,}065 \, cm^{-1}$) increasing proportionately. The band at about $31{,}015$ is a phonon sideband built upon the BMN origin. The spectra were obtained at 2 K on a 1 meter double grating Jarrell Ash spectrometer with 20 μm slits ($\sim 2 \, cm^{-1}$ resolution), with Xenon lamp excitation (and appropriate filters) and recorded via computer interfaced photon counting, with calibration and processing methods described elsewhere [5.20]. BMN is β-methylnaphthalene

trapping efficiency of naphthalene (monomer) and betamethylnaphthalene is also [5.23] within a factor of two. Thus the above order-of-magnitude C_S fluctuation is the main contribution to the experimental uncertainty (the instrumental contributions being negligible) [5.24].

We thus expect the ratio I_S/I_{total}, except for the "dark current" contribution, to be an experimental direct measure of the relative energy transfer. We also expect, for the condition of supertransfer, that this relative exciton transfer be given approximately by $\bar{P}_\infty$ (see (5.29a) and note our neglect of the second term in (5.34b)). In Fig. 5.16 we see a comparison of these experimental and theoretical relative energy transfer measures.

We notice that $\bar{P}_\infty$ is a functional of topology, and we therefore plot it in Fig. 5.16 for a number of plausible exciton interaction topologies. All of these

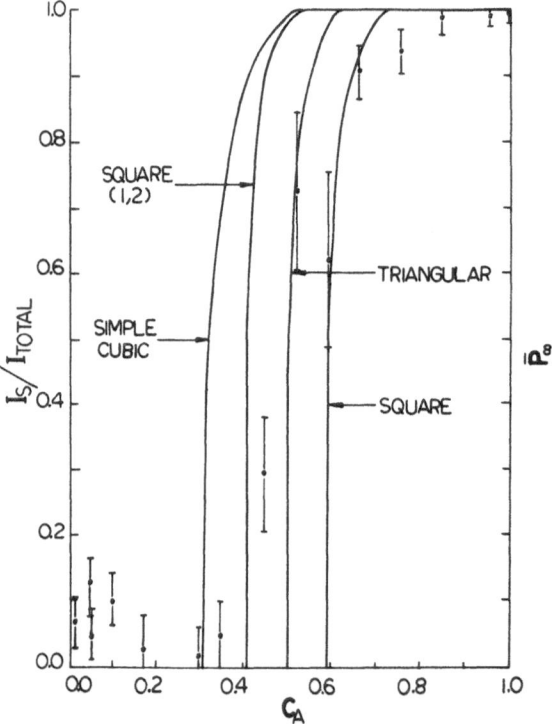

Fig. 5.16. Experimental exciton percolation: First singlet excitation of naphthalene. The experimental points are integrated fluorescence intensity ratios, where I_S is the sensor (BMN) intensity and $I_{total} = I_S + I_A$, where I_A is the naphthalene ($C_{10}H_8$) intensity (see Fig. 5.15). As $C_S \approx 10^{-3}$, $C_A \approx C_g$. For comparison we give theoretical results based on the supertransfer limit (5.29a), namely $\bar{P}_\infty$ vs C_A for various topologies. Actually, (5.29a) is only valid for $C_g \gg C_c^S$, and thus for C_g below and around percolation it should be replaced by expressions like (5.34d) or (5.34a). The maxicluster registration probabilities $\bar{P}_\infty$ are taken from Fig. 5.8 or [5.40]. Note that the experimental points given here include some refinement on those given in [5.23]. Also, note that P_∞ (not $\bar{P}_\infty$) curves were given in [5.23] (the normalization differing by a factor of C_g). These theoretical curves should actually be refined (compare Figs. 5.9, 10)

topologies have low coordination numbers (4, 6 or 8) because, as we pointed out previously, the long range interactions contribute very little to the energy transfer. The simple cubic topology refers to the limiting case where the out-of-plane exciton interactions are equal to the in-plane ones. Experimentally [5.19], however, they are smaller by at least a factor of two. The same general relative reduction in the interaction parameters happens for the in-plane non-nearest neighbors (see above). Thus we expect the square-lattice (co-ordination number 4) topology to give the lower boundary for $\bar{P}_\infty$. On the other hand, the coordination-number-8 topologies, i.e. the simple cubic and the Square (1, 2) lattice (one with next-nearest interactions equalling the nearest ones) should give roughly an upper limit for $\bar{P}_\infty$. This expectation appears to be well justified by the data in Fig. 5.16. Note, however, that more refined theoretical curves resemble the ones in Figs. 5.9, 10 (for $C_S = 0.1 \%$).

5.3.3 Conclusion and Discussion

The experimental data in Fig. 5.16 show that the naphthalene singlet exciton, at 2 K, with a relatively high sensor (supertrap) concentration and registration (trapping) efficiency, does appear to show supertransfer. This exciton transfer is consistent with the exciton percolation model. It is also consistent with reasonable exciton transfer topologies based on independent experimental information.

The supertransfer behaviour should be independent of sensor concentration C_S over some concentration region, but reduce to an exciton transport limited process (no supertransfer) at some lower C_S. This change in behaviour has been illustrated in our model computations (see Subsect. 5.2.4). It has also been borne out by our preliminary experimental results [5.26].

The unimportance of long range exciton interactions has been crucial to our specific theoretical results (the $\bar{P}_\infty$ curves in Fig. 5.8). If exciton tunneling is important, then these long range interactions can no longer be neglected. An example is given in the case of the naphthalene triplet exciton (see Sect. 5.4). Such tunneling of the singlet exciton should be most conspicuous well below the percolation concentrations (59%, 50%, 41% or 31%, depending on topology—see Fig. 5.8). The reason why tunneling is *not* important is closely related to the low temperature of the experiment ($M \gg kT$). Thus exciton migration due to naphthalene monomer-to-monomer tunneling at low C_g is likely to be quenched by the naphthalene dimers before it reaches the super-trap, any dimer-to-dimer tunneling at higher C_g is quenched by trimers, etc. Preliminary experiments [5.24] indicate tunneling contributions at temperatures where $kT \approx M$ at $C_g < C_c^S$. At even higher temperatures classical thermal climbing over the host barrier (see Subsect. 5.1.3) becomes the dominant long range energy transfer process [5.24] at low guest concentrations ($C_g < C_c^S$).

5.4 Tunneling and Superexchange

5.4.1 Clusters and Conglomerates

Terminology is a real stumbling block in emerging fields of study. Two nearest-neighbor identical guest molecules, embedded in some host, clearly form a guest cluster (pair). In a low symmetry crystal, like naphthalene, there is no difficulty imagining a pair ("cluster") made of next nearest neighbors (b-axis —see [Ref. 5.3, Fig. 5.7]), next-next nearest ones [$c+(a+b)/2$ "axis"—*ibid.*], etc. (where the "nearness" of the interactions has been measured by the center-to-center distance, which is a unique but not inevitable criterion). Extending the above approach *ad absurdum* one can define a guest "cluster" made of two guest molecules 10 lattice units apart (say along the a-axis), with the 8 in-between host molecules obviously being excluded.

Next, we are faced with explicitly considering long-range interactions. We adopt, for simplicity, the above definition of a cluster, provided that the host molecules play no part in the energy transfer except as spacers with certain dielectric tensor properties. However, in actual physical situations the host molecules may participate in the energy transfer in very specific ways: 1) by providing a (real) energy-conducting band which the guest exciton can use with the help of thermal energy (strictly speaking it has been transformed into a host exciton, but this, at finite temperatures, is often only a technicality); 2) by providing a "virtual" energy-conducting band, i.e. via quasi-resonance and superexchange (i.e. tunneling) [5.3, 27]. For either of the two cases, and especially the last one, it makes sense to include the host molecules explicitly in the energy-conducting "cluster". We have termed [5.28] such an "inhomogeneous cluster" a conglomerate.

The notion of a conglomerate has a very "ancient" forerunner [5.29], the *"Perrin-Perrin sphere of interaction"*. However, the latter was essentially envisioned as a continuous solvent sphere containing a couple of guest chromophores. Our definition emphasizes the lattice aspects [5.30] of the "solvent", and its anisotropy. It also emphasizes the topological aspects of this conglomerate, which is a haphazard agglomerate of spheres, elipsoids, discs, etc., all joined by an "interaction connectivity", and with the ability to "percolate" with increasing guest concentration or with an enhancement of the interactions. The "percolation" of conglomerates does not mean that the exciton manages to hop just once from an isolated guest site to another, but that it visits many guest sites, their minimum number depending on sensor concentration, efficiency, etc. This definition of "percolation" obviously depends on the excitation lifetime in a more drastic way than the usual case (where the range of the interactions defining a cluster also depended on this lifetime). We have called it [5.20, 23], occasionally, dynamic percolation. The arbitrariness in fixing both the sizes of the conglomerates and, therefore, their percolation concentrations, has been discussed in greater detail [5.28], and related to the method of measurement, i.e. the resolution of the spectrometer, which gives the static percolation, i.e. the concentration at which the guest spectral density-of-states becomes a quasi-continuum. We encounter here a typical case of the quantum theory, where the interaction between the measuring instrument and the system defines the "makeup" of the system, i.e. its cluster and conglomerate distributions as well as the "critical" percolation concentration.

5.4.2 The Naphthalene Triplet Exciton Experiments

Here we demonstrate that 1) dynamical exciton percolation does occur (i.e. a low concentration transition from an exciton insulator to an exciton conductor), that 2) it is very useful for the investigation of energy transfer in molecular aggregates, and that 3) it is a critical test of our current knowledge of exciton exchange and superexchange.

The trap-to-trap migration concept was introduced over a decade ago [5.27] to describe a long-range triplet exciton transfer. It was suggested that such migration could occur by a superexchange [5.19] type interaction through the host exciton band, the direct FÖRSTER type interaction [5.30] being of small importance for such long-range (non-near neighbor) triplet exciton transfer. Above we have introduced the concept of a dynamic percolation which describes the onset of free exciton flow through the guest molecules. It was shown that as a result of an increased trap-to-trap migration length in a long-lived triplet, the dynamic percolation for the triplet can occur at a much lower concentration than for the static percolation [5.28].

To experimentally observe dynamic percolation (which depends on the exciton lifetime) for a given guest (trap) a method was proposed [5.31] utilizing the doping of the binary sample with a very small concentration of a lower energy trap (supertrap) in order to monitor the exciton flow. At the onset of the dynamic percolation, the supertrap emission increases dramatically (see Sect. 5.3).

Here we present some studies on triplet energy transfer in a multi-component isotopic mixed naphthalene crystal, including the binary system $C_{10}H_8/C_{10}D_8$ with a chemically substituted supertrap. An energy denominator study of the dynamic percolation [5.20] reveals it to be highly dependent on the trap depth, thus showing the importance of the indirect, i.e. superexchange interaction through the host exciton band (see below).

The superexchange (guest-host-guest) interaction in a square lattice is given by (correcting the formula in [5.20]):

$$M_{0,n} = \Gamma_n \beta^n / \Delta^{n-1} \qquad (5.63)$$

where the second (identical) guest is n steps removed from the first via a shortest path, Γ_n is a geometrical factor described below, β is the nearest-neighbor exciton exchange interaction in a pure guest crystal and Δ is the trap depth (guest-host energy separation in an ideal mixed crystal [5.3]). In the above we neglected the superexchange interactions stemming from both non-nearest-neighbor terms and from non-shortest paths.

The geometrical factor Γ_n depends on the topology of the shortest paths between the two guests. In the relatively rare case where this path is a straight line,

$$\Gamma_n = 1 \quad \text{iff straight path}. \tag{5.64}$$

Otherwise, i.e. for zig-zag paths, it is [5.20, 32]

$$\Gamma_n = n!/[(n/2)!]^2 \quad \text{iff } n = \text{even, zig-zag path}, \tag{5.65}$$

$$\Gamma_n = n!/\{[(n-1)/2]! \, [(n+1)/2]!\} \quad \text{iff } n = \text{odd, zig-zag path}. \tag{5.66}$$

In-between values of Γ_n are obtained for non-straight, non-zig-zag paths. NIEMAN and ROBINSON [5.27] treated only the case of a straight path, which strictly applies to a linear crystal only.

To illustrate the problem we assume that in Fig. 5.2 the excitation is localized at a given instant at a guest (trap) site near one corner of the diagram and that a supertrap is located at a digonally opposite corner. Will the exciton reach the trap before it decays (radiatively or non-radiatively)? Assuming random walk, and limiting ourselves to the isolated domain given in Fig. 5.2, we note that the probability of the trap exciton reaching the supertrap depends most crucially on its ability to bridge the largest spatial gaps between guest sites. These gaps are designated as "bottlenecks" or "tunnels" with a given length. The exponential dependence of the tunneling-times upon the tunnel-length causes these maximal tunnel-lengths to account for an inordinate share of the total lifetime, compared to the smaller or "mean" tunnel lengths. Therefore, we can approximate tunneling-times with the sum of maximal tunneling-times. If a certain probable path leads the exciton to the supertrap (in time t) one can write for this path:

$$\tau \geq t = \sum_i t_i \geq \sum_j t_j. \tag{5.67}$$

Here τ is the exciton lifetime, t is the total average time of travel, i runs over all tunnels (including direct contacts of traps) and j is limited to the maximal tunnels.

The ternary system naphthalene-d_8/naphthalene-h_8/betamethylnaphthalene (BMN) was selected because of a number of theoretical and experimental advantages [5.20]. Highly purified [5.20] $C_{10}H_8/C_{10}D_8$ mixed crystals, doped with about 10^{-3} mole fraction of BMN, were excited at about 1.8 K by an appropriately filtered [5.20] xenon lamp and monitored at $1\,\text{cm}^{-1}$ resolution with photon counting (interfaced with an IBM 360/67 computer, enabling calibration, smoothing, and differential integration on a graphics terminal). For each $C_{10}H_8$ concentration, we plot (Fig. 5.17) the intensity fraction of BMN phosphorescence (0–0), the rest of the intensity being due to naphthalene-h_8 (the naphthalene-d_8 phosphorescence is quenched throughout). This BMN intensity fraction is a measure of the exciton free flow (see Sect. 5.3).

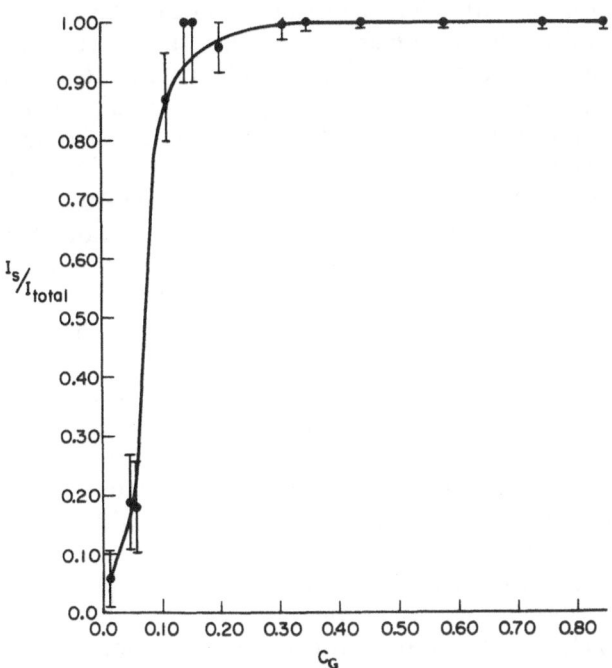

Fig. 5.17. Experimental exciton percolation: First triplet excitation of naphthalene. The experimental points are integrated phosphorescence intensity ratios, where I_s is the sensor (BMN) intensity and $I_{total} = I_s + I_A$, where I_A is the naphthalene ($C_{10}H_8$) intensity. As $C_s \approx 10^{-3}$ (mole fraction), $C_g \approx C_A$. Note that one point has been corrected in concentration, compared to [5.31]

Figure 5.17 shows that the critical concentration is 0.08 mole fraction of $C_{10}H_8$. This is far below the theoretical static site percolation concentration, based on nearest-neighbor interactions. We know that the lowest triplet exciton interactions in naphthalene are extremely short ranged [5.3] and essentially two-dimensional, limited to the ab plane; the crystal structure [5.33] yields a square lattice topology for this ab plane and for a square lattice the site percolation concentration (nearest neighbor) [5.3] is 0.59. The apparent contradiction is easily overcome if one considers exciton superexchange [5.18, 19]. With only nearest-neighbor interactions one still gets effective long-range interactions in the ab plane, enabling the $C_{10}H_8$ exciton to "jump over" (or rather tunnel through) a number of $C_{10}D_8$ sites (trap-to-trap migration [5.27] restricted to the ab plane only). The largest necessary jump ("tunnel length") for $C_A = 0.10$ is seen [Ref. 5.31, Fig. 2] to cover about 5 sites. One can say that the "dynamic" radius of the corresponding conglomerate is 5 units of nearest neighbor spacings ($5 \times 5.11\,\text{Å} = 25.55\,\text{Å}$), and the conglomerate is disk-shaped. We note that with a nearest-neighbor interaction [5.34] of $1.25\,\text{cm}^{-1}$ one gets (5.63) for a five-sites-removed superexchange pairwise interaction [5.20] about $10^{-7}\,\text{cm}^{-1}$ ($3 \times 10^3\,\text{Hz}$), giving a time constant [5.2] (tunneling-time) of about $10^{-4}\,\text{s}$. With a lifetime [5.35, 36] of 2.6 s this gives a transfer constant [5.2] of 10^4 (such conglomerates). The fact that most of the tunnel lengths are much shorter makes the exciton percolation quite feasible. On the other hand, a single 6-site tunneling may require 1% of an exciton lifetime and a 7-site tunneling requires about a lifetime. Such longer tunnel lengths are common in a 0.09 mole fraction random lattice [Ref. 5.3, Fig. 3]. We thus expect the dynamic percolation concentration to be about 0.09 mole fraction, in excellent agreement with the experimental value. We also note that SCHWÖRER and WOLF [5.36] observed a dramatic qualitative change in their 20% $C_{10}H_8/C_{10}D_8$ sample, compared with a 10% (or lower) sample, involving their ESR spectrum (the lifetime at 20% is reduced to 0.1 s). That our result (Fig. 5.17) does not reflect a singlet exciton phenomenon is shown by the latter's percolation (see Sect. 5.3) at above 30%. However, we believe this agreement to be somewhat fortuitous, in view of the relative crudeness of the model with regard to the dimensionality and nature of the exciton motion, the supertrapping efficiency, and the specifics of the exciton-phonon interactions.

Experiments were also performed [5.20] on mixed crystals with guests of 2-$DC_{10}H_7(\beta)$, 1-$DC_{10}H_7(\alpha)$ and 1,4-$D_2C_{10}H_6(\alpha_2)$ in a $C_{10}D_8$ host. In these potassium fused and zone refined crystals [5.20] the only supertraps were the lower energy isotopic impurities, i.e. $C_{10}H_8$ for the first system, $C_{10}H_8$ and β for the second one and $C_{10}H_8$, β and α for the third system. We notice that the guest-host energy separation (Δ, the trap depth or energy denominator) decreases monotonically from the first to the last [5.37]. The superexchange (5.63) interaction $M_{0,n}$ falls-off with the $(n-1)$th power of Δ. For a minimal value of $M_{0,n}$ consistent with the lifetime (which is roughly constant), the possible tunneling length n increases monotonically as Δ becomes smaller. Therefore we expect the dynamic percolation to occur at lower concentrations. This has indeed been observed [5.20], the percolation concentrations being roughly 7%, 6% and 4%, respectively.

5.4.3 Temperature Effects

First we comment that in all practical exciton conduction experiments one has a steady state, not an equilibrium state. In all experiments involving supertraps, the latter would be preferentially populated by excitons in a thermodynamic state of "equilibrium", provided kT is small compared with the guest-supertrap energy difference. In our previously described experiments on both singlet (see Sect. 5.3) and triplet (see Subsect. 5.4.2) excitons, the observation was that only a minority of the excitons ever reached the supertraps in samples of composition below the percolation concentration. One can describe such a situation as a "non-Boltzmann" system but actually it is simply a non-equilibrium system.

One common method to restore the Boltzmann "equilibrium" is to raise the temperature of the sample sufficiently so as to thermalize the excitons out of the traps into the "host" and then into the supertraps. The thermalization process is well known to follow an exponential dependence on the temperature. It should also be independent of the guest concentration, provided that all competing mechanisms are negligible.

Some temperature studies have been done [5.38] on the triplet ternary systems (see Subsect. 5.4.2). Essentially they reveal the above expected energy cascade, from shallower to deeper traps, as the temperature increases. However, at high enough temperature, "host" emission takes over. This is again expected from elementary statistical mechanics. Usually, the host has the highest concentration and thus the highest, by far, weighting factor. This is true independently of whether there is a wide or narrow host exciton bandwidth, as long as this width is small relative to the trap depth.

We notice that, occasionally, a thermalization process can be mistakenly attributed to percolation. As an example we may use the system $C_{10}D_8$, 0.35% $C_4D_4H_4(\alpha D_4)$, containing minute amounts of various supertraps (BMN, $C_{10}H_8$, αD_1, βD_1, αD_2, αD_3, etc.). At very low temperatures (1.8 K), all the emission originates from the supertraps and the guest (αD_4) seems to have "percolated" (at a concentration of 0.35%!). However, a careful temperature study reveals that thermalization of the excitons occurs down to about 2 K. This should not be totally surprising in view of the shallowness of the αD_4 trap depth [5.37] (32 cm^{-1}, i.e. about 27 cm^{-1} from the $C_{10}D_8$ exciton band-edge) and the lifetime of the excitation (order of 10 s, by interpolation of literature data [5.39]).

Is thermalization the only mechanism of energy transfer at 2 K for this experiment? The answer is negative as there has been observed a very definite concentration effect [5.20, 38] which is incompatible with thermalization as the sole factor. There is no apparent exciton percolation (quenching by the supertraps) at the lower guest concentration of 0.039% (unfortunately with the same relative supertrap concentrations [5.20]). This phenomenon clearly indicates that we have a combination of two energy transfer processes, tunneling (superexchange) and thermalization (climbing over the barrier).

Obviously, in the above described situation of both quantum-mechanical tunneling and classical thermalization, it should not be assumed that one can simply "add up" two appropriate rate constants. Instead one has to examine a situation which is similar to constructive (and sometimes destructive) interference. We assume that, for a given sample, when the tunneling length (path) is short, superexchange is efficient, but when it is long, superexchange is inefficient and thus thermalization becomes significant. We believe that this kind of exciton transfer kinetics in a random lattice can be calculated only with the help of computer simulations. We notice that here one has a very fine balance, involving kT. It appears [5.20] that our experiments are very sensitive to the absolute temperature, and that a slight variation in it, say from 1.5 K to 2 K, may drastically tip this balance...

There remains a very interesting but subtle temperature effect related to the triplet exciton percolation results (see Subsect. 5.4.2). The question can be asked: why don't the guest dimers, trimers, etc. serve as "supertraps", to compete successfully with the very dilute "official" supertrap? Certainly, at any given concentration below the site percolation concentration (59% for the square lattice) there exist certain guest clusters, with lower energy than the guest monomers and at low enough concentration, so as to exclude cluster-to-cluster tunneling, that could compete successfully with the "official" supertrap. Actually, a careful consideration of the singlet exciton experiment (see Sect. 5.3) shows that it is specifically the trapping by such dimers, trimers, etc. which prevents the BMN supertraps from trapping excitons very efficiently below percolation, and not the complete absence of superexchange (tunneling). Obviously, above percolation, the maxicluster becomes the deepest guest trap, and there is no competition with the supertraps because the maxicluster contains many of them. Returning now to the triplet case, why is not the tunneling from-guest-monomer-to-guest-monomer quenched by the dimers, trimers, etc.?

The answer to our above question is simple. The dimer, trimer, etc. excitons have the same energy as the monomer, within kT (even at 2 K)! They cannot serve, therefore, as supertraps (in contrast to the singlet case, where the energy differences are an order of magnitude larger [5.20]). However, at much lower temperatures (say 0.1 K) the triplet exciton percolation behaviour should approach that of the singlet exciton (at 1.5 K). We thus predict that at very low temperatures the triplet tunneling (superexchange) will be quenched by guest miniclusters, essentially at all concentrations up to about $C_g = 0.5$, and there will be no dynamic percolation.

Experiments at 0.1 K are still difficult. However, we have indirectly checked our hypothesis by raising the temperature (from 2 K up to 30 K) of the singlet percolation experiments (see Sect. 5.3). Our preliminary results [5.24] indicate that, for intermediate C_g concentrations (about 0.1 to 0.3), the effect of thermal detrapping of the miniclusters can be distinguished from the guest-to-host thermalization effect (which is significantly less sensitive to a variation in guest concentration). However, the two thermalization effects share a temperature (kT) domain so that at no temperature does the naphthalene singlet exciton system describe the same kind of clear-cut dynamic percolation effect exhibited by the naphthalene triplet exciton system.

5.5 A Model for the Primary Step of Photosynthesis

5.5.1 Introduction

The chemically mixed molecular crystals used for the demonstration of exciton percolation (say, naphthalene-perdeuteronaphthalene:betamethylnaphthalene) bear a striking resemblance to certain natural photosynthetic molecular aggregates (say chlorophyll a/chlorophyll b: P700). This similarity is based on: 1) similar relative exciton energy levels, with energy separations (50 to 500 cm^{-1}) larger than kT (the temperature T being very low for the alloys but ambient for the bioaggregates); 2) similar relative component concentrations (excluding the protein carriers, etc.), a typical ratio [5.5] of Host: Guest: Sensor being 0.2:0.8:0.003; 3) similar pairwise singlet exciton interactions (20 vs 300 cm^{-1}) and excitation lifetimes (70 vs 5 ns), resulting, accidentally, in a practically identical number of "steps" ($t \approx 2 \times 10^5$). This similarity led us to propose the model of a substitutionally random molecular aggregate for the light harvesting complex in photosynthetic systems, such as those in green plants.

5.5.2 The Model

Consider a set A of molecules (sites) with excitation energy E_A (with a range $\Delta E_A < kT$, where k is the Boltzmann constant and T the absolute temperature), called the energy conducting species. Consider another set B, with excitation energy E_B, where

$$E_B - E_A > kT . \tag{5.68}$$

This set is called the energy insulating species (the smear ΔE_B can be large, as long as (5.68) is obeyed). The third set S has excitation energies E_S, where

$$E_A - E_S > kT , \tag{5.69}$$

and is called the sensor or supertrap species (the smear ΔE_S can again be large, as long as (5.69) is obeyed).

The above three species form a lattice of randomly distributed substitutional sites with a given topology of interactions. The interaction topology involves the number and spatial arrangement of the physically interacting neighbors.

In our present model we limit the molar concentration C_S of the S sites to a small number, of the order of 1 % or less, such that

$$C_S \ll C_A . \tag{5.70}$$

The important parameter now is C_A or, alternatively, the ratio

$$R_A \equiv C_A/C_B . \tag{5.71}$$

We now proceed on the assumption that the energy conduction is confined to the quasi-lattice of set A and depends critically on its connectivity. This connectivity depends in turn on the concentration C_A, or alternatively, the concentration ratio R_A. The quantitative relationships are based on (5.29a).

In this supertransfer limit the energy conduction in a pure A lattice (with the same topology) is transport unlimited. Thus the energy conduction in our substitutionally random lattice is proportional to $\bar{P}_\infty^A$, the probability of an A site being inside the A maxi-cluster (the largest A cluster, see Subsec. 5.1.2). This probability is essentially zero for $0 \gtrsim C_A < C_c^S$, where C_c^S is the critical percolation concentration (i.e. 0.59 for a square lattice, 0.50 for a triangular one, 0.31 for a simple cubic one, etc.). At C_c^S the value of $\bar{P}_\infty^A$ climbs rapidly and approaches 0.99 for $C_A \approx C_c^S + 0.15$. This behavior is shown in Fig. 5.16. Thus the condition for energy transfer (energy migration) is

$$R_A > 1 \quad \text{(within a factor of 2)}. \tag{5.72}$$

To emphasize this result, we write an equivalent relationship

$$0 \gtrsim R_B \equiv \frac{C_B}{C_A} < 1 \pm 0.5. \tag{5.73}$$

5.5.3 Discussion

Our very general criterion for energy transfer, i.e. (5.72) or (5.73), is independent of whether the transfer is a coherent exciton transfer or a hopping exciton transfer (equivalent to resonant Dexter-Förster transfer). If instead of a perfectly random distribution one has partial aggregation tendencies (A with either A or B), our criterion will be somewhat affected. This should not be significant, however, as long as there is not complete segregation. Notice that any "segregation model" (or "funnel model") for photosynthesis requires some kind of surrounding of the supertrap by the A species and some partial surrounding of the latter by the B species, in some ordered or semi-ordered fashion (i.e. an "energy funnel", see Fig. 5.18). Note also that in our model an occasional thermalization of the excitation, out of the supertrap, will easily result in its trapping by a different supertrap, while this kind of "spillover" is not at all likely in the funnel model. However, the main distinction between our quasi-random model and a semi-ordered one is that in the latter there is no restriction on the ratio R_A, i.e. in the "funnel model",

$$0 \gtrsim R_A = R_B^{-1} \gtrsim \infty, \tag{5.74}$$

$$0 \gtrsim R_B = R_A^{-1} \gtrsim \infty. \tag{5.75}$$

The only time when (5.74) and (5.75) are valid in a random system is when the exciton-interactions are extremely long range. An example for this is the long-range superexchange (tunneling) interaction demonstrated by the extremely long-lived triplet states of aromatic crystals (i.e. $R_A \sim 0.1$ for the $C_{10}H_8/C_{10}D_8$ system where such percolation occurs, Sec. 5.4). We notice,

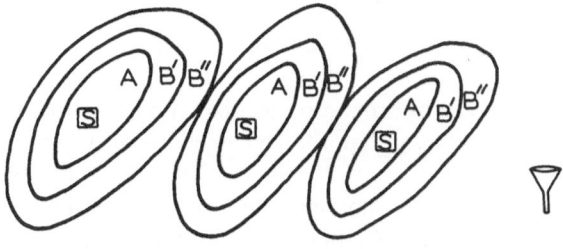

RANDOM

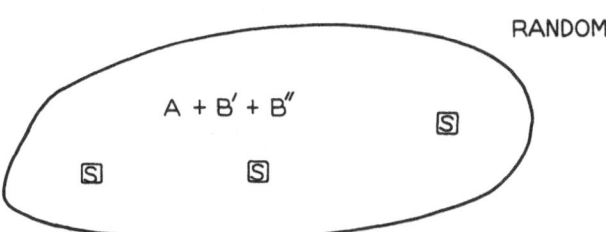

Fig. 5.18. Funnel model (top) and random lattice model (bottom) for photosynthetic system. Notice the multifunnel structure, where $E_{B''} > E_{B'} > E_A > E_S$ (see text for exact definitions; however, note that in the text no distinction was made between B' and B''). Also notice that in the random lattice model the distinction between B' and B'' is irrelevant. In the funnel model the B'' exciton with $E_{B''}$ energy falls into the B' "lattice", acquires $E_{B'}$ energy, falls into the A "lattice", acquires E_A energy and finally falls into the sensor S. In the random lattice model the B'' (or B') exciton immediately finds an A pigment and if $R_A > 1$, see (5.71), then the exciton of energy E_A will travel through the A sublattice until it finds an S site

however, that the superexchange process is efficient only for an energetically homogeneous host (i.e. $\Delta E_B \rightarrow 0$) and would be inefficient if B were a multicomponent host. On the other hand, when the excitation transfer is extremely transfer limited (i.e. very short lifetime and/or very low sensor concentration and/or low trapping efficiency), one can get into the regime where efficient energy transfer occurs only for

$$R_A \gg 1 . \tag{5.76}$$

Our computer simulations show that the above principles hold for two-dimensional lattices as small as 20×20 as well as for three-dimensional lattices with an Avogadro (or infinite) number of sites (with only minor variation due to boundary conditions). Obviously, on the other hand, for a one-dimensional-lattice with nearest neighbor interactions only, $C_c^S = 1$ and thus for such one-dimensional energy transfer,

$$R_A \rightarrow \infty , \tag{5.77}$$

i.e. no substitutional B sites are allowed (note that interstitial B species have been excluded from our discussion and should not be counted towards the R ratios!—see below).

An additional point common to our synthetic molecular alloys and to the photosynthetic systems is the requirement that the sensor (supertrap, active center), which is definitely a minority component, should be a very efficient exciton trap. This requires it to have an amalgamated phonon band (see Sec. 5.3) with the energy conducting species (A), in contrast to having localized phonons. The requirements for an amalgamated (delocalized) phonon band in synthetic alloys [5.25] boil down to a similarity in molecular structure (giving similar masses, moments of inertia, van-der-Waals packing sizes and inter-molecular force constants). This requirement would be fulfilled in the photosynthetic unit only if the reaction center (e.g. P 700 in photosystem I [5.41]) is essentially composed of the same molecule as the energy conducting species (e.g. chlorophyll a). This "phonon criterion" is unrelated to percolation or the present model and will be discussed further elsewhere.

The photosynthetic unit can be characterized with chlorophyll b as the host species having an absorbance maximum at about 650 nm, and chlorophyll a as the guest or energy conducting species with an absorbance maximum at about 680 nm. The P700 reaction center can be thought of as the sensor fulfilling the requirement (5.69) $E_A - E_S > k\,T$. These two forms of chlorophyll plus other substitutional pigments in minor amounts, plus a small amount of various carotenoids (located interstitially?) constitute the chromophore population in the photosynthetic unit, with the ratio of chlorophyll a/chlorophyll b, $R_A > 1$ for green plants [5.7]. Evidence tends to suggest that the P 700 photoactive center is composed of a dimeric aggregate of hydrated chlorophyll a [5.42]. Based on the phonon criterion, this similarity between the sensor and the A species would suggest efficient trapping of the excitation energy. This is borne out by the extremely low level of luminescence of the chlorophyll a in green plants. Obviously, the trapping of chlorophyll b excitations, mostly by chlorophyll a, is very efficient due to both the molecular similarity and the high concentration of chlorophyll a.

Finally, we cannot refrain from making an evolutional speculation. The question to be asked is: "Why are ordinary plants green?". Maybe a better question is: "Why aren't they black?" From a simple efficiency point of view it would seem to make sense for the photosynthetic unit to absorb photons throughout the solar spectrum (and especially in the photon-rich green region). Why not add a set of pigments to help do this (particularly in the green) and then let their excitation energy cascade down into the lower ("red") chlorophyll-a first excited singlet state, etc. Present pictures of photosynthesis do not seem to account well for this absence of blackness. However, our model, involving percolation, requires the condition of (5.73). All the above hypothetical "filler" pigments would be classified as B type, energy insulating species (energy insulating towards the red excitons of the A species). It simply may not be possible to both fill out the solar spectrum, the same way one fills out the tunable dye laser spectrum, and to simultaneously fullfill the requirements of (5.73). The same argument can be made for other photosystems containing "absorptivity holes" in the solar spectrum, irrespective of the "color" of the hole. We also note that a multipigment B host would not allow efficient tunneling and superexchange, in contrast to the situation described in Section 5.4.

Another interesting problem is the *ratio of antenna chlorophyll molecules per reaction center*. In green plants this is usually found [5.7], within a factor of two, to be 400:1. The evolutionary benefits of a large antenna are intuitively obvious. The interesting question is: What limits this ratio? Calculations of the sort exhibited by Figs. 5.11, 12 and 13 clearly show that at a ratio of 10^4 or higher the excitation transfer probability is significantly lower, relative to the supertransfer case (given by $\bar{P}_\infty$ on these figures). Not shown on these figures, but evident from our computations [5.45, 46] is the fact that at *ratios of 10^3 and lower* the P curve is identical with the $\bar{P}_\infty$ curve within 0.1%. This means that increasing the above ratio from say 10 to 10^3 decreases the excitation transfer very little but increasing it from about 10^3 to 10^6 leads to a catastrophic decrease in energy transfer. This result is independent of the different topologies used in our computer program (and little affected by the corrections of Figs. 9 and 10). While it does depend on the "number of steps" (n), which we assumed to be 200,000, it does not depend on it critically. So, while the latter number ($N = 200,000$, which we used for naphthalene) may easily be off by an order of magnitude for the photosynthetic units, it should not drastically affect our above conclusion, namely, that the ratio of antenna chlorophyll to reactive-center should be optimized at a value of roughly about 10^3, provided one uses our assumption of a heterogeneous (random) aggregate model and an excitation permitting about 10^5 hops.

5.6 Epilogue—What Next?

Exciton percolation is first of all an esthetically pleasing concept, connecting some beautiful and rigorous mathematics with very basic physics and applying it to chemical and biological systems. Its generality leads us to believe that, once characterized, it will soon be "discovered" under various circumstances. This has already occured in some exciton fusion studies on chemically mixed crystals (M. POPE, private communication). Moreover, it appears to us that exciton percolation is an excellent tool for the study of exciton dynamics (energy transfer). In contrast to studies on neat or slightly doped crystals, here, with heavily doped crystals, one is not too concerned about "natural" chemical impurities and/or physical defects. First, the assured presence of an efficient sensor (i.e. supertrap) at relatively high concentrations effectively neutralizes the contributions of unwanted impurity and defect traps. In addition, the presence of the "exciton insulating" host, in fairly high concentrations, introduces a potent but controlled exciton scatterer, which again overshadows the scattering contributions from defects, impurities, surfaces, etc. Finally, having available an additional parameter (C_g) makes it much easier to compare experiment and theory, as the latter should account not only for one absolute number (i. e. the exciton diffusion length in a pure crystal) but also for the qualitative and functional behaviour of the exciton transfer vs this parameter (C_g).

The apparent role that exciton percolation could well play in photosynthesis leads us to believe that a similar situation could easily be obtained for other biological systems where energy-transfer and/or communication are involved. The above described insensitivity to both defects and impurities is highly significant for such biosystems, as previously it was difficult to envision an in-vivo biosystem comparable in purity and perfection to, say, a potassium fused, vacuum treated, zone refined, ultrapure naphthalene crystal, carefully grown from the melt or from the vapor phase. The drastic property changes at the percolation concentrations, similar to those at cooperative phase transitions, could well lead to switching of energy transfer by a variety of subtle triggering mechanisms, such as, for instance, the mechanical stretching of the biopolymer leading to a change of nearest neighbor interactions. One wonders to what degree this might be related to an interaction between resonant energy transfer and proton transfer, in situations like the mechanism of enzyme action, bioenergetics, mitochondrial respiration and various regulatory steps, including possibly mental processes [5.43, 44]. Thus our model for the primary process in some photosynthetic units might have much wider implications for biomolecular aggregates.

Finally, similar to the recent applications of electron percolation in disordered semiconductors [5.10] (amorphous, glassy or polymeric) there may be technological applications of exciton percolation. Again our model for photosynthesis could serve as the basis for a sophisticated solar energy "antenna" coupled with an appropriate solar cell. Furthermore, the fact that energy transfer in molecular alloys is relatively insensitive to impurities and defects is again of utmost importance. This would minimize any effects of radiation damage (unwanted photochemistry) and enable some simple manufacturing methods of such antenna.

Acknowledgement. I would like to thank P. ARGYRAKIS, Dr. J. HOSHEN and E. M. MONBERG for help with the computations, the figures and their criticism of the manuscript. I have drawn heavily on the yet unpublished experimental and theoretical works of my above mentioned collaborators, and also on those of DRS. F. W. OCHS and P. N. PRASAD. Thanks are also due to MRS. A. DARSKY for her faithful typing of the manuscript.

This research was supported in part by N.I.H. grant NS08116, and in part by NSF grant GH-32578X, for which the author is most grateful.

References

5.1 I. B. BERLMAN: *Energy Transfer Parameters of Aromatic Compounds* (Academic Press, New York 1973)

5.2 H. C. WOLF: Prog. At. Mol. Phys. **3**, 119 (1968)

5.3 R. KOPELMAN: Excitons in Pure and Mixed Molecular Crystals, in *Excited States*, Vol. II, ed. by E. C. LIM (Academic Press, New York 1975)

5.4 D. M. HANSON: "Energy States and Energy Transfer in Molecular Crystals: A Primer" (unpublished)

5.5 K. SAUER: In *Bioenergetics of Photosynthesis*, ed. by GOVINDJEE (Academic Press, New York 1975)

5.6 R. S. KNOX: *ibid*

5.7 C. J. ARNTZEN, J.-M. BRIANTAIS: *ibid*

5.8 H.-K. HONG, R. KOPELMAN: J. Chem. Phys. **55**, 5380 (1971)

5.9 V. K. S. SHANTE, S. KIRKPATRICK: Adv. Phys. **20**, 325 (1971)

5.10 S. KIRKPATRICK: Rev. Mod. Phys. **45**, 574 (1973), effectively uses the definition $P_\infty = m'/N$

5.11 E. W. MONTROLL: J. Math. Phys. **10**, 753 (1969); J. Phys. Soc. Japan Suppl. **26**, 6 (1969) and references therein

5.12 R. P. HEMENGER, R. M. PEARLSTEIN, K. LAKATOS-LINDENBERG: J. Math. Phys. **13**, 1056 (1972)
K. LINDENBERG: J. Stat. Phys. **10**, 485 (1974)

5.13 K. LAKATOS-LINDENBERG, K. E. SHULER: J. Math. Phys. **12**, 633 (1971)

5.14 V. M. KENKRE, R. S. KNOX: Phys. Rev. B **9**, 5279 (1974)

5.15 R. P. HEMENGER, K. LAKATOS-LINDENBERG, R. B. PEARLSTEIN: J. Chem. Phys. **60**, 3271 (1974)

5.16 P. H. CHERESON, P. S. FRIEDMAN, R. KOPELMAN: J. Chem. Phys. **56**, 3716 (1972)

5.17 P. N. PRASAD, R. KOPELMAN: J. Chem. Phys. **57**, 863 (1972);
H.-K. HONG, R. KOPELMAN: J. Chem. Phys. **58**, 384 (1973)

5.18 H.-K. HONG, R. KOPELMAN: Phys. Rev. Lett. **25**, 1030 (1970)

5.19 H.-K. HONG, R. KOPELMAN: J. Chem. Phys. **55**, 724 (1971)

5.20 F. W. OCHS: Ph. D. Thesis, Univ. of Michigan (1974);
F. W. OCHS, R. KOPELMAN: J. Chem. Phys. (in press)

5.21 T. B. EL-KAREH, H. C. WOLF: Z. Naturforsch. **22**a, 1242 (1967)

5.22 S. D. COLSON, D. M. HANSON, R. KOPELMAN, G. W. ROBINSON: J. Chem. Phys. **48**, 2215 (1968)

5.22a G. J. SMALL: Private communication (1975)

5.23 R. KOPELMAN, E. M. MONBERG, F. W. OCHS, P. N. PRASAD: Phys. Rev. Lett. **34**, 1506 (1975)

5.24 E. M. MONBERG, R. KOPELMAN: unpublished

5.25 E. M. MONBERG, P. ARGYRAKIS, R. KOPELMAN: unpublished

5.26 P. ARGYRAKIS, R. KOPELMAN: Molecular Structure and Spectroscopy Symposium, The Ohio State University, Columbus, Ohio (1975), paper WG 10

5.27 G. C. NIEMAN, G. W. ROBINSON: J. Chem. Phys. **37**, 2150 (1962);
H. STERNLICHT, G. C. NIEMAN, G. W. ROBINSON: J. Chem. Phys. **38**, 1326 (1962)

5.28 H.-K. HONG, R. KOPELMAN: J. Chem. Phys. **55**, 5380 (1971)

5.29 J. PERRIN: 2me Conseil de Chimie Solvay, Bruxelles (1924); C. R. Acad. Sci. (Paris) **184**, 1097 (1927);
F. PERRIN: Ann. Chim. Physique **17**, 283 (1932)

5.30 TH. FÖRSTER: Ann. Physik (6) **2**, 55 (1948). [English translation: R. S. KNOX, Univ. of Rochester, Rochester, N. Y. 14627]

5.31 R. KOPELMAN, E. M. MONBERG, F. W. OCHS, P. N. PRASAD: J. Chem. Phys. **62**, 292 (1975)

5.32 D. P. CRAIG, M. R. PHILPOTT: Proc. Roy. Soc. A **290**, 583 (1966)

5.33 S. C. ABRAHAMS, J. M. ROBERTSON, J. G. WHITE: Acta Crystalogr. **2**, 233, 238 (1949)

5.34 D. M. HANSON: J. Chem. Phys. **52**, 3409 (1970)

5.35 M. A. EL-SAYED, M. T. WAUK, G. W. ROBINSON: Mol. Phys. **5**, 205 (1962)

5.36 M. SCHWÖRER, H. C. WOLF: Mol. Cryst. **3**, 177 (1967)

5.37 F. W. OCHS, P. N. PRASAD, R. KOPELMAN: Chem. Phys. **6**, 253 (1974)

5.38 R. KOPELMAN, F. W. OCHS, P. N. PRASAD: unpublished

5.39 R. J. WATTS, S. J. STRICKLER: J. Chem. Phys. **49**, 3867 (1968)

5.40 H. L. FRISH, J. M. HAMMERSLEY, D. J. A. WELCH: Phys. Rev. **126**, 949 (1962)

5.41 B. KOK: Biochim. Biophys. Acta **48**, 527 (1961)

5.42 F. K. FONG: Proc. Natl. Acad. Sci. USA **71**, 3692 (1974);
F. K. FONG, V. J. KOESTER: Biochim. Biophys. Acta **406**, 294 (1975);
F. K. FONG: *Theory of Molecular Relaxation: Applications in Chemistry and Biology* (Wiley-Interscience, New York 1975)

5.43 N. RESSLER: J. Theor. Biol. **23**, 425 (1969)

5.44 R. KOPELMAN: Record Chem. Prog. **31**, 211 (1970)

5.45 R. KOPELMAN: Exciton Percolation in Mixed Molecular Crystals and Aggregates: from Naphthalene to Photosynthesis. J. Phys. Chem. (in press)

5.46 J. HOSHEN, R. KOPELMAN: Exciton Percolation I. Migration Dynamics. J. Chem. Phys. (in press)

Additional References with Titles

I. I. ABRAM, R. SILBEY: Energy transfer and spectral line shapes of impurities in crystals. J. Chem. Phys. **63**, 2317 (1975)

A. P. ALEKSANDROV: Temperature dependence of the efficiency of energy transfer and nonradiative loss in an europium chelate level system. Opt. Spectrosc. (USSR) **38**, 316 (1975)

B. M. ANTIPENKO, V. B. NIKOLAEV: Role of excitation diffusion in the cooperative sensitization process. Opt. Spectrosc. **39**, 165 (1976)

A. V. ARISTOV, V. P. KOLOBKOV, P. I. KUDRYASHOV, V. S. SHEVANDIN: Quenching interaction of excited neodymium ions. Opt. Spectrosc. (USSR) **39**, 160 (1975)

F. AUZEL: Multiphonon assisted anti-Stokes and Stokes fluorescence of triply ionized rare earth earth ions. Phys. Rev. **B13**, 2809 (1976)

F. AUZEL: Multiphonon assisted anti-Stokes or Stokes fluorescence of triply ionized rare earth ions. J. Luminesc. **12/13**, 715 (1976)

M. BANCIE-GRILLOT: Cooperative luminescence of some trivalent lanthanide ions in cadmium fluoride crystals. J. Luminesc. **12/13**, 681 (1976)

T. T. BASIEV, YU. K. VORON' KO, T. G. MAMEDOV, I. A. SHCHERBAKOV: Migration of energy between Yb^{3+} ions in garnet crystals. Sov. J. Quant. Electron. **5**, 1182-7 (1975)

G. BLASSE: Energy transfer phenomena in lead sulphate. Chem. Phys. Lett. **35**, 3, 299 (1975)

A. I. BURSHTEIN, A. YU, PUSEP: Collective quenching of luminescence. Sov. Phys.-Solid State **16**, 1509

A. I. BURSHTEIN, A. YU. PUSEP: Collective quenching of luminescence. Sov. Phys.-Solid State Spectrosc. (USSR) **38**, 588 (1975)

R. R. CHANCE, A. H. MILLER, A. PROCK, R. SILBEY: Fluorescence and energy transfer near interfaces: The complete and quantitative description of the Eu^{3+}/mirror systems. J. Chem. Phys. **63**, 1589 (1975)

R. L. CONE, R. S. MELTZER: Energy transfer mechanisms in $Tb(OH)_3$ from line shapes of band-to-band exciton fluorescence. J. Chem. Phys. **62**, 3573 (1975)

H. G. DANIELMEYER: Efficiency and fluorescence quenching of stoichiometric rare earth laser materials. J. Luminesc. **12/13**, 179 (1976)

A. V. DMITRYUK, G. O. KARAPETYAN, V. I. KOSYAKOV, B. M. MAKUSHKIN, V. A. SHIROKSHIN: Cooperative luminescence in glasses activated with Yb^{3+}. Opt. Spectrosc. (USSR) **37**, 335 (1974)

J. FAVA, G. LEFLEM, J. C. BOURCET, F. GAUME-MAHN: Energy transfer between Ce^{3+} and Tb^{3+} ions in $GdAlO_3$. Mat. Res. Bull. **11**, 1 (1976)

S. I. GOLUBOV, YU. V. KONOBEEV: Theory of sensitized fluorescence in solids. Phys. Stat. Sol. **B71**, 777 (1975)

H. G. GRIMMEISS, H. TITZE: Preparation and the luminescence of antistokes-phosphors of Y_2U_3 and La_2O_3 activated by Erbirm. Z. Phys. Chem. **87**, 4/6, 208 (1973)

J. HEBER, H. MURMANN: Coherent—incoherent energy transfer in ruby and Cr^{3+}: $LaAlO_3$. J. Luminesc. **12/13**, 769 (1976)

C. HSU, R. C. POWELL: Localization and Migration of energy among Sm^{3+} ions in $CaWO_4$ crystals. Phys. Rev. Lett. **35**, 734 (1975)

R. R. JACOBS, C. B. LAYNE, M. J. WEBER, C. F. RAPP: $Ce^{3+} \rightarrow Nd^{3+}$ energy transfer in silicate glass. J. Appl. Phys. **47**, 2020 (1976)

T. KOMIYANA: Energy transfer in $Eu^{3+} - Yb^{3+}$ and $Tb^{3+} - Yb^{3+}$ system in glasses. J. Non-Crystal. Solids **18**, 107 (1975)

E. NAKAZAWA: Cooperative optical transitions of $Yb^{3+} - Yb^{3+}$ and $Gd^{3+} - Yb^{3+}$ ion pairs in $YbPO_4$ hosts. J. Luminesc. **12/13**, 675 (1976)

E. OKAMOTO, M. SEKITA, H. MASUI: Energy transfer between Er^{3+} ions in LaF_3. Phys. Rev. **B11**, 5103 (1975)

E. OKAMOTO: Resonant transfer between Er^{3+} ions in LaF_3. J. Luminesc. **12/13**, 763 (1976)

M. S. ORLOV, I. G. SAITKSLOV, A. L. STOLOV: Deactivation of excited states of Tb^{3+} ions in fluoride crystals. Sov. Phys. Solid State **17**, 1008 (1975)

V. F. PISARENKO, B. D. SUYATIN: Sensitization of luminescence of trivalent rare-earth ions (TR^{3+}) by Eu^{2+} ions in NaBr crystals. JETP Lett. **20**, 264 (1974)

J. F. POURADIER, F. AUZEL: Etude de deux transferts d'energie non radiatifs entre ions Ho^{3+} dans le fluorure d'yttrium; possibilites d'applications. J. Physique **37**, 421 (1976)

R. REISFELD, Y. ECKSTEIN: Energy transfer from Ce^{3+} to Tm^{3+} in borate and phosphate glasses. Appl. Phys. Lett. **26**, 253 (1975)

R. REISFELD, N. LIEBLICH, L. BOEHM, B. BARNETT: Energy transfer between $Bi^{3+} \rightarrow Eu^{3+} \rightarrow Sm^{3+}$ and $UO_2^{2+} + Eu^{3+}$ in oxide glasses. J. Luminesc. **12/13**, 749 (1976)

J. A. SELL, F. K. FONG: Oscillator strength determination in $LaCl_3: Pr^{3+}$ by photon upconversion. J. Chem. Phys. **62**, 4161 (1975)

S. SINGH, D. C. MILLER, J. R. POTOPOWICZ, L. K. SHICK: Emission cross section and fluorescence quenching of Nd^{3+} lanthanum penta phosphate. J. Appl. Phys. **46**, 1191 (1975)

M. J. TREADAWAY, R. C. POWELL: Energy transfer in samarium-doped calcium tungstate crystals. Phys. Rev. **B11**, 862 (1975)

R. K. WATTS: Stepwise upconversion and cooperative phenomena in fluorescent systems. In *Optical Properties of Ions in Solids*, ed. by B. DiBARTOLO, vol. 8 (Plenum Publ. Co. 1975)

R. K. WATTS: Migration and transfer of optical excitation. J. Chem. Phys. **64**, (2), 902 (1976)

M. G. ZUEV, F. A. ROZHDESTVENSKII, E. I. KRYLOV: Optical centers and excitation energy transfer in polycrystalline $LaTaO_4: Nd^{3+}$. Sov. Phys. Solid State **16**, 3, 613 (1974)

G. M. ZVEREV, I. I. KURATEV, A. M. ONISHEHENKO: Transfer of excitation energy between trivalent rare earth ions in crystals. Sov. J. Quant. Electron. **5**, 267 (1975)

Author Index

Subject Index

Applied Physics

A monthly journal

Board of Editors	**S. Amelinckx,** Mol. · **V. P. Chebotayev,** Novosibirsk **R. Gomer,** Chicago, Ill. · **H. Ibach,** Jülich **V. S. Letokhov,** Moskau · **H. K. V. Lotsch,** Heidelberg **H. J. Queisser,** Stuttgart · **F. P. Schäfer,** Göttingen **A. Seeger,** Stuttgart · **K. Shimoda,** Tokyo **T. Tamir,** Brooklyn, N.Y. · **W. T. Welford,** London **H. P. J. Wijn,** Eindhoven
Coverage	application-oriented experimental and theoretical physics: *Solid-State Physics* *Quantum Electronics* *Surface Physics* *Laser Spectroscopy* *Chemisorption* *Photophysical Chemistry* *Microwave Acoustics* *Optical Physics* *Electrophysics* *Integrated Optics*
Special Features	**rapid** publication (3–4 months) **no** page charge for **concise** reports prepublication of titles and abstracts **microfiche** edition available as well
Languages	Mostly English
Articles	original reports, and short communications review and/or tutorial papers
Manuscripts	to Springer-Verlag (Attn. H. Lotsch), P.O. Box 105 280 D-69 Heidelberg 1, F.R. Germany Place North-American orders with: Springer-Verlag New York Inc., 175 Fifth Avenue, New York. N.Y. 10010, USA

Springer-Verlag
Berlin Heidelberg New York

Topics in Current Chemistry

Fortschritte der chemischen Forschung
Managing Editor: F. L. Boschke

Springer-Verlag Berlin Heidelberg New York